THE
GREATER FLAMINGO

ALAN JOHNSON AND FRANK CÉZILLY

T & A D POYSER
London

To Luc Hoffmann, who has dedicated his life to conservation worldwide, and in memory of our close friend and colleague Heinz Hafner, whose enthusiasm for wetlands and waterfowl in the Camargue and elsewhere inspired all who knew him.

Published 2007 by T & AD Poyser, an imprint of A&C Black Publishers Ltd., 38 Soho Square, London W1D 3HB

www.acblack.com

Copyright © 2007 text by Alan Johnson and Frank Cézilly
Copyright © 2007 photographs by Alan Johnson, except where other photographers are specified.
Copyright © 2007 illustrations by Cyril Girard

The right of Alan Johnson and Frank Cézilly to be identified as the authors of this work has been asserted by them in accordance with the Copyright, Design and Patents Act 1988.

ISBN 978-0-7136-6562-8

A CIP catalogue record for this book is available from the British Library

All rights reserved. No part of this publication may be reproduced or used in any form or by any means—photographic, electronic or mechanical, including photocopying, recording, taping or information storage or retrieval systems—without permission of the publishers.

The book is produced using paper that is made from wood grown in managed sustainable forests. It is natural, renewable and recyclable. The logging and manufacturing processes conform to the environmental regulations of the country of origin.

Commissioning Editor: Nigel Redman
Project Editor: Jim Martin

Design: J&L Composition, Filey, North Yorkshire

Printed and bound in Great Britain

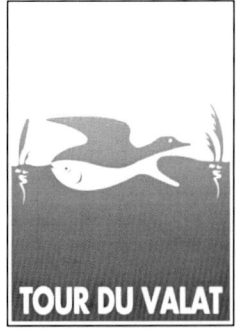

Contents

Preface	6
Acknowledgements	7
1: INTRODUCTION	9
Flamingos and man	10
Interest in flamingos: recent history	11
Focus on the Camargue	12
The flamingo specialist group	15
About this book	16
2: THE FLAMINGOS	18
Ancestry and relationships of flamingos	19
Morphology and general description	22
Life history	32
Discussion	37
3: RESEARCH	42
The pioneers and early years of discovery	42
Contemporary research—what have we learned?	45
Distribution and numbers	48
Capturing and marking flamingos	51
Capture-resighting	57
Foraging ecology	61
Breeding ecology	63
Captive flocks	66
4: DISTRIBUTION AND NUMBERS	69
Distribution	69
Numbers	74
Regional summaries	81
Population trends	85
5: MOVEMENTS	89
Flight behaviour	90
Types of movements	92
Spatial patterns	101
Some concluding remarks	107

6: Foraging ecology — 110
- Feeding apparatus — 111
- Food — 112
- Feeding behaviour and foraging methods — 113
- Foraging and drinking flights — 120

7: Mating system and mate choice — 126
- Group displays — 127
- Pair formation and copulation — 131
- Seasonal monogamy in Greater Flamingos — 133
- Age-assortative pairing — 138

8: Breeding biology — 141
- Timing of breeding — 142
- Colony establishment — 150
- Breeding stages — 153
- Breeding success — 165

9: Survival and recruitment — 173
- Causes of post-fledging mortality — 174
- Estimating survival — 177
- Factors influencing survival — 181
- Recruitment — 182

10: Conservation and management — 187
- Conservation status in the world — 188
- Threats and conflicts — 188
- Conservation actions — 199

11: Conclusions: what does the future hold? — 218
- Research in the future — 219
- Conservation issues — 223
- Epilogue — 228

12: An inventory of the more important greater flamingo breeding sites — 229
- Greater Flamingo breeding sites — 231
- West Africa — 232
- Mediterranean — 236
- Asia — 250
- East Africa — 256
- Southern Africa — 259

Appendices — 263
1: Sightings per year of flamingos PVC-banded as chicks in the Camargue 1977–2000 — 263
2: Greater Flamingo marking schemes past and present — 264

3: Characteristics of PVC leg-bands in the Mediterranean 1977–2000 265
4: Timing of annual ringing in the Camargue 266
5: Number of flamingos PVC-banded as chicks and recaptured per cohort 1977–2000 267
6: Number of pairs of Greater Flamingos breeding throughout the Old World 1972–2000 268
7: Number of pairs and distribution of Greater Flamingos breeding in West Africa 1972–2000 269
8: Number of pairs of Greater Flamingos breeding in the western Mediterranean 1972–2000 270
9: IWC counts of flamingos over the western Mediterranean 272
10: Greater Flamingo chicks fledged in the western Mediterranean 1972–2000 273
11: Number of Camargue-banded flamingos breeding in Mediterranean and West African colonies 1982–2000 275
12: Colony of first-recorded breeding of flamingos banded as chicks in the Camargue 276
13: Observations of colour-banded flamingos seen breeding at Fuente de Pedra and locatd on their foraging grounds 278
14: Mean egg-laying dates at flamingo colonies in north-west Africa, the Mediterranean and south-west Asia 279
15: The start of egg-laying in relation to date of flooding of the Etang du Fangassier, Camargue 280
16: Greater Flamingo breeding in the Camargue; egg-laying period, colony size and success 281
17: Dimensions and weights of freshly abandoned Greater Flamingo eggs from the Camargue 283
18: Stages in the growth of flamingo chicks 284
19: Duration of chick-feeding bouts 285
20: Recoveries per year after ringing of flamingos marked as chicks in the Camargue 1947–61 286
21: Scientific names of animals and plants 287

Glossary 291
References 294
Index 324

Preface

Few birds have attracted human attention as much as the flamingo. Since ancient times their exotic pink colour, their peculiar elongated body, the strange shape of their tongue and their gregarious behaviour have inspired artists, poets, explorers and even cooks. Yet, up until the middle of the 20th century, very little was known about their biology. The first serious monograph (Allen 1956) was more important for showing the gaps in our knowledge than for lifting the veil from the flamingo's mysteries. During the last few decades, a growing number of scientists have directed their attention to flamingos, and many studies have been conducted in the field and in captivity. Today, although much remains unknown, the biology of our bird is well documented. This book attempts to summarise the acquired knowledge and to show challenges for further exploration.

Alan Johnson was a carpenter when he first came to the Camargue in 1962 at the age of 21. He became so excited about the area and its birds that he decided to remain there and to devote his life to flamingos. As a carpenter he had learned to measure, to join meticulously piece to piece and to finish solidly. He has done this so well with the flamingos that over 40 years an astounding body of knowledge has been built up. By 1983 he had already earned a university doctorate, but this was only a first step in his deepening commitment to the research of the species. Today, for all who know Alan, his name is inseparable from that of the flamingo, and for all biologists the word 'flamingo' evokes inevitably Alan's name.

Frank Cézilly, Alan's associate in the writing of this book, first came to the Camargue in 1983. As a university student in biology, he was well aware of the issues and problems intriguing modern scientists, and of the methods to tackle them. In 1991, impressed by the wealth of data collected by Alan, he decided to join his team in making the flamingo research an important contribution to theoretical biology.

The aim of this book is not confined to science alone. It also intends to make an important contribution to the conservation and management of bird populations. In the 1960s, the Greater Flamingo was a threatened species in the Mediterranean. The research and management activities described in this book have enabled it to establish safe populations in this area. Those nesting in the Camargue have consolidated their numbers and nesting success, and conquered new nesting places. Thus the flamingo story can be a model for how to save an endangered species. Once again, the cooperation between science and conservation bears its fruits.

Luc Hoffmann
February 2007

Acknowledgements

Much of the data presented in this book were collected as part of a long-term population study and conservation project which has spanned three decades and three continents. The investment in time and manpower has been, to say the least, considerable. The number of people involved has been similarly impressive, and it is regrettable that all cannot be thanked individually.

The flamingo study always benefited from excellent partnership with the agricultural community. First, it would not have been possible to observe breeding flamingos at close range, to catch and ring chicks, to warden the colonies or to build new islands as the need arose, without the full cooperation of the salt company. The management of the Compagnie des Salins du Midi et des Salines de l'Est (now Les Salins) has been generous in allowing free access to their property, in particular to the Salins de Giraud, and in restoring the flamingo island at the Etang du Fangassier in 1988, and again in 1995. Their contribution to flamingo research and conservation has been of prime importance in the Mediterranean region and we would like to thank in particular the directors Messrs D. Balas, G. Boudet, A. Colas and the late A. Jullien. Secondly, despite crop losses due to flamingos foraging nocturnally in rice fields, Camargue farmers remain very tolerant and cooperative.

The World Wide Fund for Nature (WWF) and the Parc Naturel Régional de Camargue have very efficiently wardened the Fangassier colony over the last 25 years. Catching and ringing flamingos has been authorised by the Centre de Recherche sur la Biologie des Populations d'Oiseaux (CRBPO) in Paris and the French Department of Nature Conservation of the Ministry of the Environment. Many volunteers helped each year during the ringing operations. Data collection in the field has involved the assistance of a very large number of people, many on a voluntary basis. Those who have strained their eyes to read rings, wallowed in the mud making flamingo nests or catching chicks, or braved the elements to reach the flamingo tower at Fangassier during two seasons or more deserve mention: Bouzid Chalabi, Pierre-André Crochet, Nathalie Hecker, Jean-Laurent Lucchesi, Philippe Pilard, Nicolas Sadoul, Ekaterina Stotskaia, Gioia Theler, Christophe Tourenq, John Walmsley and Dianne Wilker have all devoted very long hours observing the flamingo colony in conditions which were not always optimal, sitting out storms or shivering in the cold. Many hours have been spent flying in small aircraft to census flamingos and to photograph colonies, and in particular the late Patrice Rollin's piloting and generosity were always very much appreciated. We are also most grateful to Dr Alain Tamisier from Centre National de la Recherche Scientifique (CNRS) for providing winter counts of flamingos during his flights over those parts of the Rhône delta which are less accessible on the ground.

At an international level, Wetlands International played a crucial role in stimulating contacts and cooperation among researchers and amateurs involved in flamingo censuses. Gioia Theler provided a great deal of assistance with observations in the Mediterranean and in particular in liaising with Spanish colleagues and assisting them in searching for Camargue-banded flamingos in Andalusia. At several places around the Mediterranean a handful of observers have been particularly faithful to the flamingo study over many years. Mike Smart in Tunisia and elsewhere, André Blasco and Philippe Orsini in France, Habib Dlensi in Tunisia, Luis Garcia and Manuel Rendón Martos in Spain, Alessia Atzeni, Nicola Baccetti, Vincenzo Loi and Simona Pisano in Italy, to mention just a few, have all reported hundreds of band sightings over a long period of time. The contribution by members of ornithological societies in the south of France has also been most appreciated (GRIVE, CEEP, LPO Aude, Réserve Nationale de Camargue). Their observations have contributed to our knowledge of flamingo movements in particular, but also of survival, philopatry and breeding dispersal.

Our colleagues and friends at the Tour du Valat, in particular Patrick Duncan, the late Heinz Hafner, and Jean-Paul Taris, have always been available for discussions concerning the flamingo study. Jacques Blondel, the late François Bourlière, Tim Clutton-Brock, Rudi Drent, Patrick Dugan, the late Peter Evans, John Krebs and Chris Perrins have for many years shown much interest in the flamingo study, and their evaluations and advice have always been most appreciated. A first data analysis was carried out by Graham Hirons and Rhys Green, who also gave valuable advice concerning both fieldwork and analyses. Vincent Boy has from the early days of PVC banding written the computer programs which made data handling almost a pleasure, and he carried out many of the analyses which were later refined in cooperation with Jean-Dominique Lebreton, Ruedi Nager, Roger Pradel, Giacomo Tavecchia and Anne Viallefont at the Centre d'Ecologie Fonctionelle et Evolutive at the University of Montpellier. Frank Dhermain and Kate Lessells provided particular assistance with blood sampling. We also thank Boerhringer-Manheim France (Messrs F. Boglia and J. C. Vallery) for kindly lending us technical equipment for blood chemistry analysis. François Renaud at CNRS in Montpellier helped in collection and identification of flamingo feather lice.

Many people assisted us during the preparation and final editing of this book, in particular at the Tour du Valat, Robert Bennetts, Carol Durand and Dianne Wilker, and at the University of Bourgogne, Alexandre Bauer, Pascale Laplace and François Tainturier for the final editing of the maps and tables respectively. Jevgeni Shergalin was particularly helpful in kindly providing English translations for numerous Russian-language articles on flamingos in the former USSR. Cyril Girard kindly agreed to illustrate the chapters of this book with original vignettes. We also wish to thank Andy Richford for his long-standing interest in this publication.

Finally, we wish to express our deepest gratitude to Dr Luc Hoffmann, not only for the preface to this book, but for his full support of the Tour du Valat flamingo project, which he began in the early 1950s and which remains for us a most exciting venture.

CHAPTER ONE

Introduction

Flamingos, like geese, cranes and other colonial waterbirds, are very eye-catching, whether foraging in large numbers or flying in skeins. However, they differ in many ways from other large wading birds in their preference for salty wetlands, their manner of feeding by filtering and their habit of raising their chicks in a nursery, or crèche. They are fascinating birds to observe, and have always been of great interest to birdwatchers and researchers alike. Standing among the most popular bird species, they are considered as a flagship for wetland conservation. Great progress in our understanding of the life history of flamingos has been achieved over the last twenty years, thanks to unique research and conservation efforts. The aim of the present book is to provide the reader with an updated review of the current knowledge.

It is not known precisely how many Greater Flamingos exist in the world, but reports of decreases in numbers in some areas are counter-balanced by increases in others, and from all available evidence, their overall numbers, roughly half a million, have not changed radically in recent decades. Since historical times, people have been fascinated by flamingos, even in countries where they do not occur in the wild, and their presence on particular wetlands has sometimes motivated the legal protection of an area. The species is protected throughout much of its range, as are many of the more important wetlands on which flamingos occur, but several of these areas remain unprotected and the species' future will depend on our capacity, and that of future generations, to conserve these sites.

FLAMINGOS AND MAN

The extent to which the flamingo has always held a place in the human imagination has been described in previous works (Allen 1956; Kear & Duplaix-Hall 1975; Ogilvie & Ogilvie 1986) and need not be re-explored here. The flamingo is portrayed in cave paintings at Tajo de las Figuras, Cádiz, in Andalusia, Spain, which date from the Bronze Age 5,000 years BC (Gurney 1921; Topper & Topper 1988; Mas Cornellà 2000). It also figures frequently in Carthaginian, Byzantine and Greek art and was occasionally portrayed as a hieroglyph by the early Egyptians to mean the colour red (Houlihan & Goodman 1986). Early flamingo representations are also preserved on a large number of decorated predynastic ceramic vessels from the Gerzean Period (Naqada II) of Ancient Egypt, 3500–3200 BC.

Part of man's fascination with flamingos is clearly attributable to their colourful appearance. The generic name *Phoenicopterus* is derived from the Greek and means scarlet-winged. In many languages, the common name directly refers to the flamboyance of flamingo wings (see Carro & Bernis 1968). All flamingo species are highly gregarious, with colonies of Greater Flamingos regularly exceeding 10,000 pairs and those of Lesser Flamingos sometimes exceeding 1,000,000 pairs. The spectacle of thousands of these birds taking wing cannot help but leave an impression on anyone lucky enough to see it. However, it is not colour alone that sets flamingos apart from other birds. Their behaviour is also quite distinctive, and flamingo displays provide one of the most impressive of avian spectacles. Considering the unique features of the species, above all the bill structure, Chapman & Buck (1910) even suggested that they might almost be regarded as a separate act of creation!

Their choice of habitat is also unique, because in many parts of their range flamingos inhabit highly saline and/or alkaline lakes, where conditions are generally too harsh for other species, including man. Many of these areas are remote and inhospitable to humans, which is certainly why the Lesser Flamingo's breeding colonies were not discovered and described until 1954 (Brown & Root 1971). Just three years later, the James' Flamingo, which since the beginning of the century had been presumed extinct, was rediscovered in the high Andes, where it occurs mainly at altitudes above 4,000 m (Fjeldså & Krabbe 1990).

In the past, persecution and disturbance—in the form of hunting the birds for their flesh and raiding colonies for eggs—probably had a strong effect on the numbers and distribution of flamingos, at least throughout the Mediterranean region and in parts of Asia. The Romans are reputed to have banqueted on flamingos' tongues, which they considered a delicacy (Allen 1956), while Babar the Great (1483–1530) reportedly breakfasted on their eggs, which he collected at Lake Ab-e-Istada in Afghanistan (Paludan 1959). Recipes for the species are not difficult to find in early French, Italian and Spanish culinary literature, reflecting the prevalence of flamingo hunting throughout the Mediterranean (de Marolles 1836). Flamingos were shot in the Guadalquivir Marshes in southern Spain in large numbers well into the 1960s, while French law did not give full protection to the species until as recently as 1976, although it had been illegal to shoot flamingos since 1962. Such pressures must have driven the flamingo to nest in only the most remote places.

Remarkably, this situation has now changed, at least in the western Mediterranean, and the flamingo is no longer the wary bird it formerly was. This change of behaviour has had an impact on the species' distribution, both during and outside the breeding season. Wild flamingos can currently be found feeding in close proximity to humans, or even being fed by humans, as for example in Dubai Creek in the UAE and at the ornithological park at Pont de Gau in the Camargue. The species has increased in number throughout the Mediterranean region as a result of effective protection measures and multiple conservation efforts, and, also quite remarkably, flamingos even breed now in the suburbs of a large city, at Cagliari in Sardinia. Saltpans have probably played a major role in determining the present distribution and numbers of flamingos, at least in the Mediterranean region, and they have received considerable attention in this book (Chapters 2 and 10). However, despite these successes, it seems that the species' distribution, in particular that of the Greater Flamingo, is more restricted today than was that of its ancestor species (Feduccia 1996).

INTEREST IN FLAMINGOS: RECENT HISTORY

The second half of the 20th century saw an upsurge of interest in flamingos. On one hand previously unknown breeding areas were discovered, often in remote areas, or poorly documented colonies were revisited, and much new data were gathered on flamingo numbers and distribution; on the other hand, desk or laboratory studies were undertaken, the results of which stand as empirical works which anyone with a serious interest in flamingos will have on their bookshelves. In the first instance, Etienne Gallet and Leslie Brown, for example, who studied flamingos respectively in the Camargue (Gallet 1949) and in the Rift Valley lakes of East Africa (Brown 1975), are just two of the naturalists who made valuable contributions to our knowledge of flamingo breeding biology in the wild. Other references to breeding, dating back to the end of the 19th and the first half of the 20th century, can be found in Yeates (1950). In the second period, significant contributions came from Robert Porter Allen's field trips and desk study into the life history of all species of flamingos (Allen 1956) and from the remarkable paper by Jenkin (1957) on the diet and very specialised way in which flamingos of all species feed. This latter study was so meticulous that it seems destined to remain, even despite more recent accounts (Zweers *et al.* 1995), a constant source of reference for anyone wishing to learn more about flamingos' unique manner of feeding, which we ourselves have not studied in detail. Many of these works are no longer readily available, but they are not necessarily outdated and we hasten to recommend them still to readers according to their field(s) of interest. We have widely consulted them, and other 'classic' flamingo studies, in writing the present work. However, the longest-running study of flamingos in the world, and one of the longest for any bird species, was begun in the Camargue over half a century ago.

FOCUS ON THE CAMARGUE

Luc Hoffmann was still a student when in 1947 he began chasing flamingo chicks in the Camargue in order to ring them. This was the very beginning of a study aimed at finding out more about the biology, movements and longevity of these birds in the Mediterranean region. Over the following decade, breeding was closely monitored and the ringing programme improved and intensified, until, in the early 1960s, flamingos stopped breeding in the Rhône delta. During seven long years, no chicks fledged from the Camargue and there was much concern over the Greater Flamingo's future in France and elsewhere in southern Europe.

In spite of their extensive range across southern Europe, south-west Asia and Africa, Greater Flamingos were nesting in only about 30 localities, one of the more important of these being the Camargue. This is certainly one of the best-documented colonies, since records of occasional breeding in the region date back to the mid-16th century (Quiqueran de Beaujeu 1551; Darluc 1782) and quite regular, albeit not annual, breeding had been reported since the beginning of the 20th century (Gallet 1949). The origins of the main conservation problems which had arisen in the Camargue were quite well understood because they had been identified by Gallet (1949) and confirmed by Hoffmann in the 1950s: erosion of the breeding islands, disturbance from various sources during breeding, particularly by aircraft, and predation of eggs and chicks by Yellow-legged Gulls, which were rapidly increasing in number throughout the region. Initiatives understandably moved away from research towards conservation and management.

For conservationists it was a great relief when in 1969 over 7,000 pairs of flamingos recolonised the Camargue, mostly at the Etang du Fangassier, and raised chicks for the first time since 1961. Breeding was highly successful, owing in no small part to wardening by staff of the Tour du Valat Biological Station, which Luc Hoffmann had established in the 1950s in the heart of the delta. There were high hopes that the flamingos would nest again regularly, and over the following years this optimism proved to be justified. Conservation and management efforts initiated by the Tour du Valat in the 1960s and 1970s, which included building a breeding island, paid off and gradually the Etang du Fangassier became the flamingo sanctuary it is today. Since that epic event of 1969, flamingos have nested every year in the Camargue, and from 1972 onwards only at the Etang du Fangassier. Threatened in the 1960s, the Greater Flamingo, emblem of the Camargue, made a remarkable recovery, so much so that numbers have increased beyond those projected by the action plan, and the species has become the most abundant aquatic bird breeding in the Rhône delta. This may be seen as a positive development with regard to the species' conservation, but it has also resulted in problems as the birds now cause crop damage by feeding in rice fields in spring, and they are understandably not too popular with the unlucky farmers. Flamingos are now quite accustomed to human activity, including the disturbance caused by aeroplanes, and in some places are now remarkably confiding. This change in behaviour, associated with a population increase and overcrowding at the sole breeding site in France, has driven birds to colonise new areas in the western Mediterranean in the 1990s, in some of which they now breed.

The flamingo's new-found security is due in no small degree to Luc Hoffmann's decision to warden the colony when breeding resumed in 1969. Once breeding was securely re-established, in 1977, the marking programme was reactivated (Figure 1). The advent of laminated PVC enabled us to mark birds with large coded leg-bands which allow individual recognition in the field. A band observation programme was begun both locally and internationally, to which many amateur

Figure 1. **Capture and ringing of chicks.** Between 500 and 900 flamingo chicks were captured annually for marking just prior to fledging once a year in the Camargue for over a quarter of a century. The subsequent observations of these individuals have provided the material for the study of these birds in the wild. Similar capture and marking operations, involving hundreds of people, are organised most years at breeding sites in Spain and Italy. *(J-P. Taris).*

ornithologists were soon to contribute. In 1986, colleagues in Andalusia launched a similar conservation and research project on a regular basis at the Fuente de Piedra Lagoon (Málaga), 950 km south-west of the Camargue. The success of this colony, combined with the collection of much new data, and concern for the conservation of flamingo habitats throughout the western Mediterranean and north-west Africa, led to the organisation of an international workshop on flamingos which was held at Antequera (Málaga) in 1989 (Junta de Andalucía 1991). More recently, with flamingos breeding in Sardinia and on mainland Italy, the capture-recapture study has been further expanded to include partners in Italy. At the time of writing, an extension of activities is envisaged to sites beyond the western Mediterranean which are poorly covered at present.

Multiple resightings of many thousands of flamingos have provided the basis for the study of the ecology and dynamics of the west Mediterranean sub-population of Greater Flamingos, which represents about one-fifth of the world population. Flamingos have a potentially long life expectancy, with some captive birds exceeding 60 years of age. Of those flamingos PVC-banded in 1977, one-fifth were seen back at the Fangassier colony at 20 years of age, and given our most recent estimates of adult survival, a few of those birds ringed by Luc Hoffmann in the 1950s may well still be around today!

The Camargue flamingo study has evolved as both a research and a conservation project and we have endeavoured to give adequate attention in this book to the conservation needs of these birds, while at the same time presenting the results obtained from a scientific project which we plan to pursue in the future. Management techniques which have been so successful in France and Spain could easily, where appropriate, be applied in other parts of the world. We have attempted to minimise the bias towards the Camargue but, since this area has been the theatre of such a great effort, we hope that the reader will forgive us if we have not been entirely successful in doing so.

The Camargue is one of the few places, if not the only site, in the world where flamingos breed every year, and have done so for more than 30 years now. Breeding was not so regular in the first half of the 20th century, as anyone who reads Gallet (1949) or Yeates (1950) would soon realise, and this situation is clearly anthropogenic. In addition to the effective conservation and protection of the breeding and feeding areas required to sustain a flamingo colony, annual breeding is possible because of the semi-permanent water levels of the saline lagoons which have been created and maintained by the salt industry. However, the wind of change is now blowing through this semi-agricultural industry and the Greater Flamingo's future in the Mediterranean may depend to a large degree on the survival of the historic saline complexes.

THE FLAMINGO SPECIALIST GROUP

Both individual and group initiatives undertaken during the 20th century led to a growing interest in flamingos and in particular the conservation of a species that was popular yet little studied in the wild and increasingly threatened. In 1971, the Flamingo Specialist Group was established by the former International Council for Bird Preservation (ICBP, now BirdLife International) primarily in the interests of flamingo conservation (Johnson 2000b). One of the first initiatives of the group was to undertake, in the 1970s, a practically worldwide survey of flamingos of all species (Kahl 1975b), to update knowledge on distribution and on population estimates made earlier by Brown (1959). At the same time, the group convened a Flamingo Symposium which was held at The Wildfowl Trust (now Wildfowl and Wetlands Trust) in Slimbridge (UK) in July 1973. This was the occasion for field ornithologists to unite with ethologists such as Adelheid Studer-Thiersch, whose behavioural studies have been, and still are, carried out at the Basel Zoo (Switzerland). The proceedings of this meeting were published as one of the first Poyser books (*Flamingos*, Kear & Duplaix-Hall 1975) and summarised much of what was known about all flamingo species at the time, both in captivity and in the wild. The comprehensive bibliography, along with that provided earlier by Allen (1956), will remain for a long time a source reference to all interested in flamingos at large.

There was a long wait for the second world symposium on flamingos, which finally took place in October 1998 in Miami, Florida. This three-day meeting was attended by 92 participants from 22 countries, a few of them having attended the first symposium in Slimbridge over 25 years earlier. The event was conveniently timed to follow on from the annual meeting of the Colonial Waterbird Society (now The Waterbird Society) and, just like the first symposium, it united fieldworkers, researchers and those responsible for captive flocks. The conservation and research needs of flamingos worldwide, both in the wild and in captivity, were discussed and plans were drawn up for a flamingo conservation action plan. The symposium proceedings were published as a special publication of *Waterbirds* (Baldassarre *et al.* 2000).

The Flamingo Specialist Group, which is presently one of the many specialist groups of the World Conservation Union (IUCN) Species Survival Commission (SSC) and Wetlands International, although not actually undertaking research projects, has flourished as a group of geographically widespread amateur and professional biologists who liaise through the newsletter.

ABOUT THIS BOOK

This work represents a major input in the fields of research and conservation of Greater Flamingos. It is a monograph biased to some degree by studying and managing for the interests of this Old World subspecies (or species) in an area which is heavily managed: the Rhône delta on the Mediterranean coast of southern France. We have, however, endeavoured to minimise this bias by gleaning as much information as possible from papers published on flamingos in other parts of the world, and by visiting key flamingo haunts on three continents. The first author of this book was assigned the task in 1965 of developing a conservation strategy for the flamingo in the Camargue, and later of setting up a research team. In 1978, he became Coordinator of the Flamingo Specialist Group, a position he held until 2004, and in one or other of these capacities has visited numerous key wetlands for flamingos throughout the Mediterranean region, in south-west Asia, many parts of Africa, Mexico, Argentina and Peru, and has attended meetings on four continents. The second author joined the team in 1991 for six years in order to refine data analyses, applying the most modern tools to demographic and behavioural studies.

Since the first book on flamingos (Kear & Duplaix-Hall 1975), the Greater Flamingo has bred in six new sites in the western Mediterranean, 28,000 chicks have been banded in five colonies, and there have been over a quarter of a million resightings of these birds. Over 50 scientific papers have been published, plus three doctoral theses. Here we have attempted to synthesise all current knowledge about the species up to the beginning of the 21st century. The monograph consists of 11 chapters that cover the main aspects of the biology and conservation of flamingos, including foraging ecology, mating system, breeding biology, demography, and management. We have minimised the number of mathematical tests and formulas cited in the text in favour of a lighter script than normally accompanies scientific papers.

Although flamingos seem to have bred in more places and more frequently over the past decade than they did formerly, at least in the Mediterranean region, the formation of a colony is still an ornithological event worthy of attention. We have endeavoured to include many unpublished accounts of nesting in remote areas, such as the chotts of southern Tunisia, as well as all published reports of breeding by Greater Flamingos. Findings and data which are scattered throughout both scientific journals and grey literature have been drawn together and are presented here. On the basis of these data we have compiled site sheets (Chapter 12) for those locations that we consider to be the 35 most important breeding sites, which include some localities where flamingos have been recorded breeding only once.

Much remains to be discovered about the biology and population ecology of a species as mobile and as long-lived as the Greater Flamingo. Our studies have shown that wild flamingos do not organise their lives in quite the same way as captive birds. Clearly, the life span of the species is such that the working life of one researcher is not sufficient to cover all aspects of its biology. We therefore hope that this book will stimulate interest among young scientists as well as standing as a tribute to an enormous research and conservation effort which had its genesis over half a century ago. In this respect, we are delighted that Luc Hoffmann consented to write the Preface to this book.

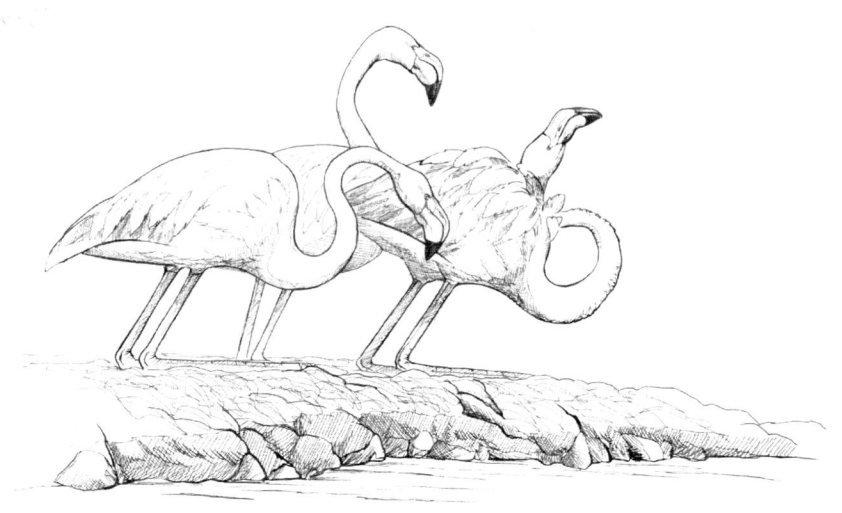

CHAPTER TWO

The flamingos

Flamingos comprise a very ancient group of birds which over the ages has attracted much attention. The group has never been satisfactorily placed in an order, nor is it unambiguously split into species. In some ways like storks, and in others like ducks, the flamingo also shares some special features with the Australian Banded Stilt. The plumage of flamingos develops over a period of years, and the resulting delay in maturity indicates that the birds have a relatively long life expectancy. In spite of their fragile appearance, flamingos are strong flyers, sometimes undertaking long journeys between breeding and foraging areas. Vocalisation is important for individual recognition between adults and chicks, and for keeping flocks together. Moult of the flight feathers can be either progressive or simultaneous as in ducks and geese. The highly specialised bill is unique among birds. The salt gland, which has so far been little studied, allows them to feed in water much saltier than other species can tolerate, and consequently they are alone in exploiting food resources in many wetlands of high salinity.

ANCESTRY AND RELATIONSHIPS OF FLAMINGOS

It is generally agreed that flamingos are amongst the oldest living birds. A detailed description of flamingo or flamingo-like fossil records can be found in Feduccia (1996). Flamingos are believed to be descendants of a very ancient group of birds whose ancestors (*Presbyornis*) can be traced back to the Cretaceous Period, 145 to 65 million years ago (Brodkorb 1963; Feduccia 1976, 1977, 1996). The oldest certain ancestor of the flamingo, however, was *Juncitarsus gracillimus*, a common colonial species inhabiting saline environments in what are now Wyoming, USA, and Germany during the Eocene Period (55 to 40 million years ago) (Feduccia 1996). The fossil of *Juncitarsus* is critical to the evolutionary record. It was a clear shorebird-flamingo mosaic, a stilt-like flamingo, intermediate in size between the extant shorebirds and the smallest of the living flamingos (Feduccia 1996). The first known flamingo, *Phoenicopterus croizeti*, occurred in Europe during the late Oligocene to early Miocene periods (40 to 20 million years ago) (Olson 1985; Feduccia 1996). It was apparently very similar to modern flamingos, with long legs and filter-feeding apparatus. Fossil evidence from Europe, North America, and Australia suggests that flamingos and their ancestors were much more widespread than they appear to be today.

PHYLOGENETIC AFFINITIES AND CLASSIFICATION

The classification of flamingos has long been the subject of considerable controversy (Sibley *et al.* 1969; Sheldon & Slikas 1997), depending on whether it is based on morphological or behavioural traits, fossil evidence of phylogeny, or by the more recent method of DNA analysis. The main question is whether similarities between flamingos and other groups of birds derive from a common ancestor or reflect convergent adaptations to similar environments. Feduccia (1996) noted that 'the flamingos... have been placed everywhere in the taxonomy and determining their phylogenetic position has been one of the most perplexing problems in avian systematics'. Up to 15 separate studies have listed the flamingos somewhere between the Anseriformes (ducks, geese and swans) and the Ciconiiformes (herons, ibises and storks). Thus, Sir Peter Scott (in Kear & Duplaix-Hall 1975) asked the question: 'Are they web-footed storks or long-legged geese— or are they just flamingos?'

Web-footed storks?

Gadow (1877, 1892) considered flamingos to be a suborder Phoenicopteri of the Ciconiiformes. Similarly, Wetmore (1960) believed that flamingos closely resembled the herons and ibises in shape and size (pelvis, sternum, pectorals, wing muscles, pterylosis, number of primaries). The hooked lower jaw is also shared with spoonbills (Cramp & Simmons 1977), although the bill is only a secondary adaptation.

Long-legged geese?

Later authors (e.g. Delacour 1961) considered flamingos to be phylogenetically close to the Anseriformes, suggesting that their long legs were merely the result of convergent evolution with other large waterbirds. Among the characteristics in support of this hypothesis are the structure of the bill and tongue, webbed feet, feeding and courtship behaviour, ground-nesting, nest-building, young resembling goslings, leaving the nest and swimming soon after hatching, creching (gathering in nurseries) of chicks, simultaneous moult to flightlessness, voice, and the feather lice which are common to certain ducks and flamingos (see also Allen 1956, Cramp & Simmons 1977).

Tall waders?

Since Scott's question, a third possibility in the classification of flamingos has arisen, based on fossil evidence, notably that of *Presbyornis* (Feduccia 1976, 1977, 1996; Olson & Feduccia 1980). Some authors concluded that flamingos and the Anatidae evolved independently from several ancient waders. Olson & Feduccia (1980) proposed that flamingos were neither long-legged geese nor bizarre storks, but were related to the Charadriiformes and derived from ancestral 'transitional shorebirds' (Feduccia 1996). They proposed classifying the flamingos as Phoenicopteridae among the Charadriiformes, next to the Recurvirostridae (stilts and avocets), which they suggested were the ancestors of flamingos. This opinion stemmed from the rather strange resemblance that flamingos have in various aspects of their biology to the Recurvirostridae and in particular to the Australian Banded Stilt. This wader has a breeding biology that is similar in several respects to that of flamingos, although its leg muscles are quite different (Burbridge & Fuller 1982). The Banded Stilt frequents salt lakes and saltpans, usually in large, dense flocks, where its main food is the brine shrimp. It breeds opportunistically in huge colonies on low islands in inland salt lakes, and was discovered breeding only in 1930. Nest densities are high, up to 18 per m², and one colony found in 1980 was estimated to have a total of 179,000 nests. Both eggs and newly hatched chicks are white, features unique among shorebirds. Furthermore, the chicks form crèches just as flamingos do. However, in spite of what might seem convincing evidence, these similarities could also have evolved through convergence.

Or are they just flamingos?

In view of this ambiguity, Storer (1971) (see also Verheyen 1959) proposed that flamingos should be considered a separate order, the Phoenicopteriformes, which he placed between the Anseriformes and the Ciconiiformes. This opinion is still largely accepted today, and is further supported by studies in molecular biology. Mainardi (1962) first observed that on the basis of immunological data, the Greater Flamingo showed links with both geese and larger wading birds. Sibley *et al.* (1969) then concluded, on the basis of the egg-white proteins and haemoglobin, that flamingos were related to both the Anseriformes and the Ciconiiformes, but were closer to the latter. Sibley & Ahlquist (1990) carried out the technique of DNA hybridisation on flamingos, the Banded Stilt, storks and ducks and confirmed that flamingos were more closely related to the Ciconiiformes than to the Anseriformes and were not related to the Charadriiformes. However, the taxonomic situation of flamingos is far from being clear-cut. First, more recent research has cast some doubts on the validity of phylogenies based on DNA hybridisation (Houde *et al.* 1995). Second, two distinct molecular phylogenies have now provided contrasted views of the relationships of flamingos to other bird species. According to Van Tuinen *et al.* (2001), the closest living relatives of flamingos are, very surprisingly, the grebes. Johnson *et al.* (2006) from their studies of lice, agree that flamingos have an ancestor which they share with grebes. Mayr (2004) also assumes this sister-group relationship and believes that flamingos evolved from a highly aquatic ancestor. Sangster (2005) further agrees to this flamingos and grebes relationship and even suggests 'Mirandornithes' (wonderful bird) as a new clade name. However, a study based on a particular locus (FGB-int 7) has concluded that the flamingos are more closely related to nighthawks, swifts and doves than to any of the above-mentioned birds (Fain & Houde 2004). It is therefore difficult to conclude exactly where in the classification of birds flamingos truly lie, especially as a debate also exists about the validity of molecular phylogenies (Feduccia 1996); thus the precise relationship of flamingos to other groups remains to be clarified. Given that there is very little consensus about the taxonomic status of flamingos, we tentatively consider them a suborder of the Ciconiiformes, although some authors (e.g. del Hoyo *et al.* 1992) still consider them to belong to a separate order designated as Phoenicopteriformes, placed between the Ciconiiformes and the Anseriformes.

Systematics

As with the classification of flamingos as a group, their division into species has not been without controversy. Until recently, most references recognised five species of flamingos belonging to three separate genera. The type species *Phoenicopterus ruber* was first described by Linnaeus in 1758, in the Bahamas. This species is the only one of the five that occurs in both the Old and the New Worlds. The Old World race was for a time considered to be a separate species, *Phoenicopterus antiquorum* (Temminck 1820), but today *Phoenicopterus ruber* is generally considered to consist of two separate subspecies, the Greater Flamingo described by Pallas in the Old World and the Caribbean (or American) Flamingo *Phoenicopterus ruber ruber* in the New World. The isolated Galapagos population of this species is also sometimes referred to as a separate subspecies, *glyphorhynchus*. The most striking feature of the latter, in marked contrast to all other flamingos, is their habit of breeding in small colonies of as few as three pairs (Valle & Coulter 1987, del Hoyo et al. 1992). The Chilean Flamingo *Phoenicopterus chilensis* (described by Molina in 1782) was formerly considered another subspecies of *P. ruber*, although it is generally recognised as a separate species today. The other species in this genus, the Lesser Flamingo *Phoenicopterus minor*, was described by Geoffroy in Senegal in 1798, and until recently was considered as belonging to a separate genus, *Phoeniconaias*. This is the only species that occurs exclusively in the Old World. The two remaining species, the James' or Puna Flamingo *Phoenicoparrus jamesi* (described by Sclater in 1886) and the Andean Flamingo *Phoenicoparrus andinus* (described by Philippi in 1854), both occur exclusively in the New World.

Even today, in spite of the use of molecular biology, the phylogenetic relations among the different flamingo species remain difficult to evaluate (Sangster 1997; Sheldon & Slikas 1997). The five species formerly separated into three genera— *Phoenicopterus*, *Phoeniconaias* and *Phoenicoparrus* (Kear & Duplaix-Hall 1975) — have been recently united by Sibley & Ahlquist (1990) under a single genus, *Phoenicopterus*. More recent results, based on DNA hybridisation (Sheldon & Slikas 1997), also suggest that the relationships among the five species are sufficiently close that they should not be separated into different genera. Ironically, this recent viewpoint was suggested decades earlier by the late Jean Delacour, who was the only participant at the Slimbridge Flamingo Symposium in 1973 who felt that all flamingos were sufficiently alike to be incorporated within the single genus *Phoenicopterus*. Finally, as this book goes to press, it is being recommended that the Greater Flamingo, just like the Chilean, should be considered a separate species from the Caribbean, on the grounds of consistent differences in plumage coloration and pattern, coloration of bill and legs and their displays and vocalisations (Knox et al. 2002). This book would then be a monography on *Phoenicopterus roseus* and not *Phoenicopterus ruber roseus*.

Although all flamingos are filter feeders, the three smaller species (Lesser, James' and Andean) have deeply keeled bills with very fine lamellae, and they feed on microscopic algae. The *Phoenicopterus* species have shallower-keeled bills lined with coarser lamellae, and their diet is mainly composed of a wide range of small or minute invertebrates which they collect either as benthos or in suspension in the water. This book is concerned with only the Old World subspecies, the Greater Flamingo, referred to here afterwards only as 'flamingo', but mention is made, where appropriate, of the other species.

MORPHOLOGY AND GENERAL DESCRIPTION

Flamingos are gregarious and live in flocks (or stands) often numbering hundreds or thousands of individuals. Their adult coloration and size make them easy to recognise, even to the inexperienced eye. Their necks and legs are longer, relative to body size, than those of any other group of birds. No other species has a bill quite like the flamingo's, and this feature alone will confirm identification of birds still in juvenile plumage. The shape and functioning of the bill and tongue, which have been described in detail by Jenkin (1957) and Zweers *et al.* (1995), are unique among birds, and the very specialised manner of feeding (see Chapter 6) is often likened to that of baleen whales. Flamingos feed by walking in water from just a few millimetres to 80 cm in depth. They can also swim buoyantly. They obtain their food from either the water column or the mud, which they may have to stir up. They have four toes and their webbed feet enable them, when necessary, to dabble like ducks and swans in order to reach the benthos. The neck is composed of 17 elongated cervical vertebrae, giving it a segmented appearance when curved. When flamingos lower their heads to feed, the bill is held pointing backwards and is practically upside down. The flamingo has the longest absorptive part of the alimentary canal, called the Meikel's tract, of all birds (Ridley 1954). It also has a nictitating membrane which protects the cornea when the bird is feeding in muddy water (Daicker *et al.* 1996). Greater Flamingos have 12 primary feathers, the outer one being very small, 27 secondaries and 14 tail feathers.

There is a popular belief that the habit of standing on one leg when at rest is something only flamingos do, although this is not true. Many species retract one leg when resting, but because they do so when concealed, or have shorter legs than flamingos, the posture may be less conspicuous. Birds do this just as much in hot weather as in cold, so this habit does not seem to have any thermoregulatory function, as some have suggested. Flamingos differ from other birds exhibiting this behaviour by having to jerk their leg and lock it in this position, which is often achieved only after several attempts. They also occasionally rest on their tarsi, young birds tending to do so more than adults. When the bird is sleeping, the neck is coiled around and the bill tucked between the scapulars.

PLUMAGE, AND WHY ADULT FLAMINGOS ARE PINK

The brightly coloured plumage of flamingos results from both the abundance of carotenoids in their natural diet and a particular efficiency in the metabolic processing of these compounds (Fox 1975). Like a few other waterbirds, such as Scarlet Ibis and Roseate Spoonbill, flamingos of all species have the capacity for biochemical oxidation of conventional plant carotenoids, the dominant being canthaxanthin. They also find important sources of carotenoid in their crustacean prey. These compounds colour not only the feathers and naked skin of the legs and face but also the blood and liver (Fox *et al.* 1967). Flamingos are particularly efficient at oxidising beta-carotene to form phoenicoxanthin and astaxanthin which are deposited in skin and plumage. A complete description of the role of carotenoids in the pigmentation of flamingos is provided by Fox (1975).

The pink coloration of adult flamingos may be of cryptic value. In some areas where flamingos feed, the water is reddened by the presence of algae and/or bacteria, and birds with a dominantly pinkish or reddish plumage are less conspicuous than would be birds in dark plumage. This is particularly true over long distances in the vast playas where the species breeds, and where the light is often bright and hazy. Chicks in their grey down may be more conspicuous than adults, but the dense crèches which the chicks form during the day resemble from a distance an island rather than a flock of birds, at least to the human eye.

Plumage development in flamingos had not been studied in detail in the field until recently. Birds banded as chicks in the Camargue, and therefore of known age and origin, were observed throughout the Mediterranean region over a period of 12 years. Coloration of plumage and bare parts was noted, primarily by the same observer, whenever the banded birds were seen at close range and in good light. Over 4,000 banded birds were assigned to the appropriate category whenever a change of plumage was observed, and observations were continued until several cohorts of banded birds acquired full plumage. Also included in the dataset were 155 ringed flamingos which died during a cold spell in southern France in January 1985.

Plumage development

Flamingos acquire their definitive adult appearance only after going through a sequence of juvenile and immature plumages (Plate 4). There is some individual variation in plumage coloration, especially in immature or sub-adult plumages. These differences, however, are so slight and insignificant that only the nine most frequently encountered plumages are illustrated and described in Table 1 (from Johnson *et al.* 1993).

Even after going through the sequence of plumages described, not all individuals manifest a very pronounced pink coloration. This coloration is not a prerequisite for breeding, but probably reflects an individual's capacity to assimilate carotenoids as pigmentation. Bright coloration, in particular of the head and neck, seems to be restricted to certain individuals. It is not necessarily linked to age since brightly coloured birds have been seen with the characteristic dark leg joints of immatures. In the Camargue, birds in bright plumage are most obvious in autumn and winter, after they have moulted and are beginning to display.

There is no obvious difference between male and female plumages, although our data indicate that more females than males had acquired adult plumage by three years of age (Figure 2). However, by the age of five years, there was no difference in the number of males and females which had still not acquired full plumage (n = 27). Some 10-year-old individuals retained traces of immature plumage, and one 14-year-old bird still had dark leg joints.

When chicks hatch they are covered in white down but this turns grey when they are a few days old. However, one of the chicks hatched in the Camargue in 1986 was clad in white down until several weeks of age, and albino chicks have been reported by Andrusenko (*in litt.*) at the colonies on Lake Tengiz (Kazakhstan).

Table 1. Colour development of plumage and soft parts in the Greater Flamingo (see Plate 4 and Figure 2). From Johnson et al. 1993).

Age (months)	a 2–5	b c.5	c c.10	d c.15	e c.20	f c.35	g c.40	h >40	i >40
base of bill	grey	grey	grey	grey or pinkish-grey	pale greyish-pink	pink	pink	deep pink	deep pink
nape	grey	off-white or grey	grey	grey	grey	white or grey	pinkish-white	pinkish-white	pink
neck	grey	grey	off-white	white	white	white	white	pinkish-white	pink
back	mottled grey-brown	mottled light grey-brown	off-white	off-white	white	white	pinkish-white	pinkish-white	pinkish-white
wing panel	grey	grey	grey	pale pink	reddish-pink	crimson	crimson	crimson	crimson
breast and belly	off-white or buff-grey	off-white or buff-grey	white	white	white	white	white to pinkish	pinkish-white	pinkish-white
legs	black or grey	black or grey	black or grey	grey	greyish-pink, scaly	pale pink	pink	deep pink	deep pink
leg-joints	black or dark grey	black or dark grey	black or dark grey	black or dark grey	darker than legs	pale pink or sombre	pink	deep pink	deep pink

Juvenile flamingos have mottled warm-buff plumage with white underparts and little or no pink in the wing-coverts. They have a grey base to the bill and either black or grey legs with darker tarso-metatarsal joints. At this stage, they are between 2.5 and 10 months of age and have not reached full size. One can be sure that they are young of the previous breeding season, even though they may be observed hundreds of kilometres from where they were hatched. The plumage of immature birds (or sub-adults) is intermediate between juvenile and full adult plumage. The iris is brown. The pink of the bill and legs is pale, and the latter have a scaly appearance. The body feathers are white, many birds having a greyish nape, and the wing-coverts are pink or reddish, not scarlet. This plumage is characteristic of birds aged 2–3 years, but some individuals may keep a trace of subdued plumage or bare parts until much later in life.

In full adult plumage, the head, neck and body are white with a pink hue, which is usually most intense on the head and neck. The primaries and most of the secondaries are black, the inner secondary feathers being pink. The wing-coverts are scarlet, the base of the bill and the legs are deep pink, and the iris is yellow.

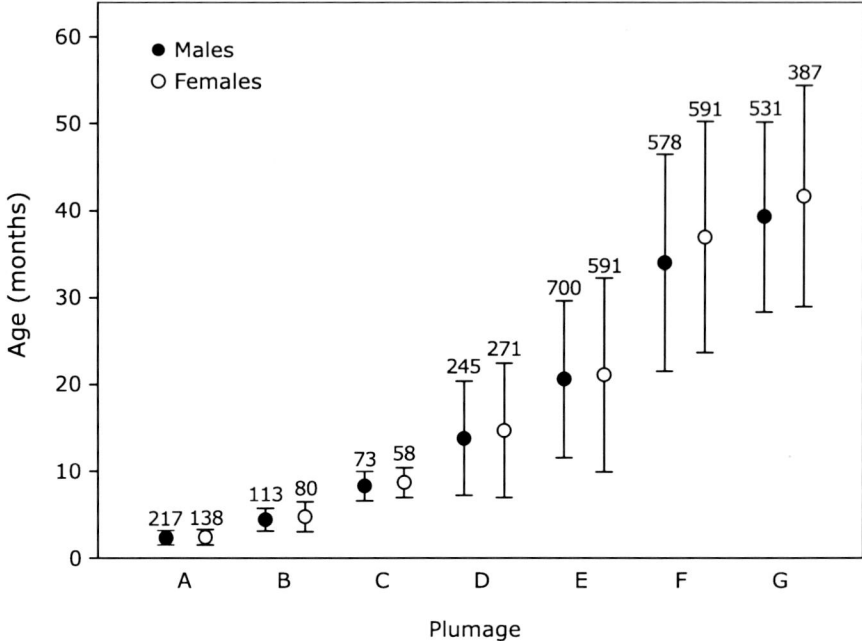

Figure 2. Comparative development of juvenile (A–C) and immature (D–G) plumages between males and females in the Greater Flamingo. Vertical bars show standard deviations. From Johnson et al. 1993.

Some adult flamingos can be seen with pale pink wing-coverts rather than scarlet. The colour of the wing-coverts is obvious when the wings are spread in display posture. This pale colour may be the remains of immature plumage; however, in captivity, this phenomenon occurs in adult birds that have fed a chick the previous season over an unusually long period of time and is thought to be due to the parents not being able to assimilate enough carotenoids to maintain their own coloration (A. Studer-Thiersch pers. comm.). The same phenomenon is commonly observed in captive flocks of Caribbean Flamingos.

Moult and flightlessness

The flamingo's first moult, as a chick, sees the original white down turn to grey at just a few days of age. This grey down is worn for about two months before being replaced by the juvenile plumage which allows the young bird to take wing at the age of 2–3 months. The colour of the contour feathers then changes progressively, as the moult of head and neck feathers, and perhaps also the scapulars, is gradual and almost continuous.

There have been no recent or in-depth studies on moult in flamingos. However, it is known to be extremely variable (Cramp & Simmons 1977), with no clear pattern emerging either on the frequency of moult during a bird's life, or on whether the birds undergo a progressive or a simultaneous moult of the flight feathers (primaries and secondaries). It is also not known how the moult relates to the breeding cycle. Since a simultaneous moult results in flightlessness for up to four weeks, this manner of moulting is restricted to areas where there is an abundance of food and a lack of disturbance. Lake Uromiyeh (Rezaiyeh) in Iran (Scott 1975) and Lake Tengiz in Kazakhstan (Demente'ev *et al.* 1951) have long been known as areas where adults become flightless and they are captured at these sites for ringing. Brown (1975) reported simultaneous moult in East Africa but gave few details. In the Camargue, many of the breeding flamingos need to fly to their foraging sites, and it is unlikely that breeding birds would moult in such a way while still feeding chicks. Yet for the past 20 years, hundreds or even thousands of full-grown flamingos have moulted their remiges simultaneously in the Etang du Fangassier during the summer (June to August). These birds are easily recognisable, since they show fear of intruders at a much greater distance than non-moulting birds, wing-flapping and revealing either gaps in their wings or a complete lack of flight feathers as they run away from the observer. Flightless adults are now caught in the Camargue each year when the chicks are rounded up for banding. This phenomenon is not new in the area, since in 1828 Crespon (*in* Allen 1956) observed 30 flamingos on the Vaccares unable to fly, and caught some of them. Nevertheless, the majority of flamingos in the Camargue maintain the ability to fly by renewing their remiges progressively.

In temperate areas, wing moult is limited to the summer. In the Camargue, we have observed birds in flight with primaries or secondaries missing from May to September, with the peak of moult in mid-July. Presumably, these flights were composed mainly of breeding birds, moving between the colony or crèche and their foraging areas. Feathers were lacking in either one or both wings but seldom symmetrically. At Lake Uromiyeh, Iran, Ashtiani (1977) observed during the summer of 1976 that non-breeding flamingos started moulting and became flightless three weeks earlier than breeding birds, which moulted only when their chicks were in the crèche. At Lake Tengiz, Andrusenko (1981) stated that it is immature and other non-breeding birds that undergo simultaneous moult of their flight feathers, and that breeding birds renew their primaries progressively. They can, however, barely fly during renewal of the outer primaries. In South Africa, Middlemiss (1961) observed that the 11 long primaries are renewed progressively, one at a time, from the proximal to the distal. We have also found this to be the case in immatures found dead in the Camargue, but have yet to observe it in adults.

Parasites

Information on both endo- and ectoparasites in flamingos is scarce. Ashford *et al.* (1994) investigated the incidence of blood parasites in 34 chicks and six adults in the Camargue and found no evidence of haematozoan infection. Several species of cestodes are known to parasitise flamingos in the Camargue (Robert & Gabrion

1991): *Flamingolepis liguloides, F. caroli, F. flamingo* (Hymenolepididae), and *Gynandrotaenia stammeri* (Progynotaeniidae). Their numbers vary widely and they tend to be distributed in different parts of the gut. The most abundant species is *F. liguloides*, and up to 4,000 adults have been found in the gut of one flamingo (Gabrion *et al.* 1982). The four species have complex life cycles and use the brine shrimp as an intermediate host. Brine shrimps become infected on ingesting parasites' eggs, which are released in the water with flamingo faeces. Flamingos eventually become infected when feeding on parasitised shrimps. A study carried out in the salt marshes of the French Mediterranean coast (Thiéry *et al.* 1990) revealed a positive relationship between the probability of infection of brine shrimp with *F. liguloides* and the local population density of flamingos.

Clay (1974, 1975) reviewed the distribution of feather lice occurring on flamingos and briefly discussed their relevance to flamingo taxonomy. We collected external parasites from 250 individuals during the banding operation in the Camargue in 1997, in an attempt to document the diversity and distribution of feather lice occurring on flamingo chicks. Five species of lice, belonging to two distinct families and four genera, were found to parasitise the Greater Flamingo (Figure 3, Table 2), with species tending to be located in distinct areas of the chick's body. For instance, the genus *Colpocephalum* was essentially found in wing feathers, whereas *Anatoecus pygaspis* was found mainly on the head, and *Anaticola phoenicopteri*, by far the most abundant species, occurred mainly on the remiges and underwing-coverts (Palma *et al.* 2002). All were already known to occur on the Greater Flamingo (Clay 1975).

BILL PATTERNS

Greater Flamingos have distinctive bill patterns, and adults have a well-defined line separating the black tip from the pink basal two-thirds of the bill. The pattern of the upper edge of the black tip differs markedly among individuals, as has been observed for Bewick's Swans (Evans 1977; Rees 1981). This feature can be used for identifying small numbers of flamingos if they can be seen face on and at close range (Figure 4) (Johnson *et al.* 1993).

Table 2. Species of feather lice known to parasitise the Greater Flamingo.

Family	Species
Menoponidae	*Colpocephalum heterosoma* Piaget
	Colpocephalum salimalii Clay
	Trinoton femoratum Piaget
Philopteridae	*Anatoecus pygaspis* Nitzsch
	Anaticola phoenicopteri Coinde

Figure 3. Feather lice. Three species of feather lice regularly found on Greater Flamingos. (a) *Anaticola* sp. (b) *Colpocephalum* sp. (c) *Anatoecus* sp.

30 *The Greater Flamingo*

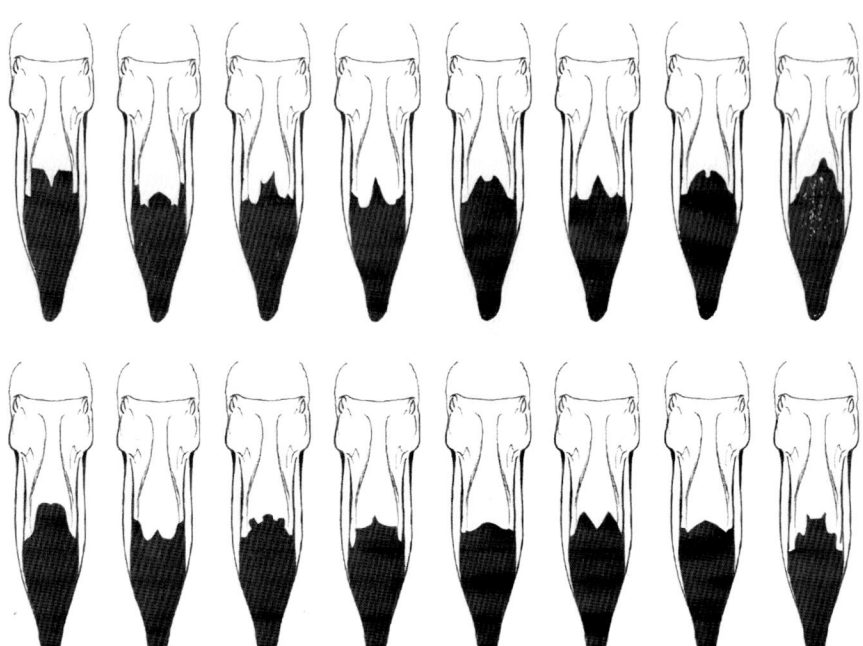

Figure 4. Vignette of bill patterns. The pattern of the upper edge of the black tip of the bill varies quite markedly between individuals, a feature which can be used for identifying small numbers of birds at close range.

Measurements

On average, males are taller and bigger than females, although size distribution overlaps between the two sexes. A large male may stand as tall as a small man, and one freshly dead bird found in the Camargue in May 1977 measured 201 cm from the tip of the bill to the toes. This individual had a wingspan of 187 cm, standard wing measurement of 457 mm, a tarsus of 390 mm, and weighed 4,450 g. These measurements exceed the range given in identification guides and handbooks (e.g. Kear & Duplaix-Hall 1975, Cramp & Simmons 1977), which were based on a limited number of specimens from southern France (Gallet 1949) and India, or museum specimens (Ali & Ripley 1978). Height is usually given as 90–155 cm and wingspan as 140–165 cm. The average wing measurement, from carpal to longest primary, is given as 428 mm (406–464 mm) for males and 380 mm (360–396 mm) for females.

Data collected in southern France from birds which died during a severe cold spell (January 1985) are particularly valuable because of the large sample size and because the birds were sexed by cloacal inspection and aged according to their plumage (Table 3). Since most, or all, died of starvation, weights were below those of healthy wild birds by about 20%, i.e. males averaged 2,727 g (range

Table 3. Measurements of flamingos found dead in southern France during the cold spell of January 1985. The values for tarsus length are slightly overestimated, by about 8% (see text for explanations).

	Juveniles	Immatures	Adults
Wing length			
Males	420.16 (±16.02) n = 62	423.55 (±14.52) n = 324	433.29 (±13.57) n = 1,111
Females	401.20 (±16.92) n = 60	393.26 (±12.28) n = 250	400.30 (±12.90) n = 669
Tarsus length			
Males	265.39 (±28.90) n = 62	325.97 (±26.31) n = 321	338.12 (±23.12) n = 1,110
Females	249.26 (±22.70) n = 61	274.73 (±19.51) n = 250	280.12 (±20.82) n = 666
Weight			
Males	2,197.28 (±259.98) n = 46	2,579.01 (±288.81) n = 221	2,727.07 (±286.96) n = 737
Females	1,868.89 (±249.24) n = 36	1,960.92 (±260.37) n = 168	2,064.64 (±240.38) n = 494

2,600–3,125 g) compared to a previous sample of 17 males which died accidentally and which averaged 3,429 g (range 2,500–4,450 g). The wing length of the winter 1985 birds was taken by the maximum chord method: flattening the primaries against a stopped rule hooked under the carpal joint, and straightening out the natural curve of the primaries by gentle pressure, then reading the length of the longest (second) primary. Tarsus length was not the orthodox measurement but was done in a simplified way, to minimise discrepancies caused by the birds' being handled by several different people. A stopped rule was hooked behind the tibia with the tarsus flexed perpendicularly and the toes bent down at 90°. This was the method employed by Richter & Bourne (1990), who sexed the adult Greater Flamingos at Detroit Zoological Park by weight and linear measurements as well as surgically. Their measurements of wing and tarsus were confined to the limits we have given, but they found no overlap between sexes. They considered wing length to be the best indicator of sex, that of females at Detroit being a maximum of 409 mm and that of males a minimum of 415 mm. Females had a maximum tarsus length of 311 mm and males a minimum of 313 mm. We have calculated that this length is on average 8% longer than true tarsus length. Males are on average taller and heavier than females by about 20%.

Both weight and length of tarsus increased between juveniles and immatures, and between immatures and adults in both males and females, as did wing length in males, indicating that flamingos continue to grow until they are in adult plumage at age 3–4 years. This contrasts with measurements taken of hand-reared chicks in Namibia (Berry & Berry 1976) and with observations of birds in Basel

Zoo (Studer-Thiersch 1986), but in these two studies, sample sizes were much smaller.

VOCALISATIONS

The flamingo's call is usually heard as a loud, rather goose-like, nasal double honk, males having a somewhat deeper voice than females. However, the call varies according to the bird's activity, whether displaying, announcing departure, alarming, in attendance at the nest or caring for the chick. Vocalisations are a particularly important component of flamingos' social behaviour, and it has been suggested that they are necessary for both mate and parent–offspring recognition (Brown 1958; Studer-Thiersch 1974). Mathevon (1997) analysed the temporal and frequency patterns of contact calls, referred to as a 'nasal double honk ka-ha' (Brown 1958; see also Studer-Thiersch 1974), and showed that vocal recognition between individuals is made possible thanks to several parameters, including call duration, slow amplitude modulation, spectrum bandwidth and the repartition of energy among harmonics. No single parameter was, however, sufficient to discriminate between individuals. This multi-parametric coding shows several acoustic convergences with other colonial species such as the Emperor Penguin or the Barnacle Goose, and is considered an adaptation to the acoustic constraints imposed by coloniality. In particular, the concentration of energy in low frequencies seems well adapted to colonial life, because low frequencies can propagate through the medium of birds' bodies better than higher frequencies (Mathevon 1997). Repetition maximises the efficiency of the transmission and the calls are usually emitted sequentially. Flamingo colonies of hundreds or thousands of pairs can be heard more than a kilometre away in favourable weather.

LIFE HISTORY

COLONIALITY

One of the most striking features of the lifestyle of flamingos is their high degree of colonialism. Colonial breeding has emerged in the evolution of several independent avian lineages, and is particularly frequent, but variable, among waterbirds (Lack 1968b; Danchin & Wagner 1997). Some waterbird species typically form large, dense colonies, whereas others are more flexible in their social organisation, nesting in large groups, loose aggregations, or even solitarily. All flamingo species show several features that are characteristic of so-called 'obligate colonial species' (Zubakin 1985; Siegel Causey & Kharitonov 1990). Except in the Galapagos, flamingos congregate to nest in extensive tightly packed groups and rarely attempt to breed solitarily. The size of the defended territory around the nest is typically small, often less than the neck length of an adult bird (Swift 1960). Breeding success also seems to depend on a minimum colony size. Strong social stimulation may be an important prerequisite for breeding activities. For instance, display activities were triggered in small captive flocks of Lesser Flamingos when mirrors provided increased visual stimulation (Pickering & Duverge 1992).

Unlike most colonial waterbirds, flamingos make no attempt to conceal their nests. Birds defaecate at the edge of the nest and do not remove eggshell remains after the chicks hatch, so nests are very conspicuous. Despite their highly gregarious behaviour, flamingos do not cooperate in defending the colony against predators, but instead show individual passive defence, and are seldom able to deter predators from eggs and chicks (Salathé 1983). About three weeks after hatching, the still-dependent young are left unattended during the day by their parents. They congregate in nursery groups, or crèches, an unusual form of social organisation, and parents come back only for short, albeit regular, visits to feed them in the evening or at night. During the day, few adults accompany the crèche. When threatened by a predator or disturbed, the chicks pack very tightly together. Features such as dense nesting, creching behaviour, absence of active defence, and absence of eggshell removal are all thought to be ancestral traits, and tend to occur in other species showing obligate colonial nesting.

The adaptive significance of coloniality in waterbirds has been widely discussed (Siegel-Causey and Kharitonov 1990; Burger and Gochfeld 1992; Rolland *et al.* 1998). Two main advantages have been proposed to explain the evolution of colonial breeding: improved exploitation of food resources and enhanced protection against predators. In flamingos, it is not clear how this combination of factors explains the evolution of colonial breeding. Evidence from the fossil record suggests that early ancestral flamingo species were breeding in very large colonies in environments subject to unpredictable rise in water levels (Feduccia 1980). Today, flamingos tend to exploit food resources in unstable environments and are opportunistic breeders in response to unpredictable fluctuations in food availability and water levels. Because they are filter-feeders and prey on small organisms, flamingos have to spend long periods foraging every day. The occasionally rapid drying out of foraging areas around nesting sites can force the birds to cover long distances to find food, and to leave their chicks unattended. In this situation the gathering of chicks into a crèche may limit vulnerability to predators. Colonial breeding may therefore have evolved because it allows chicks to join together when parents have to forage far afield.

The scarcity of suitable breeding sites, offering safety from both ground predators and flooding, might also have been an important factor in the initial evolution of colonial breeding in flamingos. The importance of flooding risk is indicated by the building of nest mounds to prevent eggs and chicks from being submerged. The susceptibility of flamingos to ground predators is attested to by observations of breeding failure when nesting is attempted in sub-optimal sites (i.e. those not surrounded by water) and may explain the extreme sensitivity of flamingos during colony site selection. Swift (1960) dismissed lack of space as a deciding factor in colony site selection because colonies often consist of tight groups of nests separated by apparently suitable, but unused, areas. However, extreme density of birds may be effective in preventing aerial predators from easily landing within the colony. For instance, Yellow-legged Gulls in flight have limited success harassing birds breeding in central locations, but are more efficient at attacking peripheral breeders from the ground (Salathé 1983). A potential disadvantage of dense packing of nests is that small inter-nest distance may prevent the birds from taking wing among other incubating birds. However, because their remote breeding sites are generally situated in open landscapes, flamingos are likely to detect approaching predators at a distance, so that the need for take-off space around the nest is reduced. Other benefits from coloniality in relation to predation risk include the 'dilution' and 'swamping' effects. With increasing colony size, the probability of an individual egg or chick being the victim of predation decreases. The same is true for chicks in crèches, which pack more tightly together when facing a predator. Colonial breeding also involves some degree of synchronisation among breeders in the date of egg-laying. If colonies are highly synchronised, predators may be overwhelmed by the number of eggs and chicks available. However, those two advantages assume that the number of predators attracted to a colony would increase less rapidly than colony size.

Once colonial breeding evolved in relation to predator avoidance and the need to minimise the risk of flooding, secondary benefits such as improved exploitation of food resources or increased opportunities for mate selection may have reinforced this behaviour. On the other hand, coloniality may have first evolved in relation to mate selection (Danchin & Wagner 1997). It has been suggested (Draulans 1988; Wagner 1993) that coloniality in birds may have evolved through the female's preference for nesting close to conspecifics in order to increase opportunities for mate selection and extra-pair copulations. However, the relevance of this hypothesis for several groups of colonial waterbirds, including flamingos, remains elusive.

Habitat

Nesting sites

Flamingos breed in dense colonies established in wetlands lying in areas of high evaporation, in arid or semi-arid regions. Salinity in such areas is far greater than in humid areas (Boyko 1966), and most nesting sites are in salt lakes, salt depressions or saltpans. Rainfall is generally sparse, irregular and unpredictable (Jenkin 1957). The birds are on or near the isotherm of 24°C during the warmest months (Allen 1956; Voous 1960), where length of day exceeds 12 h (Kear & Duplaix-Hall 1975). Colonies are established on raised ground or islands, which in saltpans are often remnants of former dykes, surrounded by water, generally shallow, which afford protection against terrestrial predators. These wetlands are usually rich in invertebrates, although in some parts of their range birds may forage far from the

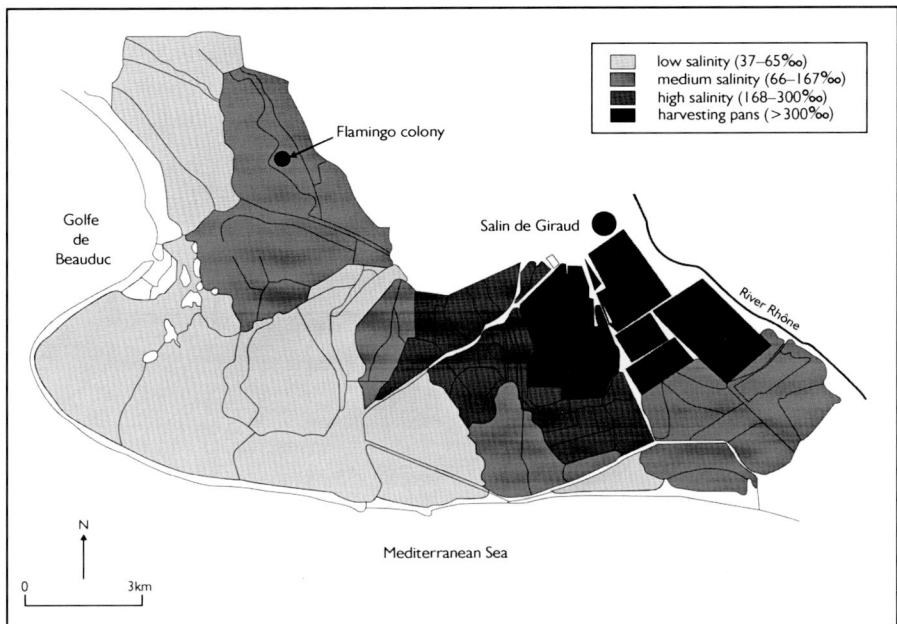

Figure 5. The Salin de Giraud, the most extensive saltpans in Europe (*c*.11,000 ha). The water, pumped from the sea at the Golfe de Beauduc, increases in salinity as it slowly flows through the system. Flamingos feed in all but the salt-harvesting pans, where there are no invertebrates.

Medium-salinity lagoons

In the range 67–167 g litre^{-1}, these evaporators are generally shallower than those of low salinity and often dry in winter. Their overlying substrate, particularly in the range 70–140 g litre^{-1}, is composed of a mat of blue-green algae. When wet, this can be very slippery and such areas are often abandoned by flamingos during periods of very strong winds. These evaporators are generally less extensive and more engineered than the low-salinity lagoons, with lower vegetation. The dykes are protected from erosion by stones. Many of the lagoons in the medium-salinity range are abandoned by flamingos in winter, but for a shorter period than the high-salinity lagoons. In those which retain some water over winter, ostracods and midge larvae occur as well as brine-shrimp cysts.

High-salinity lagoons

These ponds are characterised by their geometrical form and smaller size. The pink, almost red coloration, so obvious in many pans, particularly at the end of summer,

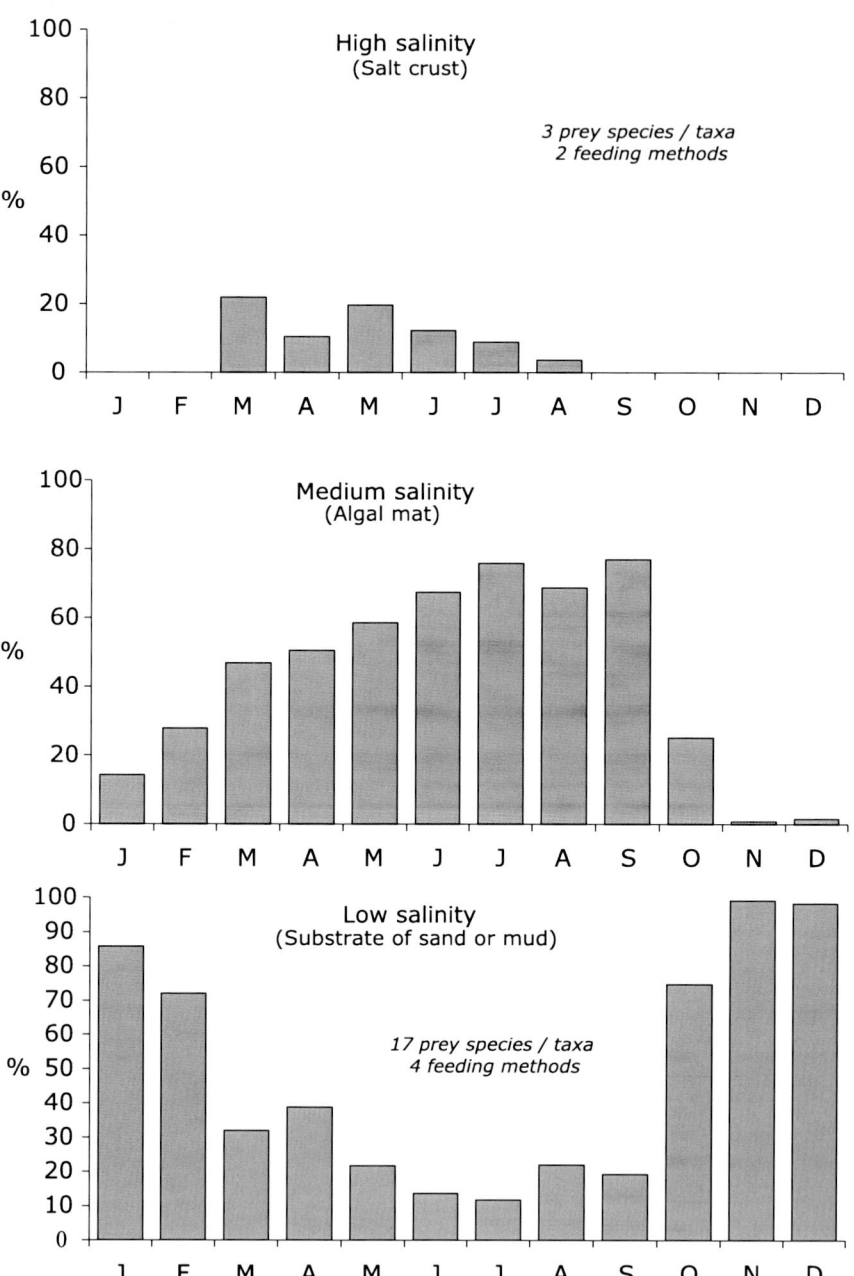

Figure 6. Pattern of occurrence of flamingos feeding in the Salin de Giraud according to the water salinity: low (37–65‰), medium (66–167‰) and high salinity (168–300‰) lagoons. The number of birds, censused at night using light intensifiers, varied from 1,200 in December to over 6,000 in June.

reflects the presence of the unicellular chlorophycean *Dunaliela salina* and/or the autotrophic bacterium *Halobacterium*. It is not caused by brine shrimps, as is sometimes popularly believed. The higher-salinity ponds in the Camargue are not used by flamingos in winter because there are no brine shrimps, the main prey here in summer. The shrimps hatch in early spring (February–March) when water temperature reaches 10°C, and they die off in autumn in October–November (Britton & Johnson 1987). Brine shrimps are therefore only available in southern France for 8–10 months of the year. Since there are no other invertebrates in the higher-salinity lagoons, these areas are abandoned by flamingos during the winter (Figure 6).

Densities of birds feeding nocturnally in the Salin de Giraud saltpans showed important temporal and spatial differences, with a maximum of 12.5 flamingos per hectare, but an average density during the yearly cycle of less than one bird per ha. Flamingos do not generally visit the harvesting beds, since the brine is too dense for even shrimps to survive.

CHAPTER THREE

Research

THE PIONEEERS AND EARLY YEARS OF DISCOVERY

Only half a century ago, relatively little was known about the breeding biology of flamingos in the wild. Colonies were established in places either still unknown to naturalists or difficult to access except by painstaking missions. However, by the end of World War II, ornithologists on three continents began venturing off the beaten track and making exciting discoveries. These are related in vivid accounts of both rewarding missions and less fruitful visits to outlying areas.

Salim Ali—the Rann of Kutch

In West India, Salim Ali (1945) visited 'Flamingo City' in the Rann of Kutch, where flamingos are known to have bred occasionally since at least 1935 (McCann in Ali 1945). He witnessed breeding in full swing and estimated that the colony held over 100,000 pairs. Ali and others revisited this site in 1960 and after many hours travelling by camel, reported even larger numbers breeding (Ali 1960; Shivrajkumar *et al.* 1960). These accounts stand as references to the two largest

Greater Flamingo colonies on record anywhere, with figures which remain quite bewildering.

Leslie Brown — the Rift Valley

In East Africa, Leslie Brown was just beginning to unveil the mystery surrounding the breeding of flamingos in the Rift Valley. He was one of the first naturalists to survey potential breeding sites using a light aircraft, which he taught himself to fly. He monitored breeding by the Greater Flamingo at Lake Elmenteita for 21 years (1951–71) and described hitherto unknown traits of the flamingo's breeding biology; for example, that adults feed only their own chicks (Brown 1958). He observed that chicks are capable of trekking long distances in very harsh conditions over soda flats (Lake Natron) and that adults can fly many kilometres to feed. Although not a researcher by profession, the observations and papers of this pioneer (e.g. Brown 1958, 1959, 1973, Brown *et al.* 1973) remain a solid scientific reference.

Domergue, Panouse, and Tixerent — the Sahara Desert

In North Africa, from 1948 to 1951, the French geologist Charles Domergue took the risk of exploring the vast and remote playas of the Sahara Desert in Tunisia, and was rewarded by discovering the breeding site of flamingos in the Chott Djerid (Domergue 1949a,b). A few years later, and at the other end of the Maghreb, in Morocco, Panouse (1958) provided the first evidence of Greater Flamingos breeding in the Iriki depression, and to the south, in Mauritania, Tixerent (in Naurois 1969a) discovered a colony on the rocky Grande Kiaone island standing in the Atlantic Ocean. Since then, unfortunately, the Moroccan site has apparently become unsuitable for waterbirds owing to upstream damming, but colonies prosper in Mauritania and offer a wonderful spectacle for those few who have witnessed flamingos breeding on islands in the sea, in an area that has since become the Banc d'Arguin National Park.

Etienne Gallet, George Yeates and Luc Hoffmann — the Camargue

In France, Etienne Gallet, an Arlesian naturalist, published his account of flamingos breeding in the Rhône delta, parts of which were still quite remote in the first half of the 20th century. His observations (Gallet 1949) ended much speculation by visiting ornithologists on the flamingo's true status in France. Gallet gave details of the size and success of the colonies established in the Rhône delta from 1914 to 1947, and he observed that flamingos degrade the islands on which they nest as well as the areas where they forage, two aspects of the flamingo's ecology which have since received much attention. His admirable book, which was translated into English and German, is a pleasure to read and is richly illustrated with remarkable black-and-white photographs.

George Yeates, a British ornithologist and photographer, visited the Camargue at the end of the 1930s (Yeates 1947) and in the 1940s (Yeates 1948, 1950). He became acquainted with Etienne Gallet and together they observed and photographed flamingos breeding. George Yeates was particularly interested in flamingos, and his books and papers relate not only his sometimes painstaking efforts to witness breeding but also acquaint the reader with much of what was known about the species half a century ago.

At the end of the 1940s, Luc Hoffmann, then an undergraduate at the University of Basel, Switzerland, also took a very active interest in the study and conservation of flamingos in the Camargue. They nested within 30 km of the Tour du Valat estate, which he was soon to acquire, and where in 1954 he founded a biological station. Luc Hoffmann continued the monitoring programme begun in the early part of the 20th century by Etienne Gallet, and he also started a ringing programme. These activities, discussed below and in Chapter 10, paved the way for the more elaborate studies which are being carried out today.

Jose Antonio Valverde — Spain

Although flamingos had for a long time been reported to occur in abundance in the marshes of the Guadalquivir in Andalusía (Chapman & Buck 1910), and also to occasionally attempt breeding there, albeit unsuccessfully (Yeates 1950; Bernis & Valverde 1954; Ferrer *et al.* 1976), this vast area of marshlands always seemed to be more important as a foraging rather than as a nesting place. One of the first reports of successful breeding by flamingos in Spain dates back to only 1963. This was the year that the late Jose Antonio Valverde, like Salim Ali and Leslie Brown a close friend of Luc Hoffmann, discovered the colony at Fuente de Piedra in the province of Málaga, 150 km east of the Guadalquivir Marshes (Valverde 1964). This inland lagoon, which was once converted into saltpans, only became favourable for breeding by flamingos after being abandoned by the salt industry. The discovery of this Spanish colony came at a time when the birds were experiencing problems in the Camargue. This was just the beginning of what was to become one of the most regular of the flamingo's breeding sites in the world, second only in importance to the Camargue. This lagoon was quite justifiably made a reserve soon afterwards, and a ceramic mosaic of some of 'Tono' Valverde's sketches of the 1963 colony adorns a wall at the entrance to the visitor centre (unveiled in his presence in January 1995).

Each decade since the mid-20th century has seen noteworthy developments in our knowledge of the flamingo's distribution and breeding. In brief, the 1950s saw the first substantiated reports of flamingos breeding in Kenya (Lake Elmenteita, Lake Natron), Tunisia (Fedjaj, Djerid), Mauritania (Kiaone Islands and Baie d'Arguin), Morocco (Iriki depression), Namibia (Etosha Pan) and Kazakhstan (Lake Tengiz). During the 1960s, colonies were discovered, or breeding was reconfirmed, in Spain (Fuente de Piedra), Tunisia (Sidi Mansour), Iran (Lake Uromiyeh), Afghanistan (Dasht-e-Nawar), Kenya (Lake Magadi) and South Africa

(St Lucia, De Hoop Vlei), and in the 1970s in Senegal (Kaolack), Spain (Doñana, Alicante), Tunisia (Sidi el Hani), Egypt (El Malaha), Turkey (Lake Tuz, Sultansazligi) and Botswana (Makgadikgadi Pans). The past two decades have also seen reports of new colonies, mostly in Europe, but these are sites newly colonised by flamingos, as opposed to newly discovered colonies.

CONTEMPORARY RESEARCH—WHAT HAVE WE LEARNED?

If the early years can be described as the years of discovery, where huge and previously unknown colonies were discovered in remote parts of the world, the past few decades (1970–2000) can be characterised as the years of description, during which our understanding of the basic life history of flamingos has unfolded. This important knowledge has become the foundation of our efforts toward conservation and management of flamingos throughout the world. In fact, a large part of this book represents a summary of this information, and, rather than anticipate ground which will be covered later in the book, we will summarise projects, including monitoring, undertaken in each region over the past 30 years, and refer the reader to the chapters containing the results of these efforts. Our division of the flamingo's range into regions (Figure 7), which we respect throughout this book, is done purely for convenience and in no way indicates boundaries or sub-populations.

West Africa: The most important countries for flamingos are Senegal and Mauritania, which have been fairly well covered by surveys in winter and during the breeding season. The size and breeding success of the colonies established on the Kiaone Islands in Mauritania, presently the most important colonies in West Africa, have been quite closely monitored (see Chapter 12, sites 3, 4), mostly by aerial surveys. Egg-laying at the colonies of the Banc d'Arguin, which lie at the southern limit of the Western Palearctic, occurs in spring, often slightly later than at some of the Mediterranean colonies to the north, whereas records of laying just to the south, in the Aftout-es-Saheli and in Senegal, which lie within the Afro-tropical region, cover a much wider period. The timing of breeding is discussed in Chapter 8, but there have been no studies of the link, if any, between this and the availability of food resources on the tidal mudflats of the Banc d'Arguin. There are no current research programmes on flamingos in West Africa, but we envision the organisation of more thorough searches for banded flamingos at the colonies in Mauritania, and the eventual initiation of banding there.

Mediterranean: This region is well covered by surveys in winter and during the breeding season (Johnson 1997b; sites 6–21), with the exception of Libya and Egypt, where no regular surveys are carried out. Research projects have been underway for many years in France (Camargue) and Spain (Fuente de Piedra) and

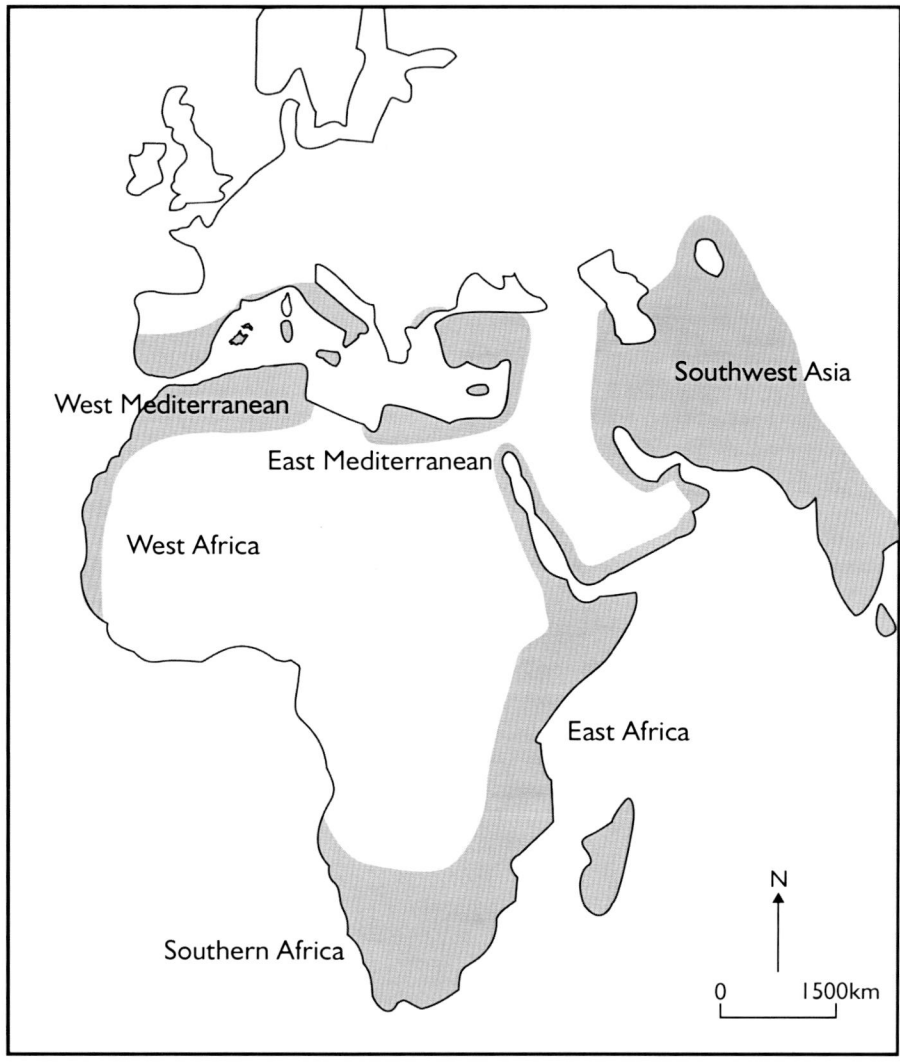

Figure 7. The greater Flamingo's world range and the geographical regions used throughout this book.

were begun more recently at breeding sites in Italy, with chicks being banded regularly or occasionally at five colonies. Data from these projects, and in particular those resulting from the Camargue study, have been widely used in this book: on dispersal (Chapter 5), foraging (Chapter 6), mating (Chapter 7), breeding behaviour (Chapter 8) and demographic parameters (Chapter 9). In Spain, the movements of several flamingos have been tracked by satellite in a study of dispersal from Fuente de Piedra and to complement the data on foraging obtained by the

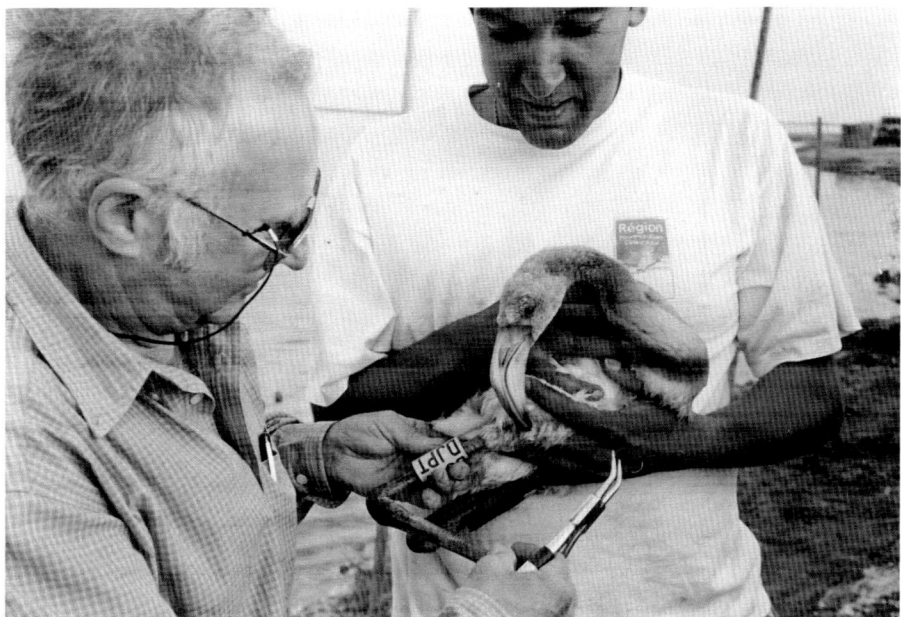

Figure 8. A coded leg band. Uniquely-coded leg-bands placed on chicks since 1977 in France, Spain and Italy allow flamingos to be identified in the field at up to 400 m. (*F. Mesléard*)

observations of banded birds (J. Amat pers. comm.), but this study is too recent for the results to be discussed here.

Asia: Although many Asian countries participate in the International Waterfowl Census (IWC) organised by Wetlands International, some of them are not well covered for logistical or political reasons, and there are also some major gaps during the breeding season, with no regular monitoring at any of the potential breeding sites and no flamingo research projects presently in place. Dates of egg-laying at the colonies in Kazakhstan, Iran and Afghanistan are discussed in Chapter 8, while observations of flamingos undertaking long-distance flights, and the recoveries of marked individuals from Lake Uromiyeh (Iran) and Lake Tengiz (Kazakhstan) shed some light on dispersal from these breeding sites and movement patterns in Asia.

East Africa: Ethiopia, Kenya and Tanzania, the most important countries for flamingos, participate in the IWC, but many wetlands which are known to sometimes harbour large numbers of flamingos are not covered because of their remoteness (i.e. Lake Natron, Lake Turkana). There are no research projects at present, and reports of breeding are very sparse and anecdotal, so that much of what is known about the breeding biology of flamingos in East Africa we owe to Leslie Brown.

Southern Africa: The more important countries for flamingos participate in the IWC surveys but, as in East Africa, there are gaps in coverage. The major sites for breeding, Etosha Pan in Namibia and Makgadikgadi Pan in Botswana, are regularly surveyed (Simmons 1996; McCulloch & Irvine 2004) and satellite tracking has been carried out at the latter site (McCulloch *et al.* 2003).

DISTRIBUTION AND NUMBERS

COUNTS AND SURVEYS

Building upon Kahl's (1975a) brief description of flamingo distribution, our knowledge has been enhanced by the more recent publications of Birdlife International on important bird areas (e.g. the Middle East, Evans 1994; Africa, Fishpool & Evans 2001), and by the series of directories edited by the World Conservation Union (IUCN) (e.g. Asian wetlands, Scott 1989) and/or United Nations Environment Programme (UNEP), IWRB (now Wetlands International) (e.g. Hughes & Hughes 1992, Scott 1995). The main source of information on the distribution and numbers of flamingos is the IWC initiated by Wetlands International, and described briefly below. The results of these surveys, made mostly during the past decade, are presented and discussed in Chapter 4.

The International Waterbird Census (IWC)

In the 1960s, Wetlands International (formerly IWRB) initiated a programme of IWC surveys carried out on specified dates each year by a network of amateur and professional ornithologists (Atkinson-Willes 1969). The IWC was launched in north-west Europe in 1967, and is the longest-running internationally coordinated biodiversity monitoring programme in the world. It was progressively extended, and currently most European and many Asian and African countries make at least one annual census, in mid-January. The IWC now covers a large part of the Greater Flamingo's world range, although regrettably the more inaccessible wetlands where flamingos occur have often been omitted from the surveys. We have ourselves participated in these waterfowl surveys. These counts provide the main source of quantitative data available (see Chapter 4) and allow us to plot the flamingo's distribution in January, during the northern winter, with increasing precision. Since 1965, all sites where flamingos occur in Mediterranean France have been surveyed in the context of the IWC, using a combination of aerial and ground surveys, and since the early 1970s most other Mediterranean countries have organised winter counts, albeit with some gaps in coverage. Our analysis of these data (see Chapter 4) called for a revision of the world population estimate of the Greater Flamingo.

The French surveys were developed into monthly counts over a period of three years (September 1977–August 1980) in order to monitor fluctuations in numbers and distribution of flamingos within the country during the yearly cycle. In addition, since 1977 there has been a survey most years of the number of birds present in May, for comparison with the number of pairs breeding (Chapter 4). These counts have also been undertaken partly on the ground (saltpans) and partly by aerial surveys.

Breeding bird surveys

Data on breeding by flamingos have been collected in a variety of ways, from random reporting at some colonies to more systematic recording at the sites which are being monitored annually (Mediterranean, West Africa, Botswana, Namibia). Records of breeding are maintained by the Flamingo Specialist Group (FSG; Wetlands International/World Conservation Union, Species Survival Commission, BirdLife International) and regularly published in the newsletter but, as with the IWC, there are gaps in coverage of the flamingo's range. All reported cases of breeding, and some unpublished data which we ourselves have collected, are given in the site sheets (Chapter 12). The more important monitoring and research programmes are indicated below.

France: The size and the breeding success of colonies in the Camargue have been monitored annually since 1947. The numbers of breeding pairs have been assessed either by visiting the colonies to count the nest mounds after breeding was completed or by aerial photography. Counts of nest mounds on the ground were made by placing a marker, such as a small pebble or a bean, on each nest mound judged by the observer to have been occupied that year. In such a manner no nests were overlooked or counted twice (Hoffmann 1959). The major disadvantage of this method, used until 1976, was that the less elaborate nests could have deteriorated and been overlooked; for this reason, it was abandoned after 1976.

Aerial photography has been used in the Camargue since 1977, when it became apparent that flamingos were gradually becoming accustomed to aircraft, and did not scare once they had laid eggs. Flights are made when all potential breeders are incubating, and before the oldest chicks begin wandering away from the nest. The plane, or helicopter, circles over the colony at an altitude of 300 m (*c.* 1,000 ft), which is high enough not to disturb the flamingos. The flight is made preferably towards the middle of the day, when there are the fewest changes of partners at the nest. Photographs (Plate 2) are taken from vertically above the colony, usually with a small telephoto lens. As the plane circles overhead, an observer in the tower hide estimates the number of non-breeding birds on the island. These birds, and the number of nests with both members of a pair present, are subtracted from the total count. The count is made later by projecting the slide onto a sheet of paper or by making an enlarged print, and then ticking each bird individually (for which an electronic dot counter is most useful). These two counting techniques both reflect the maximum numbers of pairs attempting breeding simultaneously, with no measure of turnover during the season. This is our reference to colony size used throughout this book.

The success of the Camargue colonies has been measured by counting the year's chicks just prior to fledging. This was formerly done by visual estimations from the ground; in some years, when all the chicks in the crèche were captured and ringed (1959, 1961), the exact figure was known. Since 1977 the crèche has been photographed from the air (Plate 20). At this stage, chicks are too large to be vulnerable to predation by gulls and they are generally less tightly packed than when they were small. Thus, the number counted is probably very close to the number of birds actually taking wing.

Spain: The numbers of flamingos at the Fuente de Piedra Lagoon, both feeding in the lagoon and incubating, have been monitored by monthly surveys, counts being made from vantage points on the hills around the lake, and from the hide near the colony (see Rendón-Martos *et al.* 2000). The number of chicks raised has been assessed from ground counts, from aerial photographs or, as in 1994, by capturing the whole crèche.

Some cautionary remarks

All counting methods have their strengths and weaknesses. We point out some of the weaknesses here, not to criticise the various approaches that have been used but so the reader will remember, when interpreting the resulting population estimates, that these counts are not perfect. Three areas warrant caution: (1) gaps in coverage, (2) reliability of the counts themselves, and (3) interpreting what the counts represent. Gaps in coverage are an obvious weakness of both the IWC and the FSG summaries of breeding birds. Although the geographic scope has been expanding, the gaps in any given year can be substantial, particularly for earlier counts.

The second area of concern is the reliability of counts. For example, the IWC now covers a wide geographical area at given periods, but seldom covers vast areas simultaneously. Flamingos being well known for their high mobility, it is not clear to what extent birds may have moved between counts, either escaping the census, or having been counted twice. Estimating the number of individuals can also be extremely difficult in species that form large aggregations, as flamingos do. This may be even more difficult if some counters are inexperienced. Woodworth *et al.* (1997) showed that even with experienced counters, visual censuses overestimated the number of flamingos by a factor of two, although these authors were dealing mostly with the spectacular concentrations of Lesser Flamingos for which some Rift Valley lakes are famous. Prater (1979) concluded that counters tended to slightly underestimate flocks (in flight) of between 100 and 1,000 birds and to slightly overestimate those comprising thousands of individuals. Similarly, Follestad *et al.* (1988) found that observers tended to overestimate numbers of Common Eiders when they were in large flocks of 400–2,000 birds, particularly when they were widely spread, and to underestimate smaller groups. Our own findings with monospecific flocks of Greater Flamingos on three continents, both from the air and from the ground, show that there is a tendency to underestimate numbers when birds are packed densely, as other observers have also found (Bibby *et al.* 1992; Komdeur *et al.* 1992). Some of the flamingo counts made from the air in the Camargue were carried out by A. Tamisier, who estimated that for concurrent counts of ducks, the underestimations were of the order of 20% (Dervieux *et al.* 1980).

The third area of concern is the interpretation of the counts. It is a simple, albeit tedious task to count birds on a good-quality aerial photograph of the colony. The presence, however, of either non-breeding birds or nests where both partners are present, if not taken into consideration, would give rise to an overestimation of the number of breeding pairs. Even assuming distinction is made between breeding and non-breeding birds, the resulting figure is still valid only for the day on which the photograph was taken, since the count will not include birds that attempted to breed earlier in the season but failed (unless of course they have attempted to renest) and it will not include birds that had not initiated nesting at the time the photograph was taken. This is especially important if the counts are made prior to peak occupancy of the colony. The degree of error in these counts thus depends on several factors, including the time span of laying, the proportion of the colony relaying after first (or second) clutch loss, and the number of nest mounds used by two successive pairs (for counts of mounds), in any given year. We can try to reduce these potential errors by taking into account when the counts are made and by ancillary counts of non-breeders, etc. (as is done in the Camargue), but we still need to recognise that these discrepancies exist and, in some cases, could be substantial. For example, over 50% of the nests at Fuente de Piedra can be reused within the breeding season (Rendón *et al.* 2001).

In conclusion, one should not consider existing counts, whether wintering or breeding numbers, to be an exact measure of population size. Rather, these numbers are best interpreted as providing an indication (hopefully a reasonably accurate one) of numbers. Some indications, such as the consistency of the IWC surveys from 1991 to 1998 (excepting 1996, Table 5) are encouraging, but we must still be vigilant and continue to improve our counting techniques and coverage.

CAPTURING AND MARKING FLAMINGOS

Capture and handling

Most long-term avian research projects today rely on the marking and subsequent resighting of individual birds. Many flamingos, from locations throughout the world, were marked when fledglings, or as full-grown birds when they were in moult and flightless. Techniques exist for capturing fully feathered adults, but these could result in injury to the bird(s). For instance, at the beginning of the 20th century, flamingos were caught on the Black Sea coast in large-mesh fishing nets spread across the bed of a lagoon. Hunters drove the birds towards a net which was raised once several of them were in the capture area. The unlucky flamingos' legs slipped through the meshes, preventing them from taking wing (Blussius in Bub 1991). In Kenya, flamingos have been captured for zoos by snaring, spot-lighting at night or by chasing them with a boat (Cooper 1975). This last method requires catching the flamingo by its legs as it runs across the water surface to take flight. Because of the risk of injury to the birds, prolonged disturbance, and the chances of capturing poor-quality individuals, this latter method is unacceptable to researchers and conservationists. In South America and the Caribbean, flamingos have been caught by snares (Morrison 1975; G. Baldassarre pers. comm.) and this method has also recently been used successfully without injury to birds in Spain (J. Amat pers. comm.). The snares are placed underwater and the running noose is attached to a stick by a piece of elastic, which reduces the risk of injury to the bird. Most flamingos, however, are captured as chicks by being herded into a corral, just as large mammals have been trapped in Africa. One advantage of marking chicks is that they are of known age and origin. Full-grown birds can also be captured in such a way in areas where they undergo a simultaneous moult of their flight feathers.

Mediterranean region

France: The first flamingos to be captured and ringed in the Camargue were marked as nestlings in 1947 with Paris Museum alloy rings (Hoffmann 1954; Lomont 1954b). The rings were in fact too large and slipped off their legs. In 1950 and 1952, the problem of ring loss was overcome by marking the chicks with patagial wing-tags. However, these were also found unsatisfactory because some birds were injured by the tag, and the marking process caused disturbance to the colony. In 1953 a new method of herding the crèche of chicks into a pen, or corral, was initiated (Hoffmann 1954; Bub 1991). A great advantage of this approach was that chicks were caught later in the season, when they were older and stronger. The pen was constructed on dry ground along the shore of the lagoon, which also allowed safe and efficient handling of chicks. Although somewhat refined over the years, this is the method widely used today.

The crèche, or part of it, is captured when the oldest chicks are almost fledged and no longer vulnerable to predation by gulls. This is usually at the end of July or in early August (Appendix 4), on average 106 days after the start of laying (range 91–126, n = 24), when the chicks are aged 35–77 days. They can fly at an average age of 80–81 days, although in 1978, only 103 days after the start of laying, some juveniles avoided capture by flying at a maximum age of 73 days! In contrast, in 1994, 114 days after the start of laying, many chicks, then aged 32–85 days, were considered still too small to be herded.

Preparations for the capture and marking operations begin several weeks in advance. Permission is obtained annually from the French Ministry of the Environment and from the salt company (Les Salins) who owns the land. The personnel of local conservation organisations, visiting scientists and local personalities assist the Tour du Valat staff in catching, banding, weighing and measuring several hundred chicks. On the day before the banding, all participants (150–200 people) are briefed on their roles during the operations. Herding of the crèche begins at dawn, the coolest part of the day. Beaters rapidly encircle the chicks and drive them, between convergent fences, towards two interconnecting corrals on the dyke. If any moulting adults enter the pens ahead of the main herd of chicks, they are allowed to escape by a rear gate. This gate is then closed by someone who remains hidden. Another observer is hidden at the main entrance, and when they judge that the pre-determined number of chicks has been captured, the door to the corrals is closed. The beaters instantly move to the nearest fence so that the remaining chicks, usually several thousand, can escape. They reassemble as a crèche at their preferred loafing spot, where they are rejoined shortly afterwards by those released from the pens.

The corrals are *c*. 10 m in diameter and there are 10 small pens around them. The chicks are transferred to the small pens by a team of 5–6 capturers in each corral. There is a handler assigned to each of the small pens and they pass the chicks to the porters for processing. The corrals and the pens are straw-lined to prevent the chicks from getting mud on their plumage, and the wire-netting is draped with hessian sacking in order to prevent them from injuring their bills. Separating the chicks into smaller groups reduces the risk of trampling and injury.

The chicks are handed one at a time to the teams around the pens. They are first marked with a laminated polyvinyl chloride (PVC) leg-band (Figure 8) on one tibia and a stainless-steel ring on the other. Wing and tarsus length are then measured (see description of methods in Chapter 2), as an indication of age, and the birds are weighed. In some years, blood samples and feather lice have been collected for specific studies. The last birds to be handled are generally released after less than two hours in the pens. One or two casualties are generally unavoidable each year, the price to pay for such an operation. These are birds which are either weak, or suffer injury to a wing or leg; it is exceptional for any chicks to suffer from leg paralysis, which sometimes affects long-legged birds when stressed (Young 1967).

For those who have participated, it is a memorable event, usually beginning with an impressive sunrise soon after the start of beating, and terminating with a nourishing breakfast on the beach and a bathe in the Mediterranean to wash off the mud.

Spain: The major Spanish marking scheme is carried out at Fuente de Piedra. Chicks are captured much as in the Camargue, by herding them into pens just before the oldest of them fledge, generally in June or July. When the lagoon has a high water level and many breeding flamingos, the banding operation may be delayed by some days or weeks if some adults are still incubating when the oldest

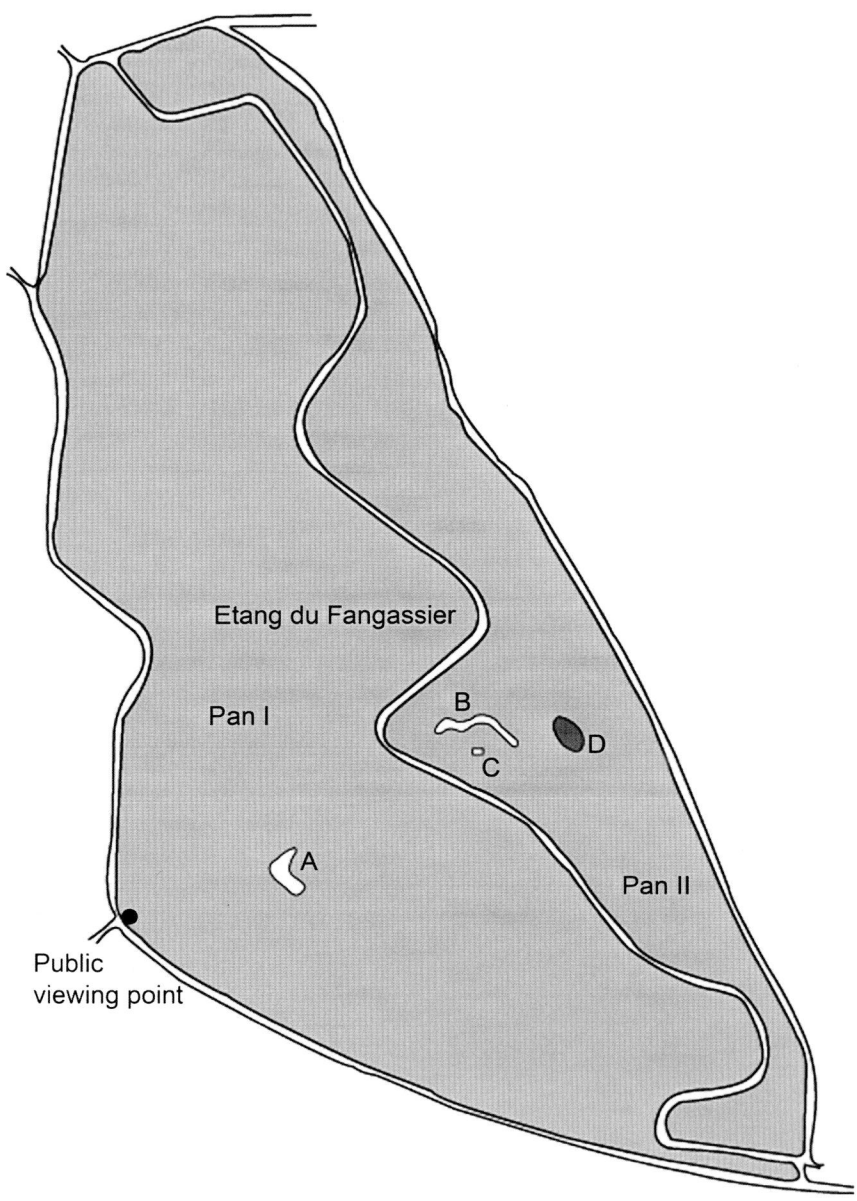

Figure 9. Flamingos have bred at the Etang du Fangassier (536 ha) every year since 1969. Island A was colonised until 1975 and Island B has been used since 1974. Each year some birds attempt to breed on the dyke separating Pans I and II, generally unsuccessfully. The tower (C), erected in 1983, allows observers to record the activities of marked birds. In July and August of most years, the nursery of chicks (D) assembles in Pan II and c. 800 fledglings are captured and marked in pens constructed on the dyke nearby.

chicks take wing. In these years, many of the young of the earlier breeders can escape capture.

Italy: In contrast to the capture operations employed in France and Spain, at Molentargius (Sardinia) and in Comacchio saltpans beaters have used boats and canoes, or have worn chest waders because of deep water or, in the case of the former site, because the lagoon is polluted by untreated waste water from the neighbouring towns.

Asia

India: Ali (1945) caught some flamingo chicks in the Rann of Kutch by running after them when they were very small, but found this method both tiring and time-consuming, and few birds were ringed.

Iran: Capture and ringing takes place at the colonies on Lake Uromiyeh in late July or early August. Three boats, each one having a driver, a capturer and a ringer, encircle the crèche and capture the swimming chicks individually. Over a period of three days, from 500 to 2,500 chicks aged *c.* 30 days were ringed in most years, from 1970 to 1990 (J. Mansoori pers. comm.), giving a grand total of 35,000 ringed birds (Behrouzi-Rad 1992). Of these, 2,250 were marked with neck collars.

Africa

Only very small numbers of flamingos have been marked anywhere in Africa, such activities being one-off operations. Brown (1973) ringed 80 Greater Flamingo chicks at Lake Magadi in Kenya in 1962; and in West Africa, Dupuy (1979) ringed 23 of the chicks raised at Kaolack saltpans in 1979. It is perhaps not surprising that these small samples have not produced any recoveries, particularly since these colonies were in quite remote areas

In South Africa, when the first and only mass breeding of flamingos took place at De Hoop Vlei (Bredasdorp district) in 1960–61 (Uys *et al.* 1963; Broekhuysen 1975), 50 chicks were captured and ringed. In Namibia, Berry (1972) reported the ringing of both Lesser and Greater Flamingos during rescue operations on Etosha Pan, 60 in 1963 and 1,500 in 1969. Some of these birds were later recovered.

RINGS AND BANDS

Mediterranean region

France: Since 1977, when the first moulded PVC bands were used, between 600 and 900 chicks, representing 7–30% of the total crèche, have been marked each year (Appendix 5). The leg-bands used in the Camargue, and subsequently at other sites in the Mediterranean, were described by Ogilvie (1972) and in Kear & Duplaix-Hall (1975). This type of band has been used since the late 1960s to mark swans and geese in northern Europe (see Rees *et al.* 1990), and cranes in North America (see Hoffman 1985), and is now widely used on waterbirds (see Flamant 1994). The bands, *c.* 1.8 mm thick, are made of laminated polyvinyl chloride (PVC), which has been manufactured in Britain since the early 1960s (see Coulson 1963). They are engraved with a 3- or 4-digit alphanumerical code repeated three times around the band (see Figure 8). The codes are engraved through the thinner coloured surface layer to expose a contrasting under layer. Each flamingo chick also receives a large Paris Museum stainless-steel ring placed on the other tibia as a more permanent marker (see below). The PVC bands weigh 5 g (3 digits) or 6 g (4 digits) and the metal rings 12.5 g, so that a band and a ring together represent less than 1% of the weight of an adult female flamingo.

Spain: Small numbers of chicks were metal-ringed at Fuente de Piedra in the 1960s and 1970s: 9 in 1964 (Bernis & Fernandez-Cruz 1965), 45 in 1969 (Fernandez-Cruz 1970) and 50 in 1977 (Blasco *et al.* 1979). In 1986, a long-term banding scheme was initiated using PVC bands (Rendón 1997) on a sample of the chicks raised each year, although in some years all of them were banded.

The chicks are marked with an Instituto Nacional para la Conservación de la Naturaleza (ICONA) metal ring on the right tibia and an orange or white PVC band on the left. The latter is engraved with a unique code allowing individual recognition in the field. A black line between the first and second digits differentiates the Spanish bands from those used in the Camargue. Breeding has occurred in 11 out of the past 15 years (1986–2000) during which time almost 10,000 chicks have been banded. The banding and resightings database is maintained by Doñana Biological Station and by the Agencia de Medioambiente de Andalucía.

Flamingos occasionally breed in the Guadalquivir Marshes (Doñana National Park) and in 1984, 173 of the chicks raised were ringed. They were marked on the tibia with a Spanish ICONA metal ring, and 153 of them were also given a red PVC band on the other tibia engraved with either one or two black lines. These bands indicate only the year and place of banding.

Italy: The Italian National Wildlife Institute (INFS) recently began a banding programme at three sites. At Orbetello (Tuscany) all 26 chicks raised in summer 1994 were captured and marked with a national museum (Bologna) metal ring on one tibia and a coded blue PVC band on the other (Dall'Antonio *et al.* 1996). In Sardinia, 404 and 383 chicks have been ringed and banded at Molentargius in 1997 and 2000 respectively (Tiana 2000), and 190 at Santa Gilla in 1999. These were all marked on one tibia with the national scheme's metal ring and on the other with a red PVC band coded for individual recognition.

Asia

Flamingos have been ringed at several sites in Asia but at only two of these are long-term projects in place. In India, for example, Ali (1945) ringed 192 chicks in the Rann of Kutch (Rajasthan) in 1945 (Bombay Natural History Society rings placed on the left tibia), but this was a one-off operation and no further ringing has taken place in the Rann of Kutch. It seems that no recoveries were ever reported (Ali & Ripley 1978). Twenty-six flamingos, presumably full-grown, were ringed at Point Calimere (Tamilnadu) in 1980–81 (Ali & Hussain 1982), but again we have no knowledge of any recoveries of these birds.

Iran: The Department of the Environment in Tehran initiated a flamingo ringing programme at Lake Uromiyeh in 1970. Large numbers of chicks and also full-grown birds, flightless during moulting, have been marked with either metal leg rings or neck collars (Argyle 1975, 1976; Scott in Kear & Duplaix-Hall 1975; Behrouzi-Rad 1992). The recovery rate of metal-ringed flamingos from Lake Uromiyeh is much lower (0.54%) than reported for birds from the Camargue (8.7%), partly perhaps because a greater proportion of Iranian birds moves to areas of low human density, where they are less likely to be reported.

Kazakhstan: A flamingo ringing scheme has been in operation in the former USSR since 1935. Both chicks and full-grown birds have been ringed, the latter being captured when flightless during their wing moult. Most were marked with just a metal ring, but in 1986, 200 individuals were also given a coded neck collar (Gavrilov 1986). Most birds have been marked at Lake Tengiz but some of the earlier ringing was done on the east coast of the Caspian Sea.

CAPTURE-RESIGHTING: ESTIMATING DEMOGRAPHIC PARAMETERS FROM MARKED INDIVIDUALS

RECOVERIES AND RESIGHTINGS

Not only does marking flamingos allow individual recognition in the field, and the study of movements and distribution of birds of known origin, but it also provides the raw information necessary to estimate essential demographic parameters such as survival rates, recruitment rates, age-specific breeding probabilities, and breeding success. Two kinds of information can be obtained from ringed or banded individuals: recoveries and resightings. Although they convey different information, both can be used to estimate mortality schedules. Recovery data are obtained from animals that have been previously ringed, at least with metal rings, and are subsequently 'recovered' dead (or severely injured). Thus, throughout the next chapters, any reference to recovery implies that the bird was found dead. In contrast, resightings correspond to observation of live banded birds. Indeed, codes engraved on flamingo bands can easily be read at 300 m through a telescope and numerous resightings from the same bird can thus be obtained.

A first distinction between recovery data and resighting data is precisely that a single bird is recovered only once (as a result of its death), whereas it can be resighted several times. A second distinction is that the probability of resighting depends upon the bird being alive, present in the site under survey and being seen by the observer, whereas the probability of recovery depends upon the bird being dead, its corpse being found and the ring being sent back to the ringing agency. Estimation of survival from recovery data has much of its origin in research on waterfowl, where hunters can quickly 'recover' large numbers of ringed birds that are harvested. This method is used today primarily for quarry species in conjunction with annual harvests. It has been applied to flamingos in the early stages of the Camargue study (e.g. Johnson 1983). However, the method has several limitations for use with flamingos. In addition, whereas each bird contributes only once to the dataset in the case of recoveries, considerably more information can be derived from resightings of the same individuals over long periods of time. This is why, following the development in the late 1980s of new statistical methods and 'user-friendly' software for the analysis of capture-mark-recapture data (Lebreton *et al.* 1992, 1993), we decided to favour the analysis of resightings over that of recoveries (Cézilly *et al.* 1996; Tavecchia *et al.* 2001). Actually, recoveries are (fortunately) much less frequent than resightings, and therefore the data basis built from resightings was also much more important than the one that could be obtained from recoveries. This is the result of banded birds being actively sought on a regular basis by the Tour du Valat flamingo team, in the Camargue and elsewhere in France, but also being increasingly reported by a network of amateur and professional ornithologists throughout the Mediterranean and in West Africa. Sightings away from the breeding localities are made wherever and whenever possible, and are reported consistently throughout the year. Informants are sent a copy of the bird's life history, in acknowledgement of their observations and to verify that data have been correctly recorded. It is also an encouragement for them to read more band codes. And this works well. Many missions to Mediterranean and West African countries have been undertaken specifically to search for banded flamingos, either breeding (Spain, Sardinia, Tunisia, Mauritania) or at other times of the year. A Flamingo Supporter Scheme, initiated in 1990, has raised money for the purchase of optical equipment and tripods which have been assigned to ornithologists throughout the Mediterranean region and West Africa, thus increasing the number of resightings of banded birds.

Today, software is advancing rapidly and new modelling techniques are now available that allow the mixing of information obtained from both recoveries and resightings in the same analysis. Thus, the road is open for new investigations making full use of the available information in the future.

DATA STORAGE AND ANALYSES

Details of resightings and recoveries of banded flamingos have been computerised since the PVC-banding programme began in 1977. Data have been stored and analysed using a computer programme written for this purpose by Vincent Boy. Thirteen variables are recorded for all birds (e.g. band code, date, site, locality, sex, status, observer, manner of recovery, precision of date) and for breeding birds in the Camargue a second programme allows nest position to be recorded. All observations of banded birds, including those made on different days in the same place or in different places on the same day, are recorded. For all recoveries the region and the manner of recovery are coded according to the European Union for Bird Ringing (EURING) international data bank (see Busse 1980). By the end of 2000, 24 years after the first chicks were PVC-banded, the number of recoveries and resightings almost reached 300,000 (Appendix 1). Over 90% of the chicks are observed at least once (or are recovered) after banding, in the crèche or elsewhere (Appendix 5), and more than 70% are seen or recovered after leaving the Fangassier. This is a very high recapture rate for a species which is not confined to a restricted area outside the breeding season. For comparison, Bewick's Swans banded in Slimbridge, Great Britain, have a recapture rate of 70–80%, and those banded in The Netherlands of 95% (Flamant 1994). More-dispersive species usually have lower recapture rates; for French-ringed White Storks, for example, it is 28% (Flamant 1994).

SURVIVAL

Analyses of flamingo survival rates are based on resightings of PVC-banded flamingos and on recoveries, described above. Survival analyses carried out thus far (Chapter 9) have been based on the resightings of flamingos in the Fangassier colony or, later in the season, with their chicks in the crèche, and exclude birds seen breeding elsewhere. A global analysis will remain a handicap until a multi-site model has been developed, since many Camargue-ringed flamingos are not philopatric, i.e. they do not subsequently return there to breed (Chapter 5). However, recoveries are too random for any emerging patterns to be completely reliable.

BAND LOSS AND ANOMALIES

The calculation of survival rates based on the resightings of banded individuals requires either that birds not lose their bands, or that some measurement of band-loss rates be taken into account, particularly with such a long-lived species as the flamingo. In addition to the PVC band, the chicks in the Camargue have been marked since 1986 with a large stainless-steel ring placed on the other tibia. The code on this ring can be read in the field at maximum distance of c. 100 m. The ring is engraved with a Paris Museum code, and intended to give some measure of the loss of PVC bands. It will withstand time much better than the aluminium or alloy rings that were used in the 1950s, and the code need only be read if a flamingo is seen to have lost its PVC band. Only six of the 7,743 flamingos banded from 1986 to 1996 have so far been recorded as having lost their band; in four other cases the PVC band had slipped down onto the tarsus, where one individual has been wearing it for 12 years!

Of over 16,000 chicks banded from 1977 to 1999, only 35 birds have so far been reported to have damaged bands, which in some cases they have worn for many years. Old bands which have been recovered do not show signs of deterioration other than the colour fading, which is of little importance. Most anomalies reported concern acorn barnacles, which can often be seen (up to six) attached to the PVC bands, but these eventually fall off.

Rees *et al.* (1990) analysed PVC Darvic and Vinylast band losses on Barnacle Geese, Bewick's and Whooper Swans. In the first species band loss was negligible at 0.35% per annum, but was higher for the swans, which wear a larger band. Although the flamingo takes the same size band as the Barnacle Goose, band loss ought to be even lower, since the flamingo's band is seldom, if ever, in contact with vegetation, or anything else which might cause wear, and apart from one or two days after banding, flamingos are seldom seen to peck at their bands.

BREEDING SUCCESS

We have estimated breeding success by dividing the number of chicks raised by the number of breeding pairs. Counting techniques have been described above (see 'Breeding bird surveys') and our measures of breeding success allow for a margin of error associated with the counting techniques, particularly of colony size.

ANNUAL VARIATION IN CHICK CONDITION

Every year during the ringing operation, each ringed chick is measured (tarsus length, wing length) and weighed. The relationship between weight and tarsus length is then used to estimate the body condition of chicks in the crèche. To this end, we model weight (W) as a function of tarsus length (T) by the allometric equation:

$$W = aT^b$$

The parameters a (intercept) and b (slope) are most easily estimated by linear regression based on logarithms:

$$\log_e(W) = \log_e(a) + b\log_e(T)$$

Residuals from the regression provide an index of body condition for each chick in a given year. The slope, a, provides an estimate of the average body condition of chicks in the crèche for a given year.

In addition to the estimation of body condition, other physiological parameters can be measured. Several haematological studies have been carried out on the biochemistry and physiology of Greater (and Chilean) Flamingos, particularly regarding the determination of normal values of blood parameters (Hawkey *et al.* 1984a; Peinado *et al.* 1992; Puerta *et al.* 1989, 1992). Such studies can be important not only for the detection of pathological states, but also to the understanding of ecological and behavioural problems (Ferrer 1990, 1992).

Blood samples were collected in the Camargue in 1992 from both adults and chicks captured during the annual marking operation. These were examined for levels of glutamic-oxalacetic transaminase (GOT) and uric acid. The former may be related to the growth rate of chicks (Puerta *et al.* 1992), and uric acid to feeding restraint and for determining nutritional condition (Okumura & Tasaki 1969; Ferrer *et al.* 1987; Garcia-Rodriguez *et al.* 1987; Ferrer 1992). Blood samples were also taken from chicks at Fuente de Piedra, in 1990, using sampling procedures similar to those used in the Camargue, and from captive birds, by Puerta *et al.* (1992).

There is no significant difference between captive and free-ranging adults in the mean values of GOT and of uric acid. The chicks, on the other hand, reveal differences in the mean values of these two parameters, which are mainly due to the low values for the Spanish birds, possibly indicating that chicks raised in the Camargue in 1992 were in better body condition than those raised in Andalusia two years earlier. The use of such parameters, together with the estimation of body condition, may prove useful in the future for documenting variations between colonies, and between years for the same colony, in the health and condition of chicks.

FORAGING ECOLOGY

The food resources of many wetlands where flamingos breed are insufficient to satisfy the needs of all the birds in the colony throughout the breeding season, and many, sometimes all, individuals at a particular site will forage in other wetlands. The first indication of this may be deduced by comparing the number of known breeding pairs in the colony with the number of birds foraging on site. Foraging flights take place mostly in the evening, and by observing these flights and their direction it is possible to estimate to which wetlands the birds may be heading. Confirmation of this requires individual recognition, which can be achieved by observing marked birds. Foraging movements have been identified by banding and dye-marking flamingos, and by radio or satellite tracking, and results are presented and discussed in Chapter 6. The different methods have their strengths and weaknesses, which we discuss below.

Banded birds: Bands proved particularly useful for identifying the feeding areas of the flamingos breeding at Fuente de Piedra in Spain. Observer effort was sufficient at the colony to identify a large number of banded birds at the nest or attending their chick, and it was similarly sufficient over parts of the feeding range in 1987–90 to allow birds to be identified at the foraging sites between sightings at the colony, within a short period (see Chapter 6). Similar observations have been made in the Camargue, although most data on the foraging areas of breeding birds there come from dye-marking.

Dye-marked birds: In order to investigate the feeding range of flamingos breeding in the Camargue, and the types of wetland used for foraging, a sample of incubating birds was marked with picric acid on the nest during four consecutive seasons (1986–89). This technique did not allow individual recognition of birds in the field but was complementary to reading band codes, which for this type of study has its disadvantages. Leg-bands cannot be seen when flamingos are feeding in deep water, and codes cannot be read at greater than *c.* 400 m, or even less when it is hazy or windy. The birds were dye-marked in May by cautiously spraying them from the boat-hide normally used by observers to enter and leave the observation tower at Fangassier (Plate 24). The floating hide was carefully positioned to within a metre of the closest flamingos incubating on the periphery of the island. Two small holes (*c.* 2 cm diameter) bored side by side in the rear of the hide allowed the observer to peep at the flamingos through one and to spray them with the picric acid through the other, the nozzle of the spray gun resting on the edge of the hole. The solution is completely harmless to the plumage. The spraying was done in calm weather, and was completed in one or, at most, two mornings' work. From 60 to 150 birds were dye-marked. Because the solution (picric-acid crystals dissolved in alcohol) was extremely volatile, only the birds incubating on or close to the edge of the island (up to *c.* 3 m) could be marked. The extent of marking varied from one flamingo to another but no effort was made to identify birds individually. The fluorescent yellow patches, which the birds extended by preening, were anywhere on the body. The colour deepened to almost gold after a day and disappeared completely after two weeks.

After dye-marking at the colony, the number of marked birds remaining on the nest was noted each day in the morning and again in the evening. The normal maximum length of attentive periods at the nest being four days, we limited our searches for the marked birds to the four days following marking. We systematically searched for off-nest birds on all wetlands in the Camargue, and also to the east of the Rhône River (Plan du Bourg), in the Petite Camargue and further west in the Languedoc. Searches were made both from the ground and from the air by a large team of observers who, at each wetland visited, noted the number of flamingos present and the number of picric-marked birds.

Over the ensuing four days, visits were made to all feeding areas within 80 km of the colony, on foot and by car to the more accessible wetlands, by pony trek into the National Nature Reserve (Vaccarès complexe), and by aerial surveys. Well-marked individuals could be seen at a distance of over one kilometre.

BREEDING ECOLOGY

Recording breeding birds

Comfortable and spacious hides have been erected for observing breeding flamingos both in the Camargue and at Fuente de Piedra. The hide at the Etang du Fangassier was built in early spring 1983, when the first cohort of PVC-banded birds began breeding in the Camargue. It was erected in March and some flamingos had already arrived in the vicinity of the breeding island. They looked on, between bouts of displaying, as the prefabricated tower of metal tubes and marine plywood took shape. It was placed 70 m to the south of the island (Figure 10, Plates 2, 24) so that observers could see the whole of the colony in favourable light. Daily observations were made for most of the daylight hours during the flamingos' seasonal occupation of the island. Observers enter and leave the tower using a floating boat-hide (Plate 24) painted matt-grey like the tower, and therefore never disturb the flamingos. These plywood-covered boats, which the observer pushes through shallow water, are moored in the bottom of the tower; the first floor is used for storage of equipment and the upper floor, 3 m above water, for observations and living quarters. Anchored by cables, this hide has withstood gales of 140 km per hour. The colony can be observed whatever the weather, but the floating hides cannot be handled when winds are in excess of force 5–6 on the Beaufort Scale. Since strong winds are frequent in the Camargue in spring, several observers have memories of being stranded in the tower for up to five days.

Banded birds are easily identified from the tower, and their behaviour recorded. They may be casually present at the colony, displaying, paired, copulating or breeding. Marker posts in front of the island at every 10° (3 m apart facing the tower) allow the position of a bird's nest to be recorded quite accurately. A bird is recorded as breeding either if its egg or later its chick is seen or, for birds nesting in the centre or at the back of the colony, if it is seen exhibiting breeding behaviour at the same nest location on at least two occasions, with at least 24 hours between observations. When the chick of a banded parent is seen, its age is estimated using a guide to size and coloration (see Chapter 2) so that the date of egg-laying can be calculated and breeding success assessed. Since most mortality occurs prior to the age of 30 days, parents are considered to have bred successfully if they are seen with a chick at least one month old.

Observations from the tower come to an end gradually each year as the chicks leave the island and move into the crèche, which usually assembles nearby. The

Figure 10. Tower hide at the Etang du Fangassier. Flamingos show no fear of the tower hide erected in early spring 1983, 70 m from the breeding island at the Etang du Fangassier. Observers, living in isolation but comfortably, are able to monitor breeding irrespective of the weather, and to record in detail the activities of banded birds visiting the Camargue colony throughout the breeding season.

crèche is observed from the neighbouring dyke(s) in the evening, during the last two hours of light, when the parents fly in to feed the chicks. At this time, observers are able to see some of the banded birds which have successfully raised a chick, among which may be some hitherto unknown breeders.

Each year, the Fangassier colony has been observed by a team of 5–6 observers on average for eight hours per day for 81 days during the egg-laying and incubation period (March–June 1983 to 2000). The crèche has been observed on average for two hours on 47 evenings (June–September 1983–2000). Over 1,000 banded birds originating from Fangassier or from Spain have been observed making breeding attempts in a single season (Appendix 11), while over 7,000 different banded birds of Camargue origin have been recorded making at least one breeding attempt during the period 1983–2000 (see Chapter 5).

Since the hide was erected in 1986, observations and recordings at Fuente de Piedra have been carried out in much the same way as those at the Etang du Fangassier, and the same protocol is applied at other colonies where observation effort has so far not been as intensive as in the Camargue or at Fuente de Piedra.

Figure 11. Chicks crèching. The chicks, 267 on the photo above, seek the shallower water where the tower hide stands. They occasionally enter the basement where some have even been fed by their parents.

Flamingos marked in the Camargue have been reported from 16 countries, 55 political or administrative regions and 700 localities (see Chapter 5). Some birds, from the earlier years of banding, now have recorded life histories exceeding 300 observations; these are the more sedentary individuals, having stayed in areas where observer effort is greatest. Others, which have been to localities where there are few ornithologists, and/or where it is difficult to approach flamingos (e.g. the Banc d'Arguin in Mauritania), may be lost to view for a period of several years, only to reappear in the Mediterranean region, in some cases after an absence of 10 years or more.

Recoveries are reported when a flamingo is captured, which in most cases means that the bird has been found weak, injured or dead. They are reported to either the Bird Museum in Paris, whose address is engraved on the metal ring, or to the French, Spanish or Italian coordinators of the different marking projects and conveyed to the holder of the database.

In Spain, a hide was built in 1986 facing the flamingo colony at Fuente de Piedra, and this has been used for observing banded birds in the colony in most years. In the beginning, the banded birds observed breeding at the lagoon were all of Camargue origin, but many of those marked with Spanish PVC bands have now reached maturity, and since 1997 their numbers have exceeded those from the Camargue.

RECORDING ATTENTIVE PERIODS

During the main incubation period, that is from one to two days after egg-laying until one to two days before hatching, flamingos change partners at the nest when the off-nest bird returns from foraging to brood. From the early years of the Camargue study, we noticed that some of the banded individuals were staying on the nest for up to four days (and presumably the intervening nights) and we began to record these attentive periods systematically for a series of pairs/nests. Those pairs/nests chosen were selected because they could be seen easily and the incubating bird identified by the observer in the tower. The nests sampled, mostly on the edge of the island facing the tower, were those where at least one of the partners was banded; in a few cases, both were. Bill patterns (see Figure 4) were sketched and used for partner identification when the incubating bird faced the tower and its tibia/band was hidden from the observer. At the time, the tower was occupied permanently, and a series of nests (from 5 to 20) was checked every hour from dawn to dusk and the presence of either the male or the female was noted. Changeovers were sometimes observed, and in other cases were known to within an hour, or took place at night.

Attentive periods were measured because it was thought that they might reflect food availability, an individual's foraging efficiency, or the distance birds travel to feed, three parameters likely to influence breeding success.

CAPTIVE FLOCKS

Flamingos have always been popular with visitors to zoos and nature parks. In the1970s, Duplaix-Hall & Kear (in Kear and Duplaix-Hall 1975) recorded 35 zoos holding almost 2,600 flamingos of all species. The number of captive flamingos had increased since the 1950s, particularly in the USA and in Europe. The first documented captive-breeding attempt occurred in Florida at the Hialeah Race Course in 1937; the first successful breeding there was in 1942. Pickering (1992) recently identified at least 30 collections in Britain alone where flamingos had bred at least once during the preceding five years. Most captive birds are maintained for exhibition purposes only, but a few flocks have been used for study. For example, breeding by Greater Flamingos at the Basel Zoo in Switzerland has been studied in detail by Studer-Thiersch (1966, 1974), whose original papers, in German, were summarised in Studer-Thiersch (1975b). In 1957, this collection, composed of Caribbean, Greater and Chilean Flamingos, was the first in Europe to start a successful breeding programme, and with one individual at least 67 years old in 2005, can boast the oldest-known flamingo of any species. Studer-Thiersch has studied many aspects of the Greater Flamingo's breeding behaviour, some of which would be difficult to observe in the field, such as the postures and simultaneous calling related to the formation of pairs. She observed that it is the female who chooses the location of the nest site and that egg-laying, which lasts from several minutes up to one hour, may take place at any time of the day. She described how flamingos, during wet weather, remove mud from the nest with their bills, sucking it up and then vigorously head-shaking outside the nest to get rid of it; in the field, only the latter part of this movement can really be seen. She also noted strong pair-bonding in captivity, something which banding in the field and subsequent observations of these birds have shown not to be the case in free-ranging flamingos (see Chapter 7). The Basel flamingos were sexed on the basis of behavioural traits, or in some cases through laparoscopy, and they were measured when captured. Tarsus length proved to be a more reliable indicator of sex than wing length, since there was less overlap between sexes. Hatching and adult–young behaviour, chick feeding and development have all been studied in detail at Basel Zoo. The composition and chemistry of the secretion which adults feed to their chicks (see Chapter 8) has also been studied at Basel.

One of the few collections in the world to hold all species of flamingos is the Wildfowl and Wetlands Trust in Slimbridge, Great Britain. Rather surprisingly, however, these birds have been very little studied. In 1989–90, the collection held a monospecific flock of 126 Greaters, which had bred for the first time in 1970 and then in 19 out of the 21 subsequent years. Individuals were not marked until 1989–90, when studies were undertaken of the ritualised displays (Lindgren & Pickering 1997) and of the breeding biology of the different species (Pickering 1992). It was observed that pinioning may impair breeding performance, since in almost two-thirds of the observed mating attempts, the male fell off the female's back. Incubation lasted on average 29.7 days (range 26–32) and was shared by both sexes, just as in the wild.

Attempts have been made by the Wildfowl and Wetlands Trust to encourage flamingos to breed by placing mirrors in their enclosure. The latter experiment was to investigate the role of visual stimuli in inducing display in Lesser Flamingos (Pickering & Duverge 1992). Bouts of display marching were prolonged on days when the mirrors were used, but the Lesser still remains the only species of flamingo never to have reproduced in captivity.

In North America, Bildstein *et al.* (1993) studied salt tolerance in the Caribbean Flamingos held in St Louis Zoo. They also observed how aggression and dominance interactions affect feeding behaviour, and quantified feeding and bill-rinsing behaviour in American and Chilean Flamingos. They compared feeding behaviour of captive birds, in St Louis and Riverbanks Zoos, with that observed in wild birds in Venezuela, and found that although free-ranging birds spent more time feeding, the mean duration of individual feeding bouts was not significantly different. Finally, Zweers *et al.* (1995) studied filter-feeding in captive flamingos in The Netherlands and produced a model confirming the findings of Jenkin (1957), and expanding on her earlier studies.

In Namibia, Berry & Berry (1976) have shown that it is possible to hand-rear chicks taken from an abandoned colony in the Etosha National Park. They successfully raised five out of seven birds which were collected either when hatching or at a maximum age of four days. The two chicks which died were collected when only one day old and had never been fed by their parents. Body weight and plumage development were monitored; in contrast to flamingos in the Northern Hemisphere, the chicks had a coral-red base to their bills until 11 weeks of age, when this changed to blue-grey. The first bird to feed naturally did so at seven weeks of age. When independent they ate an average of 1,250 g dry mass of food per day, or 9.6% of their body mass.

In captivity, it is possible to study aspects of the behaviour of flamingos which it might be difficult or impossible to observe in the field. Such studies are thus of interest and complementary to field observations of wild birds. Living in captivity, however, modifies animals' behaviour to varying degrees, and caution must be applied before assuming that captive observations are a reliable substitute for field observations. This is particularly true of flamingos, which in the wild are not only free to disperse or migrate but live in flocks much larger than those found in captivity, and in which the composition is continually changing. Pair bonds, or the lack of them, are an example (see Chapter 7).

CHAPTER FOUR

Distribution and numbers

DISTRIBUTION

Greater Flamingos occur within tropical or temperate climatic zones near the great deserts of the world, and their distribution is strongly influenced by food availability (Jenkin 1957). They congregate in or near arid localities in brackish, saline or alkaline waters in which only a few species of invertebrate prey can multiply sufficiently to meet the flamingo's needs. These wetlands are subject to seasonal drying, since they are situated in open habitats such as steppe or desert, dominated by a Mediterranean-type climate, with low precipitation—most rain falls from autumn to spring (see Di Castri *et al.* 1981)—and high rates of evaporation. During periods of drought, flamingos move to more permanent water bodies, including coastal waters (Williams & Velásquez 1997).

The flamingo's range extends from the Mediterranean region east across southwest Asia, to Kazakhstan in the north and to India and Sri Lanka in the east, along the coast of the Persian Gulf and from Ethiopia down the Rift Valley to southern Africa, including Madagascar, and spreads west to Botswana and Namibia. Flamingos are scarce in tropical West Africa, where there is little suitable habitat south of Guinea-Bissau. In the hotter parts of Africa and in the Gulf States, Pakistan and north-west India, the Greater Flamingo shares many wetlands with the Lesser Flamingo, the latter being far more abundant than the Greater in the Rift Valley of East Africa. The breeding sites of the Greater Flamingo tend to be situated on or within the isotherm of 24°C in the warmest month (Voous 1960).

WEST AFRICA

See Figure 12. Flamingos occur regularly from the northern limit of the savanna and tropical forest (Guinea-Bissau, Senegambia) in a relatively narrow belt between the Atlantic Ocean and the Sahara Desert. **Guinea-Bissau**: Bijagos Archipelago. **Senegal**: along the coast at Casamance, Sine-Saloum delta including Kaolack saltpans, wetlands at the mouth of the River Senegal (Geumbeul) and the Djoudj National Park. **Mauritania**: Chott Boul and other wetlands of the Aftout es Saheli, north to the Banc d'Arguin, the Baie d'Arguin, and up the coast in bays and wadi mouths from Nouadhibou to the Mediterranean region. They may formerly have bred (Alexander 1898), and are still frequently seen in small numbers, on the **Cape Verde Islands** (Boa Vista, Sal) (Bannerman 1953; Naurois 1969b; Naurois & Bonnaffoux 1969; Naurois 1994), but occur only occasionally in the **Canary Islands** (Bannerman 1963). They have bred in Senegal but the main breeding areas are in Mauritania, on the Banc d'Arguin. In other parts of West Africa they are confined to sites on or near the coast, where they occur locally and irregularly in **Sierra Leone**, **Liberia**, **Niger**, **Cameroon**, **Gabon** and **Congo** (Serle *et al.* 1977).

MEDITERRANEAN REGION

See Figure 13. Flamingos occur in most Mediterranean countries; the main strongholds were formerly concentrated around the western and eastern basins, and there were fewer birds around the Adriatic Sea, but numbers in the central Mediterranean have recently increased. In southern Europe the flamingo has an essentially coastal distribution, which in the west extends up the Atlantic seaboard to **Portugal** (Sado and Tagus estuaries, Castro Marim) (Guedes & Teixera 1991) with only occasional reports from further north on the Atlantic coast of **Spain**, for example at Ria de Vigo in Pontevedra (Johnson 1989a). Elsewhere in Spain, flamingos occur on most of the favourable wetlands (see Fernandez-Cruz *et al.* 1991) from the very extensive marshes and saltpans of coastal Huelva, Sevilla and Cádiz, to Fuente de Piedra (Malaga), 50 km from the coast. This last site had been the furthest inland in Europe where the species occurs regularly in large numbers, but recently flamingos have even bred at Laguna Petrola (Albacete), 100 km inland, and in La Mancha, 200 km from the Mediterranean coast, up to 160 birds were seen during the first half of 1998 (Madrigal Diaz 1999). The other Spanish sites are mostly on the Mediterranean coast, and quite widely scattered, extending from the Almeria region (Cabo de Gata) towards the north and east, through Alicante (Santa Pola and Elche-El Hondo) to the Ebro delta (Tarragona). The species occurs regularly, but in rather small numbers, on the **Balearic Islands** (Bannerman 1963).

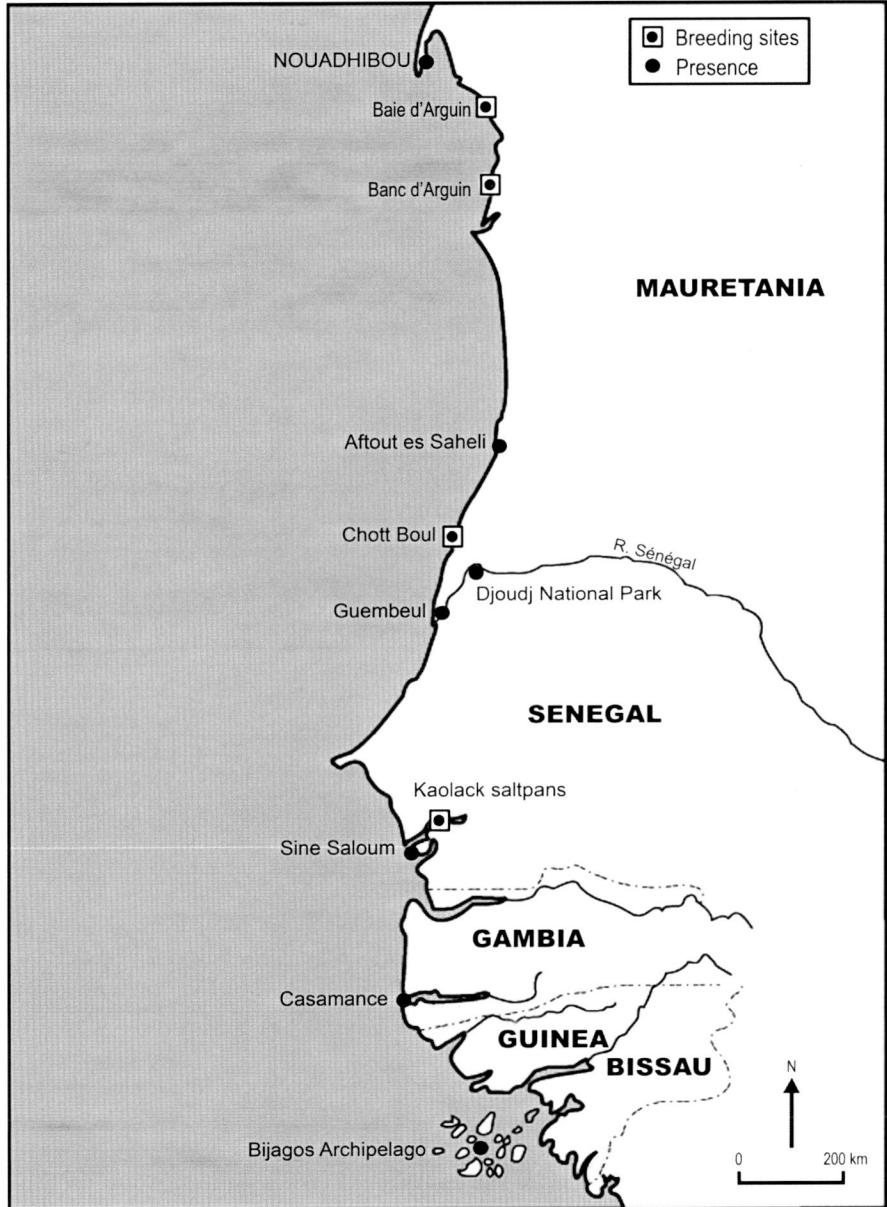

Figure 12. Important sites for Greater Flamingos in West Africa.

72 *The Greater Flamingo*

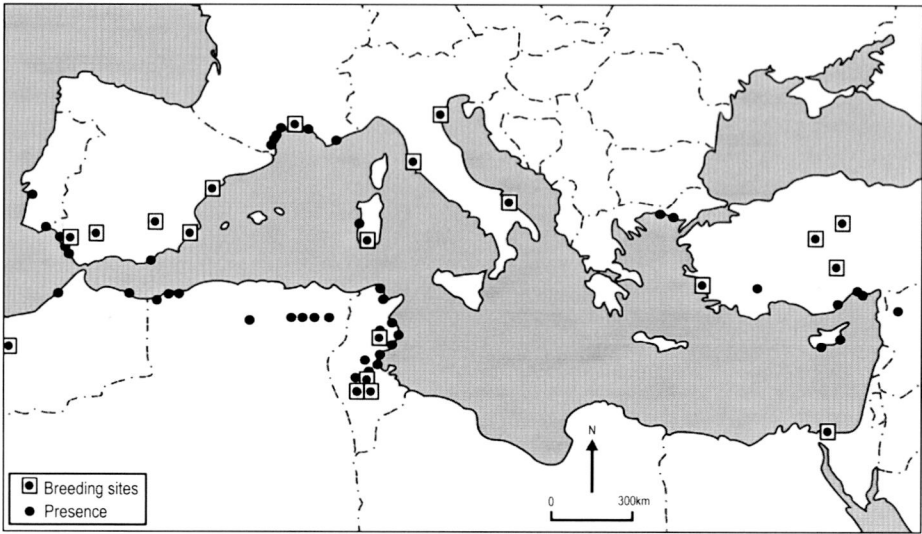

Figure 13. Important sites for Greater Flamingos in the Mediterranean region.

In **France**, practically all of the wetlands along the Mediterranean coast, from Etang de Canet (Pyrénées-orientales) to the Giens Peninsula (Var), regularly host large numbers of flamingos (Oliver 1980; Johnson in Yeatman-Berthelot 1991). Elsewhere in the Var (Orsini 1994) and Alpes-Maritimes, and on **Corsica** (Thibault 1983), the species occurs quite regularly, but only in small numbers. In **Italy**, the largest concentrations of flamingos are found in **Sardinia**, in the provinces of Cagliari and Oristano, where the species occurs throughout the year. Flamingos have recently appeared more frequently on several wetlands in Tuscany (Orbetello area), on the Adriatic coast at Apulia (Margherita di Savoia) and in Ravenna (Serra *et al.* 1997). They are scarce but regular on coastal wetlands in Sicily (Iapichino & Massa 1989).

In the Balkans, small numbers of flamingos are occasionally reported on wetlands of the Adriatic coast, from **Croatia** south to **Albania** and **Greece**. Formerly only a vagrant to Greece (Watson 1960; Bauer *et al.* 1969), the flamingo now occurs regularly and in large numbers in Macedonia and Thrace (Handrinos & Akriotis 1997) and on several islands of the Aegean Sea (**Kos, Samos, Lesvos, Limnos, Naxos** (Tsougrakis & Kardakari 1996)). On **Cyprus**, the wetlands at Larnaca and Akrotiri are very important for large numbers of flamingos on migration and in winter (Flint & Stewart 1992).

In **Turkey**, flamingos occur on wetlands of the Aegean coast, particularly near Izmir, and on inland sites in western Anatolia (Kumerloeve 1962), where they breed (Lake Tuz, Lake Seyfe) up to 300 km from the coast. They also occur, particularly in winter, around the Gulf of Adana. Further south, Jabbul Salt Lake in **Syria** is the most important wetland for flamingos in Asia Minor, sometimes host to thousands of birds (Kumerloeve 1966). There are very few reports of flamingos from **Lebanon** (Ramadan-Jaradi & Ramadan-Jaradi 1999) and **Jordan** (Vere Benson 1970) but variable numbers currently occur throughout much of the year at some locations in **Israel** (Shirihai 1996), where until 1970 the species was only a rare passage migrant and winter visitor (Paz 1987).

In North Africa, flamingos are regular visitors to the Nile delta and wetlands along the Mediterranean coast of **Egypt** (Goodman & Meininger 1989) and **Libya**, in particular in Cyrenaica, from Tocra to Benghazi. They are irregular and scarce further west in Tripolitania, from Sirte to the Tunisian border (Stanford 1954; Bundy 1976), but often abundant in the Maghreb, particularly in **Tunisia**, where they breed. Flamingos are regular visitors to **Algeria** (south Constantine region, Boughzoul and Oran (Isenmann & Moali 2000)) and **Morocco**, at wadi mouths and saltpans along both the Mediterranean and the Atlantic coasts. They are also seen inland (Zima, Merzouga and formerly Iriki), some Saharan sites being over 300 km from the coast.

Asia

Flamingos occur from the shores of the Mediterranean and Caspian Seas in the west, through the Middle East to India and Sri Lanka. In spring and summer the northerly limit of their Asian range is in Kazakhstan, at $c.$ 50°N latitude, and in winter it is at $c.$ 40°N latitude, in Turkmenistan. The main breeding areas in Asia are in Iran, north-west India and Kazakhstan, the latter being 2,500 km inland.

Throughout this vast region the following areas are host to large or very large concentrations of flamingos, the more important sites being indicated by Evans (1994) and Perennou *et al.* (1994). **Saudi Arabia**: Sabkhat-el-Fasl Lagoons (Eastern Province). **Yemen**: Aden. **Oman**: Barr al Hikman (Al Wutta and Ash Sharqiyeh regions). **UAE**: Khor Dubai, Al Wathba Lagoons (Abu Dhabi) and Ramtha Lagoons (Sharjah). **Qatar**: Al-Dhakira Mangrove and Khor al-Udeid. **Iraq**: Lake Usathe and Bahr Al Milu (Karbala), Haur Al Suwayqiyah and Haur Al Sa'adiyah (Wasit), Sudan Marshes and Haur Al Hawizeh (Maysan, Al Basrah). **Iran**: around 20 wetlands regularly hold large concentrations of flamingos, in the provinces of Azerbaijan, Mazandaran, Isfahan, Fars, Bandar Abbas, the Persian Gulf coast and Sistan/Baluchistan. **Pakistan**: regular in Sind (Ali & Ripley 1978) but also on wetlands in Baluchistan, from the Arabian Sea coast up into the northern salt lakes (Roberts 1991). **Afghanistan**: the high-altitude lakes Ab-e-Istada and Dasht-e-Nawar, where flamingos occasionally breed. Flamingos frequently occur in **Ukraine** (Osipenko, Kiev, Podolia, Kharkov region, Chernigov), **Turkmenistan** (between Chikishlyar and Gasan-Kuli), **Uzbekistan** and **Kazakhstan**, venturing north in spring and summer and breeding on the Tengiz Lakes; in winter they are to be found mainly on the southern shores of the Caspian Sea (Gistsov 1994), in Azerbaijan (Kirov Gulf), particularly in the south-east of Krasnovodsk Bay (Demente'ev *et al.* 1951; Rustamov 1994). **India**: they occur throughout much of the subcontinent but mainly in west India, east to central Uttar Pradesh and Tamil

Nadu (Grimmett *et al.* 1998). They are, however, sporadic outside of Kutch (Gujarat), absent from Bengal and Assam, and a vagrant in **Bangladesh**. **Sri Lanka**: mainly winter visitors to the northern, eastern and southern provinces, but they may breed in remote lagoons (Henry 1971; Wickramasinghi 1997).

East Africa

In East Africa, the Greater Flamingo has a similar distribution to the more abundant Lesser: from the coast of the Red Sea down to southern Africa (Brown *et al.* 1982), with major concentrations on some notable sites along the Rift Valley. The most important wetlands are found in **Ethiopia** (Lakes Shalla, Abijatta, Chitu and Debre Zeit, the Akaki area and, on the border with **Djibouti**, Lake Abbé), **Somalia**, mostly along the Indian Ocean coast but with some sightings inland (Ash & Miskell 1998), **Kenya** (on Lakes Bogoria, Nakuru, Elmenteita, Magadi, Turkana, occasionally Naivasha, Amboseli, Mombasa saltpans and **Tanzania** (Lakes Manyara, Eyasi, Burungi, Natron, Momella Lakes, with occasional breeding at Lakes Elmenteita and Natron). Greater Flamingos are widespread in **Madagascar**, where they occur on the coast, on marshes and extensive lagoons, such as Lakes Tsimanampetsotsa and Ihotry (Milon *et al.* 1973). They occur on **Aldabra** in the **Seychelles**, particularly at the eastern end of the Atoll (Penny 1974; Rainboldt *et al.* 1997), and they formerly also occurred (until 1720) on the **Ile de la Réunion** (Probst 1997).

Southern Africa

The Greater Flamingo is widespread in southern Africa (Williams & Velásquez 1997; Borello *et al.* 1998). It occurs in **Botswana** (Makgadikgadi Pans), **Malawi** (Benson & Benson 1977), **Zimbabwe, Mozambique, South Africa** (St Lucia, Bloemhofdam, Kamfersdam, Strandfontein Sewage Works), **Angola** (Dean 2000) and **Namibia** (mainly at Walvis Bay, Sandwich Harbour, Swakopmund saltpans, Cape Cross saltpans, Lake Oponono and Etosha Pan), and is a vagrant in **Zambia** (Benson *et al.* 1971). The species breeds mainly on the vast saltpans of Makgadikgadi and Etosha.

NUMBERS

World population estimates

Greater Flamingos have never been censused simultaneously throughout their entire range, although crude estimates have been made of the world population based primarily on counts during the breeding season. Brown (1959) suggested a world population of about 590,000 individuals, whereas Kahl (1975a) estimated 790,000 Greaters, with 500,000 in India alone. Following his worldwide survey for ICBP, during the period 1972–74, the author revised this estimate to 500,000 (Kahl 1975c). More recently, Wetlands International estimated the world population of Greater Flamingos to be 695,000–770,000 (Rose & Scott 1994) and later as 705,000 (Rose & Scott 1997). These estimates (Table 4) are based to some degree on Kahl's earlier figures, and are influenced by the phenomenal number of birds reported in the Rann of Kutch in 1945 (Ali 1945) and in 1960 (Ali 1960; Shivrajkumar *et al.* 1960), which recent counts indicate were probably overestimates (Johnson 2000b).

INTERNATIONAL WATERBIRD CENSUSES (IWC)

We believe that current counts of breeding birds are insufficient to project reliable world population estimates, because they exclude the non-breeding segment of the population, which, for a long-lived species such as the flamingo, can be substantial. At the present time, the IWC has probably the most reliable estimate of the world population (see Tables 5–8), but these data omit areas which in some years

Table 4. Greater Flamingo population estimates—past and present.

Region/author	Kahl 1975a*	Rose & Scott 1994	Rose & Scott 1997	Gilissen et al. 2002	Present estimate (Johnson 2000b)
India	500000	–	–	–	–
Iran	50000	–	–	–	–
Turkey	25000	–	–	–	–
Kazakhstan	20000	–	–	–	–
Afghanistan	8000	–	–	–	–
East Mediterranean, S and SW Asia		500000	500000	290000	290000
Southern Africa	75000	25000–100000	50000	65000–87000	50000
NW Sinai	5000	–	–	–	–
East Africa	50000	50000	35000	35000	35000
NW Africa	40000	–	–	–	–
France and Spain	17000	–	–	–	–
West Africa	–	40000	40000	40000	45000
West Mediterranean	–	80000	80000	100000	80000
Total world population	790000	695–770000	705000	530–552000	500000

* In Kear & Duplaix-Hall 1975.

Table 5. World population estimates. Recapitulation of Tables 6–8. Wetlands International monitoring unit 1991–98 IWC mid-January counts of Greater Flamingos throughout the Old World, and best-guess estimates for areas not covered by the IWC.

IWC figures	1991	1992	1993	1994	1995	1996	1997	1998
West Africa	11800+	17482+	5848+	19353+	15698+	28581+	60936	21253
West Med.	61302	72154	85307	92232	78600	65166	77449	80392
East Med.	363+	28963	30305+	8612	23891	26866	12478	10898
SW Asia	120054	139990	96572	128249	124751	16437	155550	195786
Southern Asia	103075	40982	59460	14291	8637	23413	5498	3083
East Africa	68042	32180	84952	52265	32370	41111	21657	20568
Southern Africa	15206	39598	18397	12270	40987	19475	24023	28323
Grand total	379842	371349	380841	327272	324934	221049	357591	360303

Best guess estimates for non-IWC areas	1991	1992	1993	1994	1995	1996	1997	1998
Banc d'Arguin	30000	30000	30000	30000	30000	30000	0	25000
Morocco	0	0	0	0	0	4000	4000	4000
Tunisia	0	0	0	0	0	15000	10000	15000
Sardinia	0	0	0	0	0	5000	0	0
Andalucia	0	0	0	0	0	0	10000	0
Egypt and Lybia	10000	10000	10000	10000	10000	10000	10000	10000
Turkey	20000	0	0	20000	20000	0	20000	20000
Iraq	3000	3000	3000	3000	3000	3000	3000	3000
Southern Asia (Pakistan, India, Sri Lanka)	0	0	0	20000	20000	20000	25000	25000
SW Asia (mainly Iran)	0	0	0	0	0	125000	0	0
East Africa (mainly Tanzania)	10000	10000	10000	10000	10000	10000	10000	10000
Southern Africa	20000	10000	20000	20000	10000	20000	20000	20000
Best-guess total	93000	63000	73000	113000	103000	242000	112000	132000
World estimate	472842	434349	453841	440272	427934	463049	469591	492303

Table 6. IWC counts of flamingos in the Mediterranean. Counts of Greater Flamingos made during the Wetlands International mid-January waterfowl censuses 1991–99 (from Rose 1992, 1995; Rose & Taylor 1993; Costa & Rufino 1997; Serra *et al.* 1997; Delany *et al.* 1999; Gilissen *et al.* 2002).

Country/year	1991	1992	1993	1994	1995	1996	1997	1998	1999
Morocco	2576	1477	4318	3612	4506	–	–	–	–
Algeria	2006	5918	2727	6126	300	5950	17011	21060	24542
Tunisia	10611	14485	9277	12808	10824	–	5438	–	6993
Portugal	1006	698	3998	3980	2915	373	668	1695	6273
Spain (incl. Balearics)	19058	21849	28577	26376	21747	24344	16895	24003	27529
France	24318	21151	27613	25983	27500	26580	23656	23070	27733
Italy (incl. Sardinia)	1727	6576	8797	13347	10808	7919	13781	10564	17537
Total west Mediterranean	61302	72154	85307	92232	78600	65166+	77449	80392	110607
Albania	–	–	1	–	271	20	4	–	–
Greece	–	–	3655	5553	4526	1508	11202	8663	7465
Cyprus	c. 343	8500	4006	c. 2500	12642	4698	1202	2230	1253
Turkey	–	20448	22210	35++	5746	20583	–	–	51755
Syria	–	–	420	510	650	–	–	–	–
Israel	20	15	13	14	56	57	70	5	139
Egypt	–	–	–	–	–	–	–	–	–
Libya	–	–	–	–	–	–	–	–	–
Total east Mediterranean	363+	28963	30305	8612+	23891	26866	12478+	10898	60612
Grand total	61665+	101117	115612	100844	102491	92032	89927	91290	171219

Table 7. IWC counts of flamingos in Asia. Counts of Greater Flamingos made during the Wetlands International mid-January waterfowl censuses 1991–99 (from Perennou & Mundkur 1991, 1992; Mundkur & Taylor 1993; Rose 1995; Lopez & Mundkur 1997; Delany et al. 1999, Gilissen et al. 2002, Li & Mundkur 2004).

Country/year	1991	1992	1993	1994	1995	1996	1997	1998	1999
Azerbaijan	13130	–	990	–	5200	5350	4150	–	–
Kuwait	–	–	6	684	–	–	–	–	–
Bahrain	–	794	–	291	–	–	–	–	–
Iran	84421	121722	88049	116031	114596	–	146646	179580	–
Oman	2575	8866	924	1287	1445	1088	1777	12994	4071
Qatar	259	404	424	550	440	395	–	–	–
Saudi Arabia	755	6626	5734	7652	2338	8066	–	–	–
Unit. Arab Emirates	2614	1491	298	1349	732	1538	2235	3092	1552
Yemen	0	87	147	405	–	–	742	–	–
Kazakhstan	0	0	–	0	–	–	–	–	–
Turkmenistan	16300	–	0	0	–	–	–	120	12
Total SW Asia	120054	139990	96572	128249	124751	16437	155550	195786	>5635
Pakistan	>50159	30300	52673	>286	>72	>118	2858	1276	1741
India	>52916	9285	6537	13867	8377	22118	40	1001	127
Sri Lanka	0	1397	250	138	188	1177	2600	806	2514
Total S Asia	103075	40982	59460	14291	8637	23413	5498	3083	4382
Grand total	223129	180972	156032	142540	133388	39850	161048	198869	>10017

No counts for Iraq, no count for Iran in 1996 and 1999, and Turkmenistan 1992, 1995–97. Poor coverage of Pakistan 1994–99. In 1991 a further 25,000 flamingos were seen in India but not identified to species.

Distribution and numbers 79

Table 8. IWC counts of flamingos in Africa. Counts of Greater Flamingos made during the Wetlands International mid-January waterfowl censuses in 1991–2000 in West Africa, East Africa and Southern Africa (from Perennou 1991,1992; Taylor 1993; Taylor & Rose 1994; Dodman & Taylor 1995, 1996; Dodman et al. 1997, 1999, Dodman & Diagana 2003).

	1991	1992	1993	1994	1995	1996	1997	1998	1999	2000
Cameroon	0	–	0	0	1	0	0	0	–	–
Sierra Leone	–	0	–	403	–	–	–	–	–	–
Niger	–	0	0	0	1	0	0	0	–	–
Guinea	–	–	–	–	–	–	–	–	–	125
Guinea Bissau	–	–	–	–	–	–	54	31	100	98
Senegal	11800	17482	4698	17970	2540	12118	21276	18823	17564	30336
The Gambia	–	–	25	–	–	–	0	0	–	–
Mauritania	–	–	1125	980	13156	16463	39606	2399+	17165	72408
Total West Africa	11800	17482	5848	19353	15698	28581	60936	21253+	34880	102967
Eritrea	–	–	–	–	–	–	–	–	–	30
Sudan	–	–	156	–	–	–	130	–	–	–
Ethiopia	57453	15688	59548	35387	19599	24543	5473	2307	3632	1488
Uganda	–	0	–	–	2	–	0	–	–	–
Kenya	10589	16492	25248	12378	11657	16250	15454	15383	7876	19115
Tanzania	–	–	–	4500	1112	318	600	2878	–	–
Total East Africa	68042	32180	84952	52265	32370	41111	21657	20568	11508	20633
Malawi	0	24	1	–	28	–	0	0	48	0
Zambia	0	–	0	–	1	1	14	12	–	–
Zimbabwe	–	–	7	–	2	–	0	1	–	–
Mozambique	–	–	–	–	–	–	546	209	–	–
Botswana	111	7515	170	877	1205	969	998	394	1044	1287
South Africa	559	17742	–	7878	24158	2	2888	2665	9833	5340
Madagascar	–	–	121	264	669	3412	0	479	–	–
Namibia	14536	14317	18098	3251	14924	15091	19577	24563	24946	304
Total S Africa	15206	39598	18397	12270	40987	19475	24023	28323	35871	6931
Grand total	95048	89260	109197	83888	89055	89167	106616	70144+	82259	130531

may have had a substantial number of flamingos, and which are too important to simply ignore. Consequently, we have formulated educated approximations for these areas (Table 5) based on the best-available ancillary evidence. Our annual approximations for areas not surveyed during the period 1991–98 range from 63,000 to 242,000 birds. We have provided a basis for these estimations in the regional summaries below.

COUNTS OF BREEDING NUMBERS

The greatest number of pairs reported breeding worldwide in any given year (Figure 14, Appendix 6) was just over 97,000 in 1998, followed by 64,000 in 1991, and 57,000 in both 1986 and 1988. In two of these years, 1986 and 1991, there were no reports of breeding from East Africa and southern Africa, and in 1988 no reports from Asia. Better coverage of the area by the IWC and more surveys during the breeding season may confirm whether colonies of the size indicated by Ali (1960) and Shivrajkumar et al. (1960) still exist in Asia.

By combining these rough approximations and the IWC surveys, we estimate the world population as 500,000 birds (Table 4; Johnson 2000b). Although this estimate is admittedly crude, we believe that it is more realistic, and better substantiated, than estimates using earlier counts from Asia that were probably inflated. There seems little justification for assuming that there are 500,000 Greater Flamingos in Asia alone, since neither the IWC counts nor the data available on breeding over the past 25 years support such high numbers. We have perhaps been a little modest with our estimates for non-IWC areas but we have been generous with our estimate of the world total of 500,000 flamingos, since we fell short of this figure by between 7,000 (1998) and 72,000 (1995) birds.

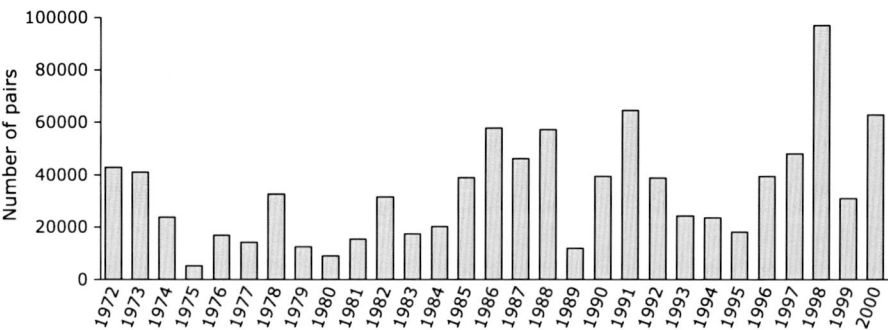

Figure 14. The number of pairs of Greater Flamingos recorded breeding throughout the Old World, 1972–2000. See Appendix 6.

REGIONAL SUMMARIES

WEST AFRICA

Wintering numbers: The main concentrations of flamingos have always been found on the wetlands near the mouth of the River Senegal, including the Djoudj National Park, and north along the coast of Mauritania (Aftout es Saheli) to the intertidal flats of the Banc d'Arguin National Park. It is important to note that this latter area was not included in the IWC survey (Table 8) until 1997. There were at least 32,000 flamingos wintering on the Banc d'Arguin in 1971–72 (Pététin & Trotignon 1972), 60,000 in 1978–79 (Trotignon & Trotignon 1981) and 52,000 in 1979–80 (Trotignon *et al.* 1980). For the period 1990–94 Gowthorpe *et al.* (1996) considered the population of flamingos in the park to be around 25,000–30,000, and the 1997 census (Dodman *et al.* 1997) revealed a similar figure of 35,287 birds. We have therefore taken 30,000 as our 'best estimate' of the number of flamingos present in the park during the period 1991–96, prior to its inclusion in the IWC. This results in less inter-annual variation in our estimates of numbers in West Africa, and suggests a normal wintering population in the order of 40,000–60,000 flamingos.

Breeding numbers: Flamingos have been confirmed breeding in West Africa (Figure 15, Appendices 6, 7) in 22 out of the past 29 years (1972–2000). The number of breeding pairs has varied greatly, with a maximum of almost 17,000 in 1982.

MEDITERRANEAN REGION

Wintering numbers: Most countries bordering the Mediterranean participate in the IWC (Table 6) but there are some important gaps in some years, which we

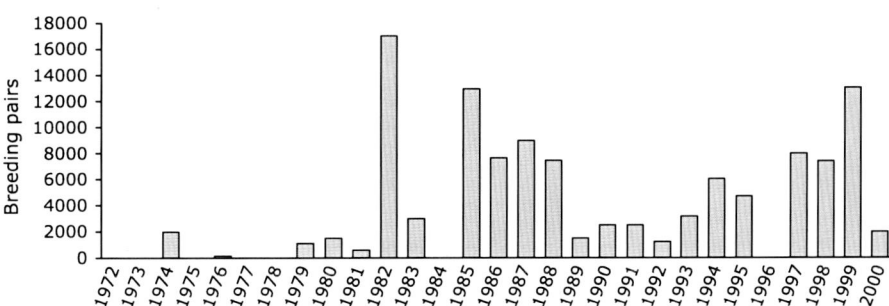

Figure 15. The number of pairs of Greater Flamingos recorded breeding in West Africa, 1972–2000. See Appendix 7.

discuss below. **Tunisia**: the five counts made in 1991–95 revealed an average of 11,600 flamingos in Tunisia. In each of these years, however, sites which are known to be important for flamingos were not counted. In January 2000, over 27,000 flamingos were counted throughout the country (H. Azafzaf, Tunisian Ornithological Group *in litt.*), and we therefore consider 15,000 to be a reasonable best-guess estimate for numbers in the country in mid-winter during the three years when few wetlands were visited or no counts at all were made (1996–98). **Egypt** and **Libya**: the two winter censuses carried out in Egypt revealed very different numbers of flamingos, with 7,500 in 1979/80 and over 20,000 in 1989/90 (Walmsley in Goodman & Meininger 1989). No recent count data are available for Libya, where Booth (in Stanford 1954) reported that 150–2,000 were regular from August to April. On the basis of these observations, we have retained a figure of 10,000 flamingos in Egypt and Libya for the period 1991–98. **Turkey**: three IWC counts (in 1992, 1993, 1996) revealed an average of just over 20,000 flamingos in winter. We have, therefore, taken this as our best guess for years when no counts were made, or when they were incomplete (1991, 1994, 1995, 1997, 1998).

The flamingos wintering in the western Mediterranean were censused on several occasions during the 1970s and 1980s (Figure 16, Appendix 9), when numbers were similar to those reported during the more recent IWC counts. If we allow for omissions in the IWC coverage, there are probably about 80,000 flamingos regularly wintering around the western basin of the Mediterranean. In the eastern Mediterranean, four censuses during the 1990s revealed 24,000–30,000 flamingos wintering principally in Turkey, Greece and Cyprus. By averaging these censuses and taking into account an additional 10,000 flamingos probably wintering in Libya and Egypt (see Goodman & Meininger 1989), we can estimate that about 37,000 flamingos winter around the eastern basin of the Mediterranean. Thus, we believe that around 117,000 flamingos probably winter in the whole Mediterranean region.

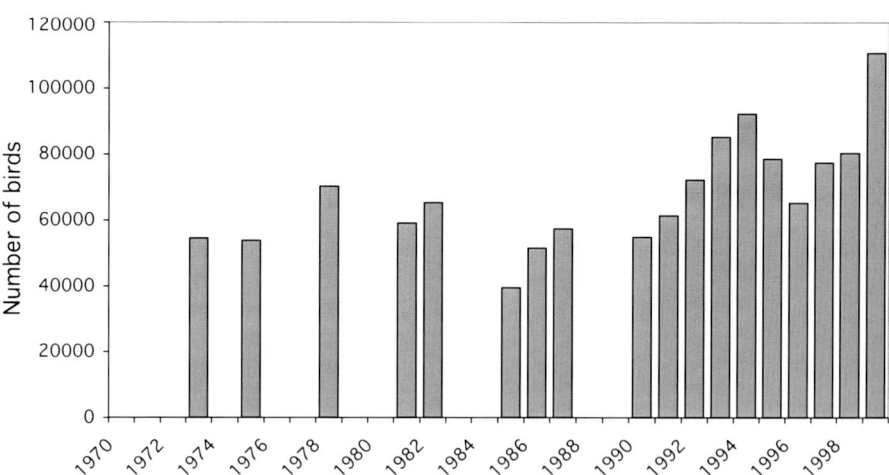

Figure 16. IWC counts of flamingos over the western Mediterranean. See Appendix 9.

Breeding numbers: Breeding throughout the Mediterranean region is generally well documented, though some colonies in North Africa may have gone undetected. In 1991, a year of complete coverage of the region from Spain to Turkey, seven colonies were established, totalling over 43,000 breeding pairs. This breeding season was preceded by good autumn to spring rainfall, which allowed breeding in temporary sites in Tunisia, Spain and Turkey. In 1998, also a year with high water levels, these figures were surpassed; in the western Mediterranean alone there were over 44,000 breeding pairs (Figure 17, Appendix 8). It appears therefore that colonies are established annually at one or more of about 15 breeding sites in the Mediterranean and from 20,000–50,000 pairs of flamingos breed.

ASIA

Wintering numbers: With the exception of Iraq and the year 1996, there has been good coverage of south-west Asia by the IWC. Counts have revealed 96,000–196,000 flamingos in the region (Table 7), with the largest numbers occurring in Iran. Although Iraq has not been counted in recent years, relatively few (just above 3,000) birds were counted in the country in two out of five censuses between 1967 and 1979 (Perennou *et al.* 1994). In southern Asia, the large variation in numbers reported from India and Pakistan may be due to gaps in coverage, particularly for the Rann of Kutch in Gujarat. Perennou *et al.* (1994) recognise these gaps and assume an additional 20,000 birds in uncounted areas. The maximum number of birds ever counted over the whole of Asia was 223,000 in 1991. Our best-guess estimates for non-IWC areas in south-west Asia are 3,000 flamingos in Iraq in 1991–98 and 125,000 in Iran in 1996, and for non-IWC areas in southern Asia they are 20,000 in 1994–96 and 25,000 in 1997–98 in Pakistan, India and Sri Lanka.

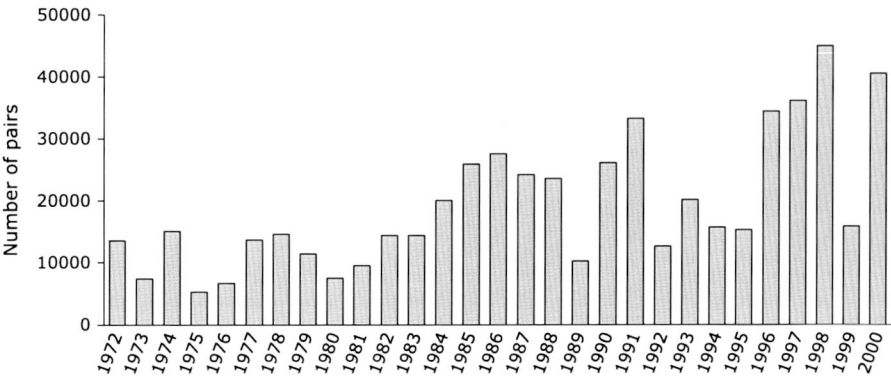

Figure 17. The number of pairs of Greater Flamingos recorded breeding in western Mediterranean, 1972–2000. See Appendix 8.

Breeding numbers: Lake Uromiyeh has been the most important breeding site in Asia in recent years, with large colonies being established there quite regularly from the 1970s to 2000. Lake Tengiz in Kazakhstan is probably of similar importance, although data on breeding numbers (site 23 Chapter 12) are lacking. Similar data gaps exist for colonies in Afghanistan and the Rann of Kutch (India). Despite these data gaps, it is clear that numbers of birds in the Asian flamingo colonies (Appendix 6) are far lower than previously reported for this continent, and the IWC counts also tend to indicate the presence of fewer flamingos in Asia than earlier evaluations suggested, although more recently Jadhav & Parasharya (2004) reported 28,333 Greater Flamingos on some of the wetlands in Gujarat state in January 2003.

EAST AFRICA

Wintering numbers: There are several gaps and weaknesses in coverage of this region (Table 8): (1) Tanzanian wetlands were not counted prior to 1994, (2) Lake Natron (Tanzania/Kenya) has never been censused during the IWC surveys, (3) Lake Magadi (Kenya) was not censused in 1994, and (4) counts at Lake Abijatta (Ethiopia) are extrapolations from sample counts, thus differing from the totals given in the summary tables of the African Waterfowl Censuses. These gaps may at least partially explain the variation from 20,000 to 85,000 birds reported for East Africa. Our best estimate for non-IWC areas is 10,000 flamingos (1991–98) based on a complete lack of counts at Lake Natron and on no counts elsewhere in Tanzania in 1991–93, no count at Lake Magadi (Kenya) in 1994, and poor coverage of Ethiopian wetlands during 1995–98. Over the past seven years, at least 30,000 and perhaps as many as 100,000 Greater Flamingos have wintered in East Africa (Katondo (1997) refers to over 101,000 Greater Flamingos on Lake Eyasi alone in January 1995, which for unknown reasons are excluded from the IWC totals for Tanzania).

Breeding numbers: Breeding-season counts since 1972 have never exceeded 2,500 pairs. However, only aerial surveys provide adequate coverage of areas such as Lake Natron, and even then it is difficult to distinguish Greater Flamingos from the far-more-abundant Lessers (see Woodworth *et al.* 1997). It would, therefore, not be surprising to discover that the figures given in Appendix 6 underestimate both the numbers of breeding pairs and the frequency of breeding in East Africa.

SOUTHERN AFRICA

Wintering numbers: The IWC surveys have rarely included thorough coverage of some wetlands (e.g. the Makgadikgadi Pans in Botswana and Etosha Pan in Namibia) which can host very large numbers of flamingos when flooded. For example, Parker (in Borello *et al.* 1998) reported an estimated 300,000 Greaters on

Sua Pan in November 1974. During the IWC counts (Table 8), from 7,000 to 41,000 birds were reported. However, Simmons (1996) estimated 47,427 Greater Flamingos in southern Africa during January in 1993 and 1994. Thus, our best guesses for the number of flamingos overlooked in southern Africa during the IWC are 10,000 birds in 1992 and 1995 and 20,000 birds in the other years of survey.

Breeding numbers: Although breeding in southern Africa seems to be irregular, when colonies are established they can be large, for example at Etosha Pan (Namibia) 27,000 birds in 1971 (Berry 1972), and at Makgadakgadi Pan (Botswana) 18,000 pairs in 1978 (Robertson & Johnson 1979), 25,000 pairs in 1988 (Hancock 1990), 36,000 birds in 1993 (Liversedge in Simmons 1996) and *c.* 24,000 pairs in 1999–2000 (McCulloch & Irvine 2004).

POPULATION TRENDS

The lack of precision, and notably the gaps in coverage of some important Asian and African sites, prevents any reliable assessment of global population trends. Neither should our proposed world population estimate (Table 4) be interpreted as indicating a decrease in the numbers of flamingos in the world, compared to previous estimates, since we believe that in fact it represents a more accurate estimate using the best data available at present. Population trends can only be detected using reliable long-term quantitative data gathered over many years. For wintering numbers the most reliable long-term data are from the IWC, which has counted the number of birds wintering in France since 1965 (Figure 18, Appendix 9). Coverage of the whole of the western Mediterranean (Figure 16, Appendix 9) has been sufficient to compute numbers throughout the region in 18 of these 36 years.

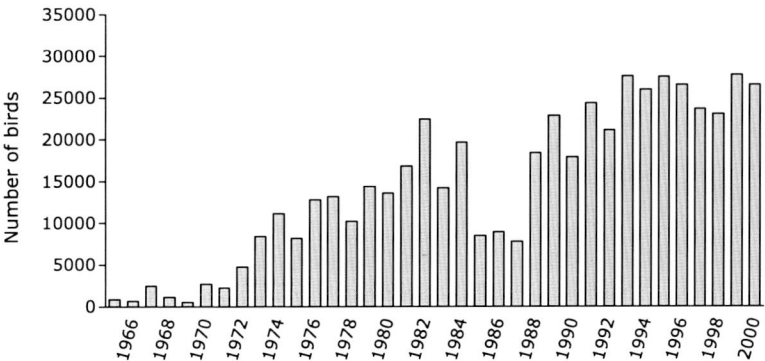

Figure 18. IWC counts of flamingos in France, 1965–2000. See Appendix 9.

For breeding population trends, the number of breeding pairs has been estimated in the Camargue for 54 years (Figure 19, Appendix 8, Table 13) (Johnson 1997b, 1999), in Spain for 35 years (Rendón-Martos & Johnson 1996), and over the whole western Mediterranean for over 29 years (Johnson 1997b) (Figure 17, Appendix 8). There are also sufficient records for West Africa (Figure 15, Appendix 7) for the past 29 years. We therefore restrict the following analyses to these data sources, recognising that even these trends may be confounded if there have been concurrent shifts in distribution among regions.

TRENDS IN THE NUMBER OF WINTERING BIRDS

The number of flamingos wintering in France (Figure 18, Appendix 9) has steadily increased (Spearman rank-order correlation coefficient, rs = 0.9102, n = 36, p < 0.001) from the early 1970s until the present. Only during the severely cold winters of 1985–87 were numbers much lower, owing to the death of many birds during the first of these three winters, and to emigration. Over the past decade (1990–99) the number of birds censused in January in southern France has been more than seven times that of the first 10 years of winter monitoring (1965–74). Throughout the western Mediterranean, however, the IWC figures (Figure 16, Appendix 9) indicate a moderate but significant increase (rs = 0.6154, n = 18, p = 0.013). This may be because some of the birds that breed in the western Mediterranean winter outside the region, probably in West Africa. It could also imply an increase in mortality, but it is unlikely that mortality alone could explain the dramatic differences observed. However, it emphasises the need to evaluate flamingo population dynamics at large geographic scales which correspond to the biological realities of the population.

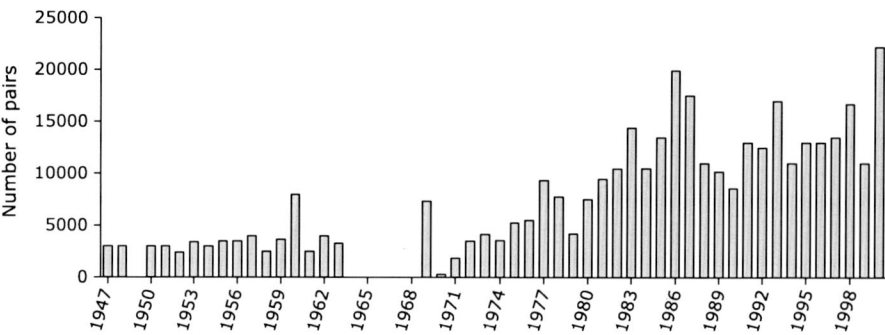

Figure 19. The number of pairs of Greater flamingos recorded breeding in the Camargue, 1947–2000. See Appendix 16.

TRENDS IN THE NUMBER OF BREEDING BIRDS

In contrast to numbers of wintering birds, the number of flamingos breeding in the western Mediterranean has increased more or less steadily over the past three decades (Figure 17, Appendix 8, rs = 0.7207, n = 29, p < 0.001). More than 30,000 pairs have bred in six of the past 12 years (1989–2000), whereas during the decade 1972–81, the three most successful years achieved only half this number. This reflects a general increase in the number of flamingos breeding in France (Figure 19, Appendix 16, rs = 0.8215, n = 54, p = 0.001; Johnson 1994, 1999) and Spain (Rendón-Martos & Johnson 1996; Rendón 1997; Johnson 1997a), and a more recent expansion of breeding in Italy (Brichetti & Cherubini 1997; Johnson 1997a).

Flamingos tend to be more faithful to breeding sites which are permanent or semi-permanent, as in the Camargue and, to a lesser extent, in Spain and Italy. This fidelity to secure breeding places has led to an increase in the number of breeding pairs. In the Camargue, only about 10,000 pairs are able to breed each year on the specially provided island, but more have bred at Fangassier since 1983, by nesting in the depressions between nest mounds, by taking over nests of the earlier breeders as they are vacated by the chicks moving into the crèche, or by extending the colony onto a neighbouring dyke and/or island. This last strategy had little success until recently.

Tunisia contributes substantially to the number of breeding birds in the western Mediterranean, but only during years of sufficient rainfall (Johnson 1997b, 1999). In contrast to the western Mediterranean, in West Africa the number of birds breeding (Figure 15, Appendix 7) has not shown a consistent increase (rs = –0.0344, n = 24, p = 0.546), although there have been periods of higher numbers. Gowthorpe et al. (1996) reported a decline in the number of flamingos occurring on the Banc d'Arguin of West Africa, but the numbers of breeding birds reported in West Africa are not entirely consistent with this claim. Although numbers have certainly varied considerably, breeding has probably been more consistent. While the current number of breeding birds may be lower than in a few exceptional years during the 1980s, there has been an increasing trend during the 1990s and numbers are certainly higher and more stable than they were during the 1970s (Appendix 7).

Corresponding with increases in the breeding population size throughout the western Mediterranean, there has also been a recent increase in the number of colony sites occupied (Figure 20, Appendix 8). During the period 1972–92, from one to three colonies were established in any given year, while during the period 1993–2000, between four and eight sites, six of which were never used before 1993, have been occupied simultaneously.

The increase in the size of the breeding population has been accompanied by an extension of the species' range, so that flamingos now occur regularly and sometimes in abundance on wetlands where they were scarce or absent two or three decades ago, particularly in the central and eastern parts of the Mediterranean. For

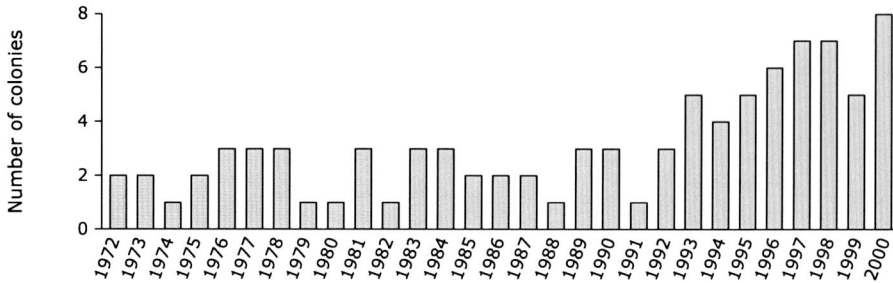

Figure 20. The number of flamingo colonies established in the western Mediterranean, 1972–2000. See Appendix 8.

example, during midwinter waterfowl counts in Italy, and in particular in Sardinia, flamingos were recorded on 43 wetlands in 1991, on 47 in 1992, on 119 in 1993, on 154 in 1994 and on no fewer than 217 in 1995 (Serra *et al.* 1997). If left undisturbed, flamingos now exploit wetlands in close proximity to human presence, including airports.

Brown *et al.* (1982) suggested that the Greater Flamingo might be increasing in southern Africa due to increased man-made habitats (such as saltworks, sewage lagoons or large dams), but perhaps decreasing in the East African Rift Valley due to interference from Great White Pelicans. Simmons (1996) more recently reported a decrease of Greater Flamingos across Africa since the mid 1970s, in particular in southern Africa. Since his report, however, there has been massive breeding again in Etosha, in March 1997, by 15,000–20,000 Greaters (R. Simmons pers. comm.) and in Makgadikgadi in 1999–2000 by over 23,000 pairs (McCulloch & Irvine 2004). This may warrant a revision of the trend reported by Simmons (1996); it also emphasises the need to consider trends over long timescales, and to consider changes in distribution in relation to a variable environment.

CHAPTER FIVE

Movements

Understanding the dispersion of birds (i.e. their movements and the resulting spatial distribution) is essential from both an ecological and a conservation perspective. From a demographic perspective, all population change is a result of births, deaths, emigration and immigration. The last two are definitively movement, and the first two are strongly influenced by movement (Parker & Stuart 1976). From a genetics perspective, movement is an essential element of gene flow. From a conservation perspective, in addition to demography and genetics, understanding how and why birds move between habitats, and their resulting distribution, is essential for setting up reserves, protecting critical habitats, and targeting conservation efforts. In this chapter we explore the movements of Greater Flamingos, starting with the mechanics and behaviour of flight. We then explore several types of movement, some elements of the resulting spatial patterns, and offer some concluding thoughts.

FLIGHT BEHAVIOUR

Flamingos often need to run across the surface of the water for several metres in order to become airborne, although when facing the wind they can rise into the air almost as soon as they open their wings. Once they are on the wing they fly with neck and legs extended, and flocks form skeins or V-formations. With a tail-wind birds generally fly high, but against a strong head-wind they often fly just above the surface of the water, flocks rising in undulating lines as they pass over land. Oddly, birds will sometimes preen in flight; they stop beating their wings and swing their head around to preen the back feathers momentarily. An experienced observer can tell from the manner of flight whether the flocks are moving between local feeding areas and/or breeding sites, dispersing, displaying (see Chapter 7), or have been disturbed by aircraft, which some still fear. When disturbed they do not immediately form skeins, but fly in dense flocks like ducks or waders, all birds changing direction simultaneously and with surprising rapidity. At these times, the observer will appreciate the strength and agility of flamingos on the wing. Flamingos do not soar but beat their wings continuously, only gliding to alight. During level flight, the head is maintained on a fixed plane by a snaking, compensatory movement of the neck, the body rising on the downward wingbeat and falling on the upward. This is quite visible when birds are flying singly or in pairs.

Long-distance flights are undertaken mostly at night (Brown in Kear & Duplaix-Hall 1975; Curry-Lindahl 1981; Johnson 1989a; Shirihai 1996; Williams & Velásquez 1997; Rendón-Martos *et al.* 2000; McCulloch *et al.* 2003). For example, during post-breeding and post-fledging dispersal, many birds have been observed leaving the Camargue and heading south across the Mediterranean in the late afternoon, shortly before sunset. Indeed, the position of the sun on the horizon appears to be highly informative for nocturnal migrants (Sandberg 1991 in Berthold 1996; Terrill 1991 in Berthold 1996). One flamingo (code CAL) seen near Aigues-Mortes (Gard), west of the Camargue, on the afternoon of 13 September 1978, was observed the following morning near Oristano in Sardinia, 550 km to the SSE. They frequently take advantage of favourable winds for their trans-Mediterranean flights (Johnson 1989a). The mistral, a wind which blows from the NNW down the Rhône Valley, favours birds heading toward Sardinia and/or Tunisia, destinations lying conveniently to the SSE. This wind sometimes prevails during the whole sea crossing; it not only increases the birds' flying (ground) speed but also assures good visibility, at least at the start of their flight, and presumably thus also favours navigation. If by using favourable winds birds increase flying speed by an average 30% (Liechti & Bruderer 1998), then a flight from the Camargue to Tunisia, a distance of 800–850 km, could be accomplished by flamingos in 10–11 h as opposed to 14–15 h in calm weather. Several studies have demonstrated the importance of tail-winds for migration, for example for Greater Canada Geese (Wege & Raveling 1983), Bewick's Swans (Evans 1979), and Common Cranes (Alerstam & Bauer 1973).

The flying speed of flamingos has been estimated by several authors to be around 50–60 km/h (Allen 1956; Rooth 1965; Brown in Kear & Duplaix-Hall 1975; Brown *et al.* 1982), although more precise data have recently been gathered by B. Bruderer and A. Boldt of the Swiss Ornithological Institute, Sempach. They radar-tracked four flamingos in low-level flight (at 100 m) on Mallorca (Spain). These birds were probably not undertaking a long-distance flight, and were flying against the wind at a ground speed of 13.7 m/s (49.3 km/h) (Bruderer & Boldt 2001). They have twice recorded flamingos on radar over southern Israel, four and five individuals migrating at heights of 1,020 m and 1,290 m above ground respectively. In one instance they were flying at an air speed of 19.1 m/s (68.8 km/h) assisted by a light tail-wind (16.1 km/h) and in the other at 14.9 m/s (53.6 km/h) with a slightly favourable side-wind of 11.5 km/h. These authors refer to 15.9–15.2 m/s (57.7–54.7 km/h) as realistic migratory speeds for the Greater Flamingo. For comparison with other species, the air speed of Whooper Swans has been measured at 60 km/h, Common Cranes at 67 km/h, Bean Geese at 72 km/h and Common Eiders at 74 km/h (Alerstam 1990). The ground speed of Greater Canada Geese on migration in North America has been measured at up to 96 km/h, which the birds sustained over a distance of at least 24 km (Bellrose & Crompton 1981).

The height at which flamingos undertake long-distance flights seems to depend largely on whether they are crossing land or sea, and on the wind strength and direction (Brooke & Birkhead 1991). Flocks departing from France over the Mediterranean have been judged by observers to be flying at a maximum height of *c.* 250 m, usually much lower, at least until lost to the observer on the horizon. Incoming flights in spring from the south-east, seen arriving both in the Aude, south France, against a head-wind (Y. Kayser pers. comm.), and in the Camargue, have all been judged by the observers to be at less than 50 m altitude, and some birds were just above the waves. Occasional observations have also been made of low-flying birds at sea between Italy and the North African coast (Henderson 1921) and across the Strait of Gibraltar (Verner 1909). This latter narrow stretch of ocean, however, seems also to be crossed by flamingos flying very high, as reported by Chapman & Buck (1910). In the Caribbean, Rooth (1965) observed Caribbean Flamingos departing from Bonaire in the late afternoon low over the waves, presumably heading for the coast of Venezuela, around 100 km to the south.

When crossing landmasses, flamingos fly higher. At Eilat, in Israel, a group of 85 birds was observed by Y. Kayser and E. Didner (pers. comm.) on 25 March 1996. The birds arrived in the evening from the south, flying quite low along the coast of the Gulf of Aqaba. When they reached Eilat, they circled for approximately half an hour, gaining altitude before moving off at a much greater height to the north-east (in the direction of Iraq and Iran). Shirihai (1996) referred to flamingos among the species of birds which have been tracked by radar and identified by telescope flying at an altitude of 2,000–6,000 m above sea-level over the arid, central parts of the country. Bruderer & Boldt (2001) also tracked four birds flying SSW over the Arava valley (–150 m) on 29 September 1991 at 10.27 hrs at an altitude of 1020 m and five birds above the Negev Desert (450 m) on 2 April 1992 at 10.00 hrs heading NNE at 1287 m. In Spain, B. Bruderer and F. Liechti (Swiss Ornithological Institute) detected a flamingo on their radar screen flying at an altitude of between 4210 m and 4786 m over Malaga (in Rendón-Martos 1996).

Flocks of flamingos observed departing from the Camargue vary in size from some tens up to several hundred birds. Most are composed of 20–60 birds, and very few are of more than 200 (Figure 21). Kokhlov (1995) reported 10 flights of flamingos along the east coast of the Caspian Sea in October–November, numbering from 4 to 350 birds, and Varshavskiy *et al.* (1977) reported flocks of 150–180 birds, more often in the range 30–60 birds. Similarly, in Cyprus we have seen several flocks of between 110 and 210 birds, but Flint & Stewart (1992) mentioned some of up to 1,000. All of the flights we have seen in the Mediterranean region have been composed of either adults, or adults and juveniles, and it seems that juveniles seldom undertake long-distance flights on their own, although, as we shall describe, they apparently disperse to a greater extent than older birds, and at times wander on their own. Nothing is known of the relationships, if any, among members of flocks.

TYPES OF MOVEMENTS

Animals, and birds in particular, exhibit several different types of movements, which are discussed below (except for local foraging flights, which are reported in the next chapter). At first glance, the way that we categorise different types of movements may seem unusual. For example, we consider 'post-fledging' dispersal and 'natal' dispersal separately. They both deal with dispersal of a juvenile from its

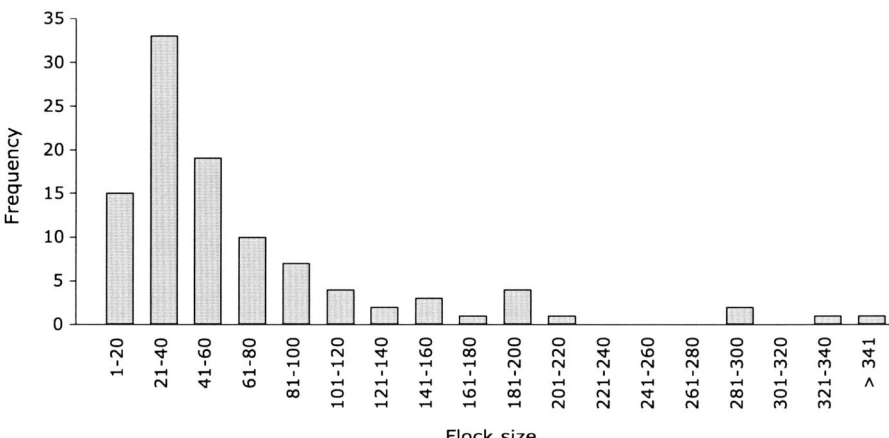

Figure 21. Size of flamingo flocks departing from the Camargue on post-nuptial, trans-Mediterranean flights.

natal site, but the process of dispersal has a beginning, a middle, and an end. The beginning, or departure from a site, probably implies, in the case of post-fledging dispersal, an innate releasing mechanism, triggered by environmental conditions. The middle may consist of exploration, particularly in the case of juveniles, or be the journey of adults following a regular migratory flyway. The end is the process of habitat selection, which determines where birds settle. This could be the habitat where they choose to stop briefly, remain during non-breeding, or where they ultimately breed. In addition, the underlying causes and/or consequences of dispersal may be demographic (i.e. influencing the probability of survival or reproduction) or genetic (i.e. influencing gene flow). Thus, the departure of a juvenile from its natal site (post-fledging dispersal) may have certain reasons or consequences from a demographic perspective that are independent of where they ultimately breed; but from a genetic standpoint, the dispersal from their natal site to their breeding site (i.e. natal dispersal) is what affects gene flow. In the following presentation, we have tried to group movements relative to function, and have separated some groups that at first glance may appear similar but for which the underlying causes or consequences may be quite different.

POST-FLEDGING DISPERSAL

Soon after they are able to fly, juveniles often disperse, and may be seen only a few days later hundreds of kilometres away from their natal colony. In the Northern Hemisphere, birds typically disperse from the more northerly colonies (Camargue (Fig. 5.2) and Lake Uromiyeh (Fig. 5.3)) in a general southerly direction. From the more centrally situated colonies (Fuente de Piedra, Molentargius), they disperse in all directions, and from the more southerly colonies (Tunisia, Mauritania) they probably disperse in a predominantly northerly direction.

Bauer & Glutz von Blotzheim (1966) stated that it is mostly young flamingos which undertake sporadic movements. Scott (1975), summarising movements from Lake Uromiyeh, likewise noted that the younger flamingos undertake an extensive dispersal, generally southwards, rather than following well-defined migration routes. Behrouzi-Rad (1992), analysing ring recoveries from Lake Uromiyeh, also reported that juveniles are more likely than adults to undertake long-distance journeys. These findings are supported to some extent in the Western Palearctic by recoveries of marked birds. Many of the flamingos banded in the Camargue which have been sighted in West Africa, a distance of 3,500 km, are known to have flown there during their first autumn. Indeed, 34 of 57 banded flamingos observed in Senegal from 1983 to 1992 were first seen there as juveniles. A large proportion of these banded birds (46) later returned to the Mediterranean region, where most of them apparently remained, since up to 1998 only one of them had been resighted in West Africa despite relatively good observer effort in Senegal during the period 1989–92. These observations tend to be in accordance with Lack's (1968a) statement regarding partial migrants that birds seemingly

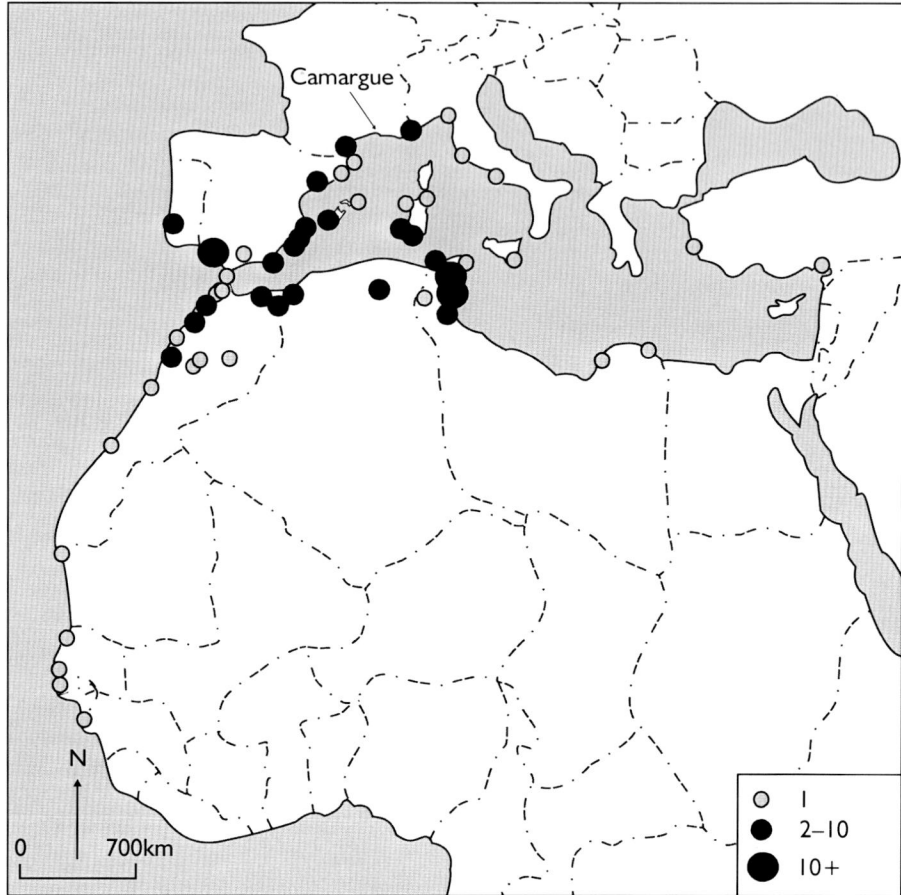

Figure 22. Geographical location of recoveries outside southern France of flamingos ringed as chicks in the Camargue in 1947–61 (Museum Paris alloy rings).

become less dispersive as they grow older and tend to spend the non-breeding season closer to their eventual breeding site. Similarly, young Northern Gannets are known to move much greater distances than adults, and they remain in their wintering areas over summer (Nelson 1978).

It is most unusual for flamingos to disperse from the Camargue to the north, but in 1998 a group of six juveniles appeared on 1 September on Lake Geneva (Vaud) Switzerland, 383 km north-east of the Camargue (Winkler 1999). One of them (code DCJT) had been banded in the Camargue a month earlier and had been seen in the crèche only nine days before being sighted in Switzerland. These birds moved 65 km further north on 2 September, to Lake Neuchâtel, 440 km NNE of the Camargue, where they stayed until 5 November, much to the excitement of local

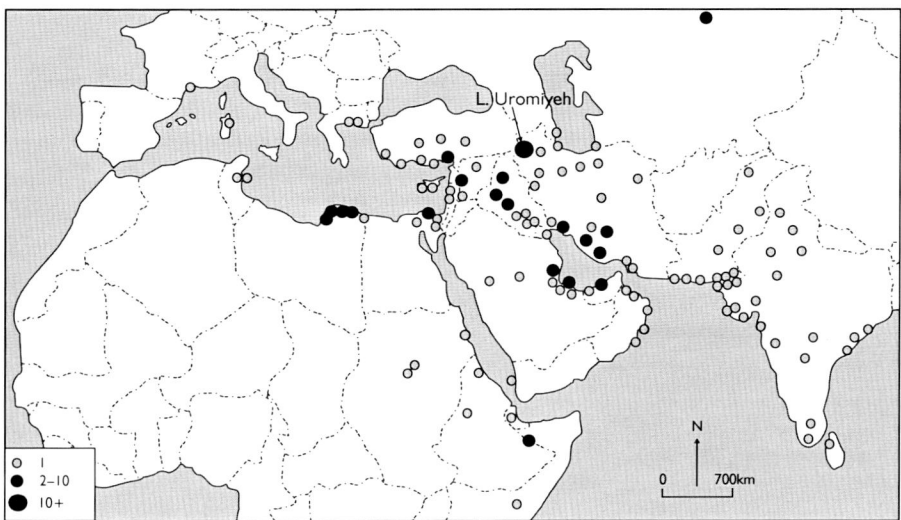

Figure 23. Geographical location of recoveries of Greater Flamingos ringed at Lake Uromiyeh, Iran.

birdwatchers. On 10 November the banded bird was observed in France (Hautes-Alpes) 150 km to the south of Lake Neuchâtel, where it stayed until 1 December. It was seen again the following February back on the south coast of France. Just one year earlier a group of 10 juveniles was observed on 31 August 1997 at St Martin-en-Bresse (Saône et Loire), 389 km north of Camargue. These birds were probably also from the Camargue but there was no banded individual to prove this. Such types of movements are referred to as intermittent or intermediate migratory movements (Berthold 1996).

In addition to observations of marked individuals, the appearance of juvenile flamingos in areas where the species does not breed, or before the locally bred young are on the wing, illustrates post-fledging dispersal (Johnson 1989a). Such first-year juveniles are easily distinguishable from locally bred young by their larger size and paler plumage. These 'visitors' have been observed in the Camargue (Johnson 1973, 1989a), in Andalusia (Johnson 1989a; Fernandez-Cruz et al. 1991), and on the Banc d'Arguin in Mauritania (Cézilly et al. 1994a). It is likely that in the case of the Camargue and Andalusia these birds originated from more southerly sites where laying takes place earlier (Johnson 1973, 1989a), but in the case of the recently fledged juveniles observed on the Banc d'Arguin their origin is less clear. We suspect that they originated either from Chott Boul, southern Mauritania, or from a hitherto unknown site in Saharan Africa.

Post-fledging dispersal may also influence which sites birds use in later years outside the breeding season and many birds are faithful to wetlands which they visited during their first or second year of life (Green et al. 1989).

Natal dispersal

Flamingos banded as chicks in the Camargue (Figure 24) and at Fuente de Piedra have been observed breeding at their natal colony and in several other colonies in the Mediterranean and in southern Mauritania. Birds banded at these sites have also been reported in attendance at the colonies in the Banc d'Arguin (Mauritania) and in central Anatolia (Turkey), where we suspect they were breeding. Appendices 11 and 12 summarise the distribution of Camargue-banded birds recorded breeding at colonies in the Mediterranean and West Africa. These raw data give an insight into natal dispersion, although observer effort and colony size have varied considerably among sites. No intensive effort has yet been made to detect banded birds from the western Mediterranean in Asiatic colonies, Tunisia or West Africa, both because of the remoteness of the breeding sites and because of the infrequency of breeding.

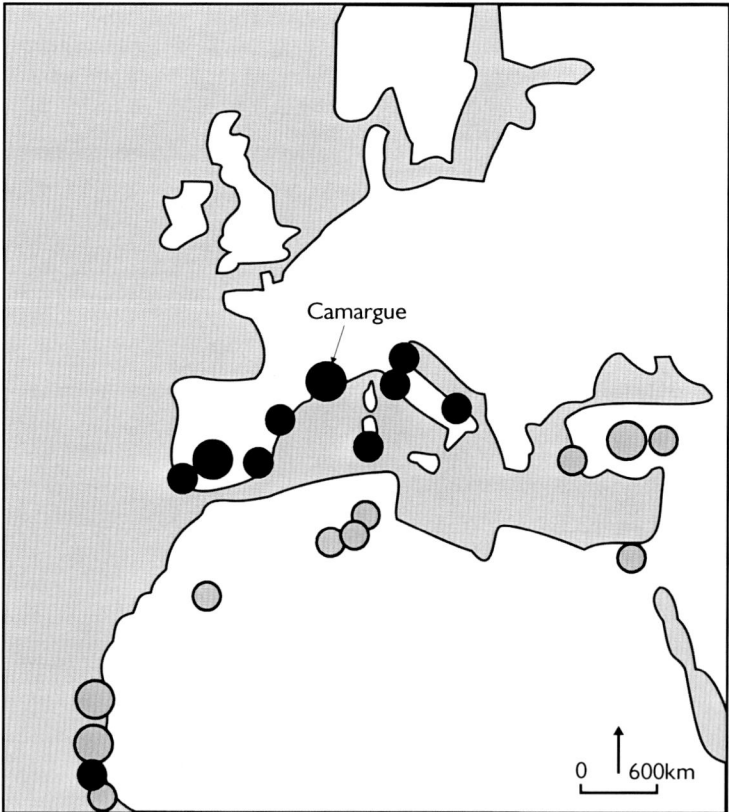

Figure 24. Flamingo colonies in the Mediterranean and West Africa and those where birds banded as chicks in the Camargue have been recorded breeding. See Appendices 11 and 12.

Natal dispersal between the Camargue and Spain was first studied for the 1986–92 period (Nager *et al.* 1996). However, these observations were limited to movements between the two colonies, rather than to dispersal per se. The two colonies differ in that breeding took place every year in the Camargue, whereas breeding at Fuente de Piedra occurred only in years of abundant rainfall prior to breeding. During the study period, the majority of natal dispersers (95.1%) were observed in 1988 and 1990, when 23.7% and 68.7% respectively of young flamingos hatched in the Camargue recruited into the colony at Fuente de Piedra, following major breeding failures at the Fangassier colony in 1987 and 1989. During the other years, young flamingos were very philopatric, only 0.4% of the females and 1.8% of the males having dispersed. However, higher rates of natal dispersal were observed in years when there was low access to the Camargue breeding colony (estimated from the ratio of potential recruits to the actual number of sites occupied by breeding birds) than in years with more favourable accessibility to breeding. Overall, natal dispersal did not differ between males and females and was not related to age of the flamingos. The average age of the Camargue-banded birds breeding for the first time at Fuente de Piedra was lower than that for those breeding first at their natal colony in the Camargue (Nager *et al.* 1996). We present here (Appendix 12) an updated account of the colony of first breeding for Camargue-banded birds over a larger area and a longer period of time.

Both years of high dispersal rates were preceded by large-scale disturbances at the Camargue colony; in 1987 a child's balloon was blown onto the breeding island at hatching (Johnson in Boutin & Chérain 1989) and in 1989 a strike by workers in the saltpans delayed flooding of the Fangassier lagoon. That same year, flamingos were further disturbed by an aggressive Black Swan (Johnson in Boutin *et al.* 1991; Cézilly *et al.* 1995; see Chapter 10). It is also common for immature birds to visit the colony in the Camargue, and if such birds were present in the Camargue during these disturbances, it may have discouraged them from nesting there. However, it is not known whether any of the 'dispersed' birds were present during the disturbances, and we can only speculate on the connection between these events and the higher dispersal rates the following year.

BREEDING DISPERSAL

Breeding-site fidelity may be influenced by previous breeding experience and is often positively correlated with age and breeding success. Some of the flamingos which bred in the Camargue in 1989 returned to breed in later years, while others moved to Fuente de Piedra. Those which later bred at Fuente de Piedra did not differ in age from the site-faithful flamingos, but they had made fewer breeding attempts in previous years, suggesting that they may have been less experienced than philopatric birds. Alternatively, dispersing birds may have had a stronger tendency to disperse, and may have bred at colonies other than Fuente de Piedra in years when they were not detected.

Of 4,229 Camargue-banded flamingos recorded breeding at least twice during 1986–92, we observed 174 (4.1%) that changed colonies after their first known breeding attempt. This, however, can be only an underestimation of the true rate of breeding dispersal because of low observer coverage and intermittent breeding at other colonies over the same time period. Camargue-banded flamingos having once bred in the Camargue dispersed less often (2.94%) than those having bred at Fuente de Piedra (8.14%) (Nager et al. 1996). Again this may reflect the relative instability of the principal Spanish breeding site.

As with natal dispersal, the highest proportion of breeding dispersal occurred in either 1988 or 1990, years following major disturbances at the colony. Mainly because of these disturbances, breeding success was very low in 1987 (12.6% of nests produced young) but was high in 1989 (70% of nests produced young). Of the 88 birds that dispersed from the Camargue to Fuente de Piedra, 83 (94.3%) did so in either 1988 or 1990. In these two years, a higher proportion of breeders dispersed from the Camargue to Fuente de Piedra (12.3%) than during the remaining years (0.4%). In contrast, the rate of dispersal from Fuente de Piedra to the Camargue was no higher in years following dry winters, when no breeding occurred in Andalusia (7.9%), than in the remaining years.

In 1989, most of the dispersing flamingos (13 out of 15) emigrated from the Camargue after their first or second breeding attempt. Comparing their breeding success in 1989 with that of the site-faithful birds, which also bred then for the first or second time, it was found that only 15.4% of the dispersing flamingos raised a chick in 1989 whereas 39.5% of the site-faithful birds did so. The study by Nager et al. (1996) found no differences between the sexes in the proportion of birds showing breeding dispersal, either from the Camargue to Fuente de Piedra or vice versa.

Although disturbance, and perhaps the consequent lower breeding success, is not the only factor causing breeding dispersal, studies in the Camargue and at Fuente de Piedra have shown this to be an important determinant at the former site. The lower breeding-site fidelity among 'Spanish' birds was not due to a higher immigration into the Camargue in years when there was no breeding at Fuente de Piedra.

INTER-COLONY MOVEMENTS

A total of 145 banded flamingos observed in the Camargue before the end of each June from 1986 to 1994 had been seen earlier the same season at Fuente de Piedra (950 km to the south-west). This represents an average 10% of the Camargue-banded flamingos identified at Fuente de Piedra during February–May of each year. Twelve individuals had attempted breeding at Fuente de Piedra prior to departure, and 23 attempted breeding on arrival in the Camargue, with only one individual suspected of having attempted nesting at both sites. In contrast, no birds that had been in the Camargue during early spring of these years were observed at Fuente de Piedra during the breeding season. Colonies are established at Fuente de Piedra earlier in the year than is the case in the Camargue, and the lagoon usually becomes less attractive as the season advances, whereas much the opposite is the case for the Camargue. There is also a negative correlation between March water levels at Fuente de Piedra and the proportion of banded birds first observed there and then shortly afterwards in the Camargue ($r = -0.85$, $n = 6$ $p = 0.031$). Given the time difference in breeding commencement between the two colonies, some birds may sample the environment (e.g. food or water levels) at Fuente de Piedra before deciding whether to breed there or to move, as many do, to the Camargue, where conditions are more predictable. For flamingos moving north from Fuente de Piedra to Camargue the same spring, the median date between sightings at each colony is 23 April.

During simultaneous observations at Molentargius (Sardinia) and in the Camargue, from February to June each year between 1993 and 1997, we recorded 93 banded flamingos of Camargue origin visiting both colonies during the same season. The pattern of movements between the Camargue and Sardinia differs from that observed between the Camargue and Fuente de Piedra in two respects. Firstly, only 19 individuals moved north from Sardinia to the Camargue, an oversea crossing of 550 km, whereas 74 individuals moved in the reverse direction, some of these being failed breeders. The median date between resightings at the two colonies the same season was 16 April for birds arriving in the Camargue and 14 May for those arriving in Sardinia (these differences are statistically highly significant). Molentargius, in contrast to Fuente de Piedra, is a saltpan with a permanent water level, and flamingos breed there at about the same time of the year as they do in the Camargue.

'MIGRATION' AND WANDERINGS

Only in the more northerly parts of their range do flamingos undertake movements which may be termed truly migratory, with birds regularly moving twice a year between geographically separated breeding sites and wintering areas. Flamingos which nest in Kazakhstan winter on the shores of the Caspian Sea when the breeding areas are under ice and snow. In other parts of their distribution, some flamingos may undertake long-distance movements but they remain within the confines of the species' year-round range. Some flamingos banded in the Camargue have life histories indicating annual crossings of the Mediterranean, but they are few in number. In addition, we know from multiple resightings of banded birds that, at the end of summer and in autumn, perhaps also at other times of the year, flamingos cross the Mediterranean in both directions, leaving southern France for North Africa and vice versa. Flights in either direction may involve birds on post-nuptial or post-fledging dispersal, and such movements may also be weather-related (Berthold 1993), either in response to diminishing food resources, or, in the more southerly sites, as temporary wetlands dry out. At times birds fly great distances outside their normal range, as in autumn 1906 and again in November 1907, when both adult and juvenile flamingos appeared in considerable numbers in central Siberia, at locations approximately 2,500 km beyond the normal range (Allen 1956). There are also two records of flocks of flamingos in Primorskiy Kray, on the coast of the Sea of Japan, the most recent being of 20 birds in October 1984 (Litvinenko & Shibaev 1999). In true migrants, such movements are referred to as 'reverse migration' as birds migrate on a route 180° from the normal.

ENVIRONMENTALLY INDUCED DISPERSAL

Temporary wetlands

Flamingos are opportunists and their normal seasonal movements may be subject to change by irregular weather conditions. Favourable water levels in the temporary wetlands which the birds frequent when they are flooded allow the flamingos to expand their range following periods of heavy rain. Simmons *et al.* (1999) believe that the ability of flamingos and other wetland birds to find and exploit ephemeral wetlands in arid landscapes is due in part to their behaviour of following massive thunder heads, as some raptors are known to do. Flint & Stewart (1993) state that flamingos in Cyprus usually appear on the salt lakes at Larnaca (Figure 28) and Akrotiri within one to two days of flooding, suggesting regular passage overhead. In the Camargue, small groups or pairs of flamingos fly over the Fangassier lagoon in February–March, seemingly to see whether the site is suitable for breeding.

Extreme climatic events

Mass movements of flamingos, in response to extreme climatological phenomena other than drought, occur only in the more northerly parts of the species' range. They are most infrequent but the two examples reported below, the first in response to severe cold and the second to gale-force winds, occurred quite recently in southern France.

Severely cold winters: Spells of severely cold weather occur periodically in the northern limit of the species' normal winter range: France, Greece, southern Turkey and around the Caspian Sea (Azerbaijan, Turkmenistan, Iran). Flamingos can survive without food for several days, but if lagoons remain frozen for longer periods the birds may die of starvation. Confronted by severe cold, some flamingos stay at their usual wintering sites and others try to escape. It is a time when food is scarce, however, and the weather not necessarily favourable for a long-distance flight. At times, escape movements occur from both the Caspian Sea (Demente'ev *et al.* 1951) and from the Camargue. One such major event in France occurred during our studies. In January 1985, during the last spell of severely cold weather in southern France in the 20th century, some birds reacted to the onset of freezing conditions by fleeing south-west along the coast from the Camargue to the Languedoc-Roussillon region, as revealed by an increase in the number of flamingos in this region, but it was also exceptionally cold there. Other birds attempted to escape over the Mediterranean and some birds reached Corsica, a distance of at least 250 km, where flamingos were reported in 21 localities, the largest flock numbering 150–200 birds (Johnson 1989a).

Gales: Unusually strong south-easterly winds (up to 126 km/h) blew across France on 7 to 8 November 1982 (Johnson 1989a). During this gale, and shortly afterwards, flamingos were reported from several sites some tens or even hundreds of kilometres inland to the north and north-west of their normal distribution (Hafner *et al.* 1985). During this time, the Languedoc lagoons, which had been used by thousands of foraging flamingos, were flooded by the sea. Birds displaced by the flooding may have simply been blown inland, unable to fly against gale-force winds. For example, one banded flamingo (code BZU), was observed in the Languedoc on 15 October 1982 and was seen shortly after the storm, on 14 and 20 November, in the Drome department, 180 km north of the Camargue. This bird returned to the Camargue and bred six months later.

SPATIAL PATTERNS

Brown *et al.* (1982) suggested that there were discrete populations of flamingos centred in four areas: 1) Europe and the Middle East migrating in the winter to North and East Africa, 2) Mauritania and Senegal, 3) East Africa Rift Valley, and 4) southern Africa. Our own findings do not entirely agree with this, and although many flamingos undoubtedly do move only within the above-mentioned divisions, these are not discrete populations. For example, ring recoveries and band resightings now show that movements between the Camargue and West Africa, and vice versa, are not uncommon, yet we ourselves at the beginning of our studies believed that exchange of birds between these two regions was negligible. We now know this not to be the case and we suspect there to be an exchange of birds between other divisions indicated by Brown *et al.* (1982). Therefore, although we agree in principle with Brown *et al.* that there are distinct regions within which there appear to be greater exchanges of birds, we suggest, in light of more recent data, revising the concept of segregated populations to recognise the likelihood of some level of exchange between the more distant areas. The current evidence, therefore, indicates three and possibly four major sub-populations, at least partially influenced by distance, but not the same as those proposed by Brown *et al.* (1982). These sub-populations are as follows: (1) west Mediterranean and north-west Africa, (2) east Mediterranean and south-west Asia, and (3) East and southern Africa, although the latter may be better described as two separate sub-populations.

SUB-POPULATIONS AND FLYWAYS

West Mediterranean and north-west Africa

Observations of birds flying along the coast, and numerous recoveries and resightings of marked flamingos, indicate that the Atlantic coastline of Morocco is a regular flyway for flamingos moving between West Africa (Guinea-Bissau, Senegambia, Mauritania) and the Mediterranean region, and vice versa (Valverde 1957; Heim de Balsac & Mayaud 1962). This route is part of the East Atlantic flyway, which is used twice a year by myriad birds which breed in the Palearctic region and winter in areas bordering the Sahara Desert, or further south. Some flamingos, having wintered in West Africa, fly north in early spring to breed in the Mediterranean region, but there is also movement north later in the year, coinciding with post-breeding dispersal from the West African colonies. Flamingos usually move south along this flyway in late summer and autumn, when they leave their European breeding sites to spend the non-breeding season in West Africa.

> *From Camargue to Senegal*
> Flamingo code AVBD was banded in the Camargue on 9 August 1989 and remained in the crèche until at least 16 September. It was observed on 19 January 1990 near the mouth of the River Senegal, where it stayed until at least March 1993 (34 resightings).
>
> *. . .and back to Camargue*
> Flamingo code AAJC, banded in the Camargue in July 1985, was resighted four times in Senegal between 12 May and 6 June 1990. It was observed at Fuente de Piedra (Spain) on 25 July 1990 and then in France from 16 October 1990 onwards.

Flamingos also occasionally fly inland from the Atlantic coast in southern Morocco, along the valley of the Oued Dra on the southern edge of the Antiatlas Mountains (A. P. Robin pers. comm.). They visit or pass through the Iriki depression (where they formerly bred), the Dayet Merzouga and, in Algeria, the Daiet Tiour (Dupuy 1969). This latter wetland is on the edge of the Saharan Atlas, 650 km east of the Atlantic coast and 550 km south of the Mediterranean Sea. From here birds may fly north to the Mediterranean coast or east along the northern edge of the Sahara Desert to wetlands in Algeria and/or Tunisia (see for example Brosset 1959).

The Strait of Gibraltar, best known to ornithologists for the spectacular migration of raptors and other landbirds, is also crossed by flamingos moving between Europe and Africa (Chapman & Buck 1910; Pineau & Giraud-Audine 1979; Telleria 1981). There are no favourable wetlands immediately along the coast on either side of the Strait, so flamingos crossing at this point are probably moving between the Merja Zerga or Larache in Morocco and the Bay of Cádiz or wetlands of the lower Guadalquivir in Spain, a flight of 100–200 km. Flamingos probably also cross the western Mediterranean further east when moving between Andalusia and eastern Morocco, Algeria or Tunisia.

In the Mediterranean region, an extensive network of flight lines links the multitude of wetlands situated around the Mediterranean (Figure 25). At least three major flyways are apparent, two of which are in the western Mediterranean.

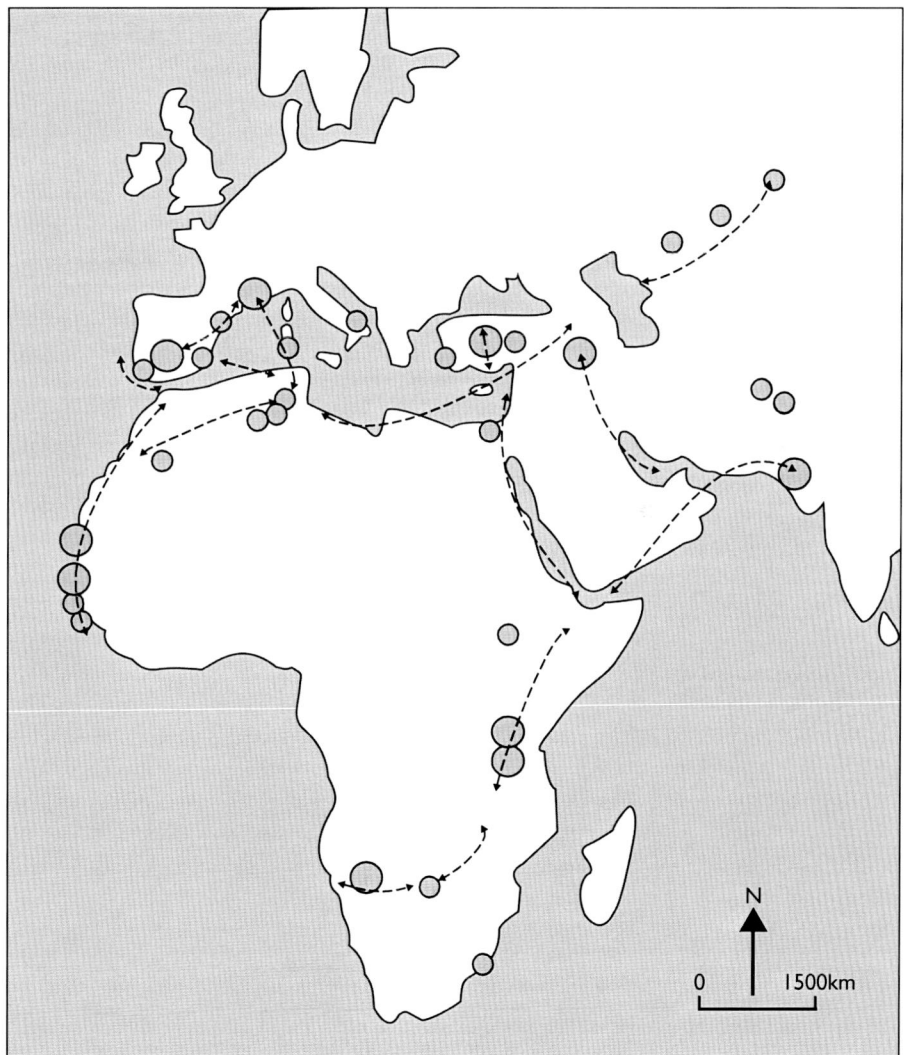

Figure 25. Some major flyways of the Greater Flamingo.

1. The Atlantic coastal flyway to and from West Africa continues from Morocco into southern Portugal and along the Mediterranean coast of the Iberian Peninsula into south-west Spain, and from south-west Spain either east along the Mediterranean coast to the Almeria region or inland from wetlands bordering the Gulf of Cádiz to the Fuente de Piedra Lagoon. From Fuente de Piedra or Almeria, birds move up the Mediterranean coast into southern France. Flamingos observed in spring arriving over the sea into Aude (Y. Kayser pers. comm.) could originate from the Ebro delta, the most northerly site where the species regularly occurs in Catalonia. From Almeria, some flamingos also cross the Mediterranean to North Africa and probably Sardinia, where many banded birds from Fuente de Piedra have been resighted.
2. Between southern France, particularly the Camargue, and Sardinia and Tunisia, favourable winds in both directions assist birds with the sea crossings along this flyway to and from the Rhône delta. From August to October, the Arles beach, west of the river mouth, is a good place to observe flamingos on post-nuptial flights, shortly before sunset on days of mistral. Flamingos travel this flyway in a southerly direction on post-breeding/post-fledging dispersal and in a northerly direction on spring migration. In addition, there have been numerous resightings of banded flamingos flying north across the Mediterranean in summer and autumn in a type of 'reverse migration', possibly by birds in response to the drying out of feeding areas in North Africa and/or Sardinia.

> During post-fledging dispersal, flamingo code MATH, banded at Molentargius, Sardinia, on 21 July 1997, was observed in the Camargue on 8 August the same year. Another bird from the same cohort, band code MAJC, was resighted in Tunisia at least four times from 12 August.

East Mediterranean and south-west Asia

In the east Mediterranean, a major flyway passes over the island of Cyprus (Figure 25). This is a predominantly ENE to WSW flyway, confirmed by sightings of birds heading south or south-west in autumn and in a generally north-easterly direction in spring (Flint & Stewart 1992). Lake Larnaca is a favourite location for seeing flamingos departing in the evening in January and February. Many of the flamingos using this flyway are probably moving between their winter quarters, on Cyprus or along the coast of North Africa between Egypt and Morocco, and the breeding localities in Iran (Lake Uromiyeh) and Kazakhstan (Lake Tengiz). There have been recoveries in Cyprus and North Africa of both Iranian-ringed (Flint & Stewart 1992) and Russian-ringed flamingos; while other birds must originate from Anatolia to the north, birds from Iran or Kazakhstan may fly via Turkey and/or Syria before reaching Cyprus.

Several banded flamingos from the Camargue and Fuente de Piedra have been resighted in north-eastern Greece. Some of these birds had previously been observed in Sardinia and/or Tunisia, suggesting east–west movements in the Mediterranean along the coast of North Africa, rather than over northern Italy and western Greece (see Johnson 1990). Resightings of banded birds from the Camargue also indicate movements between Cyprus, Turkey (western Anatolia) and Greece, but so far there is no evidence of birds crossing western Greece or mainland Italy, although such movements may well occur.

Multiple sightings throughout the Mediterranean
Flamingo code ALNF, banded in the Camargue in 1987, was resighted near Izmir in Turkey, before (June 1991) and again after (January 1992) being sighted in Thrace, Greece (August 1991). It later visited Sardinia (July 1994), returned again to western Turkey (February 1995) and finally was resighted in Foggia, Italy, several times between November 1998 and June 2000.

In south-west Asia, flamingos which visit the lakes of central Kazakhstan in spring and summer, and any yearlings fledged there, move to the south and the south-west in late summer and autumn. Some spend the winter on the shores of the Caspian Sea, in Turkmenistan (Karabogaz Bay) and in Azerbaijan and northern Iran, while others travel much further to destinations in the Mediterranean, around the Arabian Sea and probably in north-east Africa as well, perhaps meeting at the junction of three continents with birds from Lake Uromiyeh. Spring and autumn migration along both coasts of the Caspian Sea has been observed. Further west, birds have been reported north of Baku in Dagestan, flying north in May and south again in September. On the eastern shore of the Caspian Sea they move north in April and early May (Krasnovodsk Gulf) and return south in October and November (Khokhlov 1995). After leaving the Caspian Sea, spring migrants have been reported crossing the Ustyurt Plateau, some birds stopping in the northern part of the Aral Sea (Ametov 1981). The destination of these birds apparently depends on the amount of flooding in the lakes of central Kazakhstan. While it is clear that there are major flyways of flamingos around the Caspian Sea and north-east to central Kazakhstan, a distance of around 1,000 km, it is also clear, from an abundance of Russian literature, that flamingos disperse widely in Kazakhstan and Uzbekistan and at times fly as far as Siberia, 3,000 km or more from the Caspian Sea (Lukovtsev 1990). There are even two records of flocks in Primorskiy Kray, on the coast of the Sea of Japan, the most recent being of 20 birds in October 1984 (Litvinenko & Shibaev 1999). Flamingos formerly appeared in autumn on the southern part of the Aral Sea, at the mouth and along the lower reaches of the Amu Daria, and they still occur occasionally in the Altay region of eastern Uzbekistan. Flamingos appear very sporadically at widely separated locations in the south-west of the former USSR. The natural influence of precipitation rates on the level of many water bodies is contrasted by the gradual drying out of other wetlands through human interference, the case of the Aral Sea being the best known.

Flamingos winter in, and pass through, countries bordering the Arabian Peninsula (Bundy and Warr 1980; Phillips 1982; Mohamed 1991; Platt 1994; Porter *et al.* 1996) and recoveries of ringed birds from Lake Uromiyeh indicate that both the Red Sea and the Persian Gulf coasts are flyways used by flamingos originating from Iran (Figure 25). Behrouzi-Rad (1992) states that adults from Lake Uromiyeh undertake only rather short flights to the south-west, not usually leaving the country, but that young birds in their first autumn travel over long distances.

There is also movement of flamingos along the Arabian Sea coast from India and Pakistan (Ali & Ripley 1978; Roberts 1991) to Djibouti (Welch & Welch 1984), Somalia (Ash & Miskell 1998), Sudan and Ethiopia. The recoveries of Iranian-ringed flamingos in these countries confirm movements between the Palearctic region and East Africa as far south as *c.* 10°N, as suggested by Brown *et al.* (1982). Migrating birds from the north-east reach the Rift Valley after crossing the Arabian Peninsula or flying along the coast of the Red Sea (Curry-Lindahl 1981).

East and southern Africa

Reports of movements of flamingos throughout these two regions refer mainly to the more abundant Lesser Flamingo. However, since the two species cohabit, we suspect that they use the same flyways, either when moving within these regions or, eventually, between these regions.

In East Africa, some of the flamingos which breed in Kenya and Tanzania may be resident within East Africa (see Curry-Lindahl 1981, Brown *et al.* 1982), with movements up and down the Rift Valley occurring according to foraging conditions at the different lakes and in relation to breeding. In southern Africa, Williams & Velásquez (1997) refer to movements in the late afternoon along the Namibian coast and inland after dark towards Etosha Pan, a distance of *c.* 500 km, where the birds arrive before daybreak. Movements between Etosha Pan and the Atlantic coast are confirmed by two recoveries of newly fledged juveniles ringed at Etosha in June 1969, one recovered two weeks later at Walvis Bay, 500 km to the SSW and the other one month later near Möwe Bay, 405 km to the west (Berry 1972; Elliott & Jarvis 1973).

The number of flamingos occurring at times on Etosha Pan seems to far exceed the numbers which occur on the Atlantic coast, and Berry (1972) thought it likely that there was movement between Etosha Pan, Lake Ngami and the Makgadikgadi Pans in Botswana, as well as between south-western and East Africa. Pans in Bushmanland provide stopover sites en route to the Makgadikgadi Pans (Williams & Velásquez 1997). A bird ringed at Walvis Bay in June 1994 was recovered in March 1997 near Ghazni in Botswana, 766 km inland to the NNE (Underhill *et al.* 1999). Borello *et al.* (1998) report considerable movements of flamingos east from Botswana at *c.* 21°–22°S latitude. Birds believed to be heading for coastal wetlands in Mozambique are colliding at night with newly erected north–south power-lines in the south of Zimbabwe. Sightings in Zambia and Malawi suggest movements along the Luangwa and Rift Valleys, thereby linking southern and eastern Africa, including Madagascar. These two regions may also be linked, however, by movements along the Indian Ocean coastline, as suggested by Underhill *et al.* (1999). Finally, one of the fledglings ringed at Bredasdorp, South Africa, in January 1961 was recovered a year later at Brandvlei, 515 km to the north (Underhill *et al.* 1999). More recently, data obtained from satellite tracking of three Greater Flamingos have established that birds leaving the Makgadikgadi saltpans in Botswana can fly either west to Namibia or south to South Africa (McCulloch *et al.* 2003).

North-west Europe — feral flamingos?

Since the 1980s, flamingos have occurred in north-west Europe, in countries bordering the North Sea (northern France, The Netherlands, Germany and Denmark), 1,000 km north of the Mediterranean (Treep 2000). Most of these birds are either Chilean or Caribbean Flamingos, escaped from captivity, but there are also some Greater Flamingos in these groups. The origin of these birds is not known, but until large numbers of juveniles are seen, or there is a resighting or recovery of a bird banded in the wild, we think that these flamingos should also be considered feral. The most northerly records of truly wild birds in western Europe, at least in recent years, have involved movements of less than 500 km from the Mediterranean coast.

SOME CONCLUDING REMARKS

We have used the term 'migration' in a very wide sense to indicate large-scale movements of groups of individuals from one spatial unit to another (see Baker 1978, Berthold 1996). Although the existence of flyways used by flamingos seems obvious, they do not seem to depend on particular landscape features as is sometimes observed in other bird species (Berthold 1996). Over most parts of their range flamingos do not limit themselves to regular temporal and spatial 'migrations', nor are their movements confined to certain flyways, some of which we have mentioned here. Flamingos are dispersive, erratic, migratory, partially migratory, irruptive, sedentary, and so on; movement patterns differ not only from one sub-population to another, according to geographical location, but also individually, as demonstrated by multiple resightings of birds banded at Mediterranean colonies. Although band resightings never reveal full life-history traits, they nevertheless illustrate the different strategies adopted by flamingos.

Some banded birds from the Camargue, observed over 300 times during 21 years, have never been seen outside France. Others, about two-thirds of the resighted Camargue flamingos, have been observed in other countries. Thousands of life histories of marked birds reveal many years of fidelity to the same breeding site and, outside the breeding season, to the same foraging areas, whether near the

Table 9. Recoveries and resightings of flamingos marked as chicks in Iran, France, Spain and Sardinia. Recoveries a) of flamingos ringed in Iran, b) of flamingos ringed in Camargue with Museum Paris alloy and steel rings 1947–99. Resightings of live flamingos banded with PVC leg-bands (code read in field) marked c) in France, d) in Spain and e) in Sardinia. Data from S. Sadeghi-Zadegan for Iran and from the respective ringing centres in France, Spain and Italy.

	a	b	c	d	e
Algeria	–	41	17	32	1
Azerbaijan	2	–	–	–	–
Bahrain	4	–	–	–	–
Balearic Is.	–	4	11	–	–
Corsica	–	1	13	–	–
Cyprus	3	–	9	6	–
Djibouti	1	–	–	–	–
Dubai	4	–	–	–	–
Egypt	10	3	5	–	–
Ethiopia	4	–	–	–	–
France	1	536	245425	8078	88
Greece	2	–	37	12	–
Guinea Bissau	–	–	6	4	–
India	23	–	–	–	–
Iran	47	–	–	–	–
Iraq	19	–	–	–	–
Israel	3	–	–	–	–
Italy (excl. Sardinia)	–	5	509	2387	21
Kazakhstan	8	–	–	–	–
Kuwait	1	–	–	–	–
Libya	13	8	7	1	–
Mali	–	1	–	–	–
Mauritania	–	4	41	25	1
Morocco	1	39	369	868	5
Oman	5	–	–	–	–
Pakistan	21	–	–	–	–
Portugal	–	14	267	958	5
Qatar	8	–	–	–	–
Sardinia	1	83	3029	see Italy	103
Saudi Arabia	3	–	–	–	–
Senegal	–	4	119	502	–
Sicily	–	3	32	see Italy	2
Somalia	3	–	–	–	–
Spain	–	240	39827	39773	32
Sri Lanka	1	–	–	–	–
Sudan	3	–	–	–	–
Syria	4	–	–	–	–
Tunisia	4	102	2504	1535	70
Turkey	22	3	20	4	–
Turkmenistan	1	–	–	–	–
USSR	6	–	–	–	–

colony or hundreds of kilometres away. The distance travelled by individual flamingos from the same cohort during their lifetimes varies tremendously, some birds having never been observed further from their natal colony than the distance to the nearest favourable feeding areas; whereas others regularly cross the Mediterranean.

Recoveries and resightings of marked flamingos have demonstrated that an occasional or regular exchange of birds takes place between almost all neighbouring regions where the species occurs in the Northern Hemisphere. There is thus gene flow between sub-populations to which the more sedentary individuals belong, as observed in the Camargue, although there is no proof yet of movements between eastern and southern Africa, nor between western Africa and the Rift Valley. The recovery in Mali (Table 9), 900 km inland from the Atlantic coast, of a flamingo banded in the Camargue may be aberrant, or it may be an indication that birds sometimes cross Africa south of the Sahara. Indeed, it is still unknown whether the small population of Lesser Flamingos occurring in Mauritania, Senegal and Guinea-Bissau is sustained by breeding in West Africa, of which only one, unsuccessful attempt has so far been recorded (Naurois 1965), or whether these birds originate from eastern (or southern) Africa.

CHAPTER SIX

Foraging ecology

Although flamingos feed primarily on very small animals, their foraging ecology is probably closer to that of grazing species such as geese than to that of conventional predators like herons. In order to meet their daily food requirements, flamingos must spend considerable time filtering water to capture various invertebrate prey and other food items. One of their most distinctive characteristics is the filter-feeding bill (Figure 26), which is unique among birds. Although the anatomy and function of the filter, and the various foraging techniques used by flamingos in the wild have been extensively described by previous authors (Jenkin 1957; Rooth 1965; Zweers *et al.* 1995; Mascitti & Kravetz 2002), the selection of feeding habitat by flamingos has so far received little attention. In addition, the precise impact of foraging flamingos on the invertebrate communities of various wetlands has been largely ignored (but see Hurlbert *et al.* 1986, Glassom & Branch 1997a). In this chapter we summarise current knowledge of the flamingo's foraging ecology and introduce new observations on movements between breeding sites and feeding habitats in the Camargue and elsewhere.

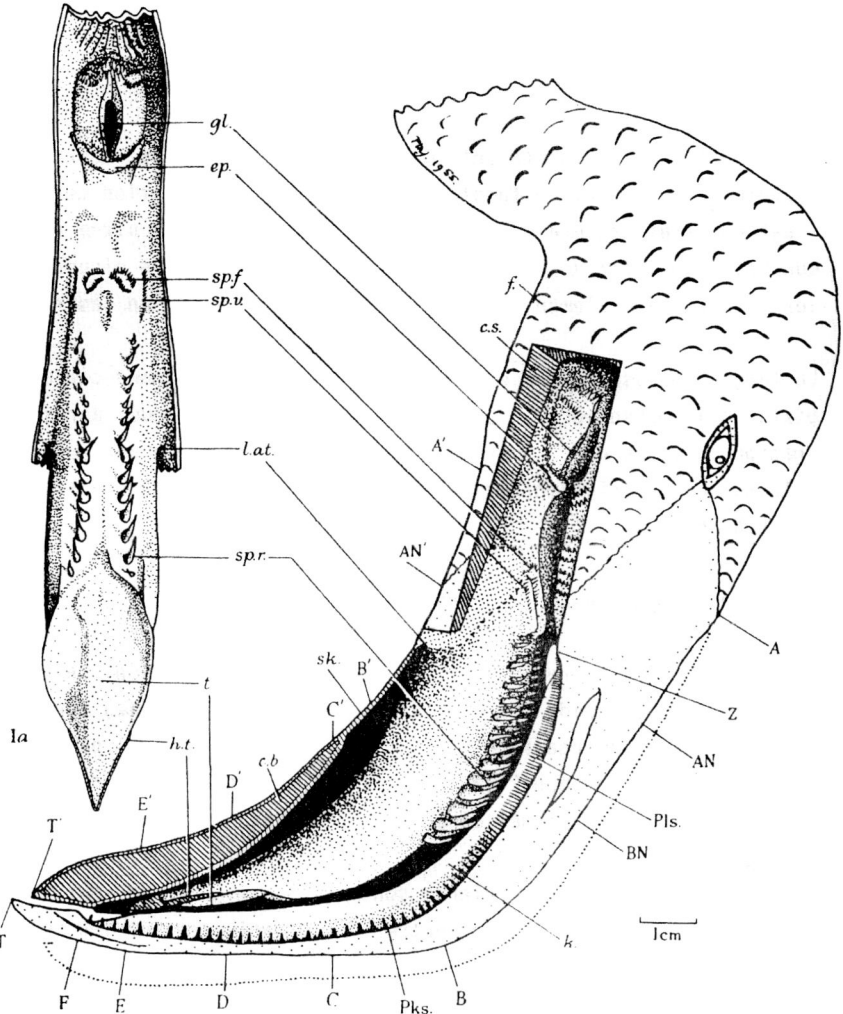

Figure 26. The Greater Flamingo's bill and tongue, a unique and complex mechanism designed for filtering organisms from mud and water. Reprinted from Jenkin 1957.

FEEDING APPARATUS

The anatomy of the flamingo's bill and tongue and the method by which the birds obtain their food have been extensively described by Jenkin (1957), Zweers *et al.* (1995) and Mascitti & Kravetz (2002). Our description is based to a large extent on their works, to which we refer the reader for a more detailed account of this fundamental aspect of the flamingo's ecology. The flamingo has a thick, fleshy tongue (Figure 26), which is housed in the large trough-like lower mandible, or jaw, which is attached to the skull. The tip, or distal part, of the tongue is flattened with raised edges and it is fringed on the upper surface by two rows of spines which curve backwards towards the throat. There are also two fans of minute spines at the base of the tongue. Strong muscles at the base allow the flamingo to activate its tongue in a rapid back and forth piston-like movement four times per second, creating suction. The upper mandible (which becomes the lower during feeding) is rather lid-like, and 'hinged' to the skull. The edges are lined with hook-like filters—coarse towards the tip of the bill and gradually becoming finer towards the gape—which prevent the bill from closing completely. The inner surface of the upper mandible is lined with plates of lamellae directed inwards and backwards towards the tongue. During feeding the flattened distal one-third of the down-curved bill is generally held horizontally. Being thin it allows the flamingo to feed in water only a few millimetres in depth. According to Jenkin (1957), flamingos are able to choose the size of the organisms they swallow, but not the type. However, they sometimes feed using their bills as forceps. This enables them to catch larger prey which they can either see or feel, since the bill has many tactile organs and also, according to Zweers *et al.* (1995), taste buds which may allow the flamingo to discriminate between food items.

Filter-feeding is clearly the flamingo's most commonly used foraging method. The flamingo opens its bill to a narrow gape of *c.* 5 mm at the tip, and sweeps it from side to side, while sucking water containing potential food items into the bill through the outer lamellae. The water crosses the platelets of lamellae lining the upper mandible, which retain the food as the bill is closed again to seal the filter. With each inward movement of the tongue, the water is expelled near the base of the bill. This can be seen in the field when flamingos are feeding in shallow water, and can also be heard at close range on a calm day. When flamingos feed on mud, the process is apparently reversed. The outer lamellae prevent the coarser particles from entering the bill while the finer mud is swallowed and the organic content extracted. At close range, when a feeding flamingo raises its head to swallow, the bolus of food accumulated in the bill can be seen as a bulge moving down its neck (Zweers *et al.* 1995).

FOOD

Flamingos feed primarily on aquatic invertebrates and their eggs and larvae. The range of species and taxa which forms the basis of the Greater Flamingo's diet is very wide. A few taxa which are widespread in the world, and at times abundant throughout the flamingos' range, form an important part of the species' diet. Among these are fairy shrimp, brine shrimp and brine-fly larvae in hypersaline waters, the larvae of midges in fresh, brackish and saline waters, and bivalve molluscs on the tidal mudflats of the Atlantic coast in West Africa (Mauritania). Most food items are collected by flamingos when they are filter-feeding, but some of the larger prey may be located by sight, for example fish, ragworms, soldier crabs (Macnae 1960), or by touch, for example bivalve molluscs.

Flamingos are not restricted to feeding on invertebrates, but also feed on the seeds of aquatic plants (see Madon 1932, Abdulali 1964), including rice, and sometimes eat mud to extract its organic content. According to Jenkin (1957), 80% of the grit found in Greater Flamingo stomachs measures more than 0.5 mm in diameter. From a monospecific invertebrate community in hypersaline lagoons, the variety of larvae increases considerably as the water salinity decreases. Particularly prevalent are insects—beetles (Coleoptera), dipteran flies and the smaller hemipteran bugs and dragonflies and damselflies (Odonata)—and molluscs (Mollusca) and arthropods (Crustacea and Arachnida). Even in saline waters of the Camargue, more than 15 species occur and are probably consumed by flamingos— polychaete worm (ragworm), tubificid worms, gastropods, crustaceans (copepods, amphipods, ostracods, phyllopods and decapods) and insect larvae (non-biting midges (Chironomidae), shore flies (Ephydridae), dolichopodid flies) and the smaller Coleoptera. Flamingos have a very varied diet since the distribution and density of the different prey species differ from one wetland to another, and in temperate regions may be subject to seasonal cycles. Flamingos will take advantage of any abundance/population explosion of these invertebrates. This is particularly obvious, for example, with brine shrimp in the saltpans of the Camargue. These phyllopods occur in profusion during the summer months in water with a salinity exceeding 80 g/l. Their eggs, or cysts, are also taken (see MacDonald 1980), particularly when they are concentrated, as often occurs along the downwind shoreline. Although brine shrimp may sustain a large proportion of the flamingos breeding in some places, as they do in the Camargue, it would be erroneous to believe that flamingos feed only on these crustaceans. During the breeding season many birds also feed in brackish or fresh water where there are no brine shrimp, and in winter there are regularly over 20,000 flamingos in southern France, all feeding on other food items.

FEEDING BEHAVIOUR AND FORAGING METHODS

Flamingos sometimes swim when feeding in deeper water, but they usually forage in water from 5 to 50 cm in depth (Figure 27). They forage gregariously, in flocks of up to several thousand birds. Little is known about the way flamingos are attracted to and distribute themselves over foraging grounds. Arengo & Baldassare (1995) suggested that Caribbean Flamingos in Mexico exploit food resources in accordance with the ideal free distribution model (Fretwell & Lucas 1970), that is to say that foraging individuals are distributed over foraging patches in direct

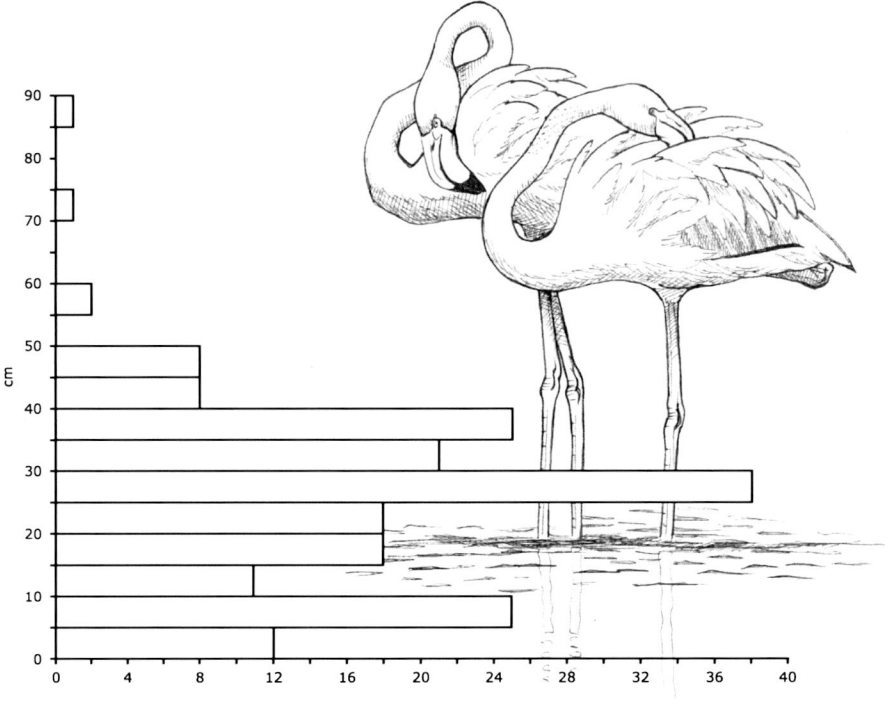

Figure 27. Flamingos sometimes swim to feed but they most frequently forage in water from 5–50 cm in depth.

proportion to local food density. Not only does flock density vary according to food resources available, but feeding methods do too. Birds usually feed independently within a flock, but communal feeding, with the birds driving their prey before them, does occasionally occur. On Lake Larnaca in Cyprus, long undulating lines of birds three or four deep (Figure 28) have been seen moving forward, probably filtering brine shrimp from the water column. The leading birds presumably displace many shrimps which they themselves will not catch, but which are then available to the birds behind them. Other species of birds occasionally take advantage of this flamingo feeding strategy, feeding alongside them (Figure 29), such as Slender-billed Gulls (Walmsley 1991). A recent study of Caribbean Flamingos (Arengo & Baldassarre 2002) suggested that groups of foraging individuals may signal location of food to conspecifics and may also provide some information on food quality through feeding behaviour.

Little is known about the precise impact of predation by Greater Flamingos on the invertebrate fauna. Experiments performed in Namibia with enclosures

Foraging ecology 115

Figure 28. Larnaca salt lake, Cyprus. Several thousand flamingos form a long undulating line as they forage, probably on brine shrimp and their cysts.

(Glassom & Branch 1997a) suggest, however, that this impact can be strong. Inside enclosures preventing predation by flamingos, macrofaunal numbers (mainly polychaete worms and amphipods) increased by approximately 3 times the density of control plots subtidally and 1.5 times the control density intertidally. Unfortunately, similar experiments have not been carried out in other parts of the species' range, such that the precise impact of flamingos on local invertebrate fauna among the various habitats they exploit remains undocumented.

Indeed, although flamingos are specialised filter-feeders, they are very versatile and can obtain food in several ways. The different foraging methods have been described by Rooth (1965) and we have expanded upon his observations of Caribbean flamingos (Figure 30), since these methods are basically the same as those

Figure 29. Other species of birds may feed on invertebrates such as brine shrimp displaced in the wake of flamingos as they up-end and thrust themselves forwards.

Figure 30. Some of the different methods of feeding. (a) walking and filtering or grubbing, (b) up-ending, (c) stamping, (d) stamping 'marking time', (e) skimming, (f) walking 'leaving tracks'.

used by the Old World race. Flamingos adapt their feeding to the substrate of the lagoon, to the potential prey items available in the mud and/or water, and to water depth. When several different methods of feeding are used by flamingos on the same wetland, the birds are probably foraging for different prey. The methods are described here in order of frequency of use in the Mediterranean region:

Walking and either filtering in water column or 'grubbing': Birds walk slowly forwards in water which varies in depth from a few millimetres to belly-depth (50–70 cm). The head is immersed for periods of 5–25 seconds. This widely used method allows birds in hypersaline lagoons to filter brine shrimp from the water column or, in less salty waters, to 'grub' for benthic invertebrates. During 'grubbing', relatively large particles of food are grasped with the beak tip and transported through the mouth by a catch-and-throw mechanism (Zweers *et al.* 1995). In deeper water this type of feeding leads to up-ending.

Up-ending: If the water is slightly too deep for a walking flamingo to reach the bottom, it will swim and up-end like a swan in order to reach the benthos or lower layers of water. According to Rooth (1965) females can feed this way in a maximum of 90 cm of water and males in depths of up to 105 cm, i.e. some 30 cm deeper than when walking. Rooth (1965) refers to this method as 'grubbing'. Since he observed this behaviour in areas where the water was not much deeper than 1 m, he believed that the birds were bottom-feeding. The birds need to paddle like ducks in order to maintain a vertical position, and this may move them forwards through the water in search of new food supplies. On a calm day the splashing of their feet can be heard at a considerable distance.

Stamping: This method of feeding is unique to flamingos. While stamping, and sinking into the mud or sand, the birds gradually turn in a circle around their bill, normally describing at least one complete circle. Not only is the behaviour itself quite striking to observe, but in places where the water recedes, the bed of a lagoon may be covered with saucer-like depressions; Gallet (1949) referred to these as 'feeding cones' (Figure 31) and Glassom & Branch (1997a) as 'wheelies'. The depressions measured at a wetland near Montpellier (O. Pineau pers. comm.) varied in size from 63–96 cm diameter (average 80 cm, n = 119). There were 70–83 depressions per 100 m^2 (n = 8), and they covered 39% of the marsh bed. Flamingos use this method only in soft substrates, since they are either feeding on the invertebrates/larvae buried in the sand or mud, or they are actually eating the mud that they sieve from the centre of the depression. This method of foraging may seem a lot of effort for an apparently small reward, but it is commonly used year-round. When seen from the air, the exposed depressions in the substrate look remarkably like flamingo nests.

Stamping—'marking time': The flamingo stamps with its feet, lifting one leg and then the other in rather rapid succession. The heel joint is continually extended and

Figure 31. **Feeding cones**. Also referred to as 'rings' or 'wheelies', feeding cones can be seen when water recedes from areas where flamingos have been foraging by 'stamping' in the sand or mud in search of benthic food items.

retracted and the bill is held underwater, just above the surface of the mud. The bird may move backwards slightly as it captures invertebrates which it has brought into suspension. Rooth (1965) found that Caribbean Flamingos used this method when feeding on salt-fly larvae. Prolonged stamping by large numbers of birds can affect the ecology of a wetland by reducing or eliminating certain submerged macrophytes (Montes & Bernues 1990), by increasing nutrient release from the sediment into the water column (Comin *et al.* 1997), and by reducing the macrofauna, as observed in southern Africa (Glassom & Branch 1997a,b) and in Guadalquivir Marshes, where Chapman, as long ago as 1884 (in Jenkin 1957), said 'they tear up grasses and water plants'.

Skimming: When food items are concentrated at the surface of the water, flamingos will collect them by skimming with their bills in a side-to-side scything movement, with only the flattened distal part of the upper mandible in contact with the water. They do this when swimming or walking, sometimes standing and sieving as food drifts by. This is how Lesser Flamingos feed on blue-green algae. Greater Flamingos, when feeding in this fashion, often do so in areas where there are brine shrimp, particularly where brine-shrimp cysts are floating on the surface. The downwind shoreline is often covered with a 'scum' of shrimp cysts (which to the untrained eye may appear to be fine sand).

Walking 'leaving tracks': The flamingo walks slowly forwards, the head immersed, and snakes its neck from side to side. The bill is trailed through the mud, which the flamingo is eating, and this leaves a track 1 cm or so deep and some 2–3 m long. In the Camargue, flamingos feeding in this way are widely dispersed and apparently use this method mainly in winter, suggesting that they feed on organic ooze only when their more normal food is in short supply. Similar behaviour has been reported for the Caribbean Flamingo on Bonaire (Rooth 1982).

Heron-like 'running': This is a method only very rarely used on both sides of the Atlantic. It is restricted to areas of shallow water where birds feed individually or in small groups. They are not filtering or grubbing, but seizing prey which they detect visually. These may be small fish trapped in a pool after water has receded or, as observed in the Camargue, ragworms, possibly as they come to the surface of the mud or sand of a lagoon to reproduce. The flamingo dashes forwards with the neck fully extended and lunges to catch whatever prey it detects, the bill being used as a pair of forceps and not held upside down. Macnae (1960) reported flamingos feeding in a similar manner on the intertidal flats of Inhaca Island, Mozambique. He observed that they were stooping and seizing soldier crabs which they stalked at low tide.

VIGILANCE WHILE FORAGING

Foraging with the head lowered is incompatible with the detection of approaching predators or other sorts of disturbance. Therefore, flamingos regularly interrupt their current feeding activity to raise their head and briefly scan the environment. Flock foraging may allow birds to decrease individual investment in vigilance while maintaining a constant level of predator detection through corporate vigilance. Indeed, Beauchamp & McNeil (2003) have shown that vigilance decreases with increasing group size in a similar way both during day and during night in Caribbean Flamingos. However, they also observed that the size of foraging flocks was reduced at night, and that flamingos increased vigilance during dark nights, suggesting that the amount of time they can allocate to foraging is limited by low light levels. Vigilance in Greater Flamingos has not yet been systematically studied (but see Beauchamp 2005).

CIRCADIAN RHYTHMS

Flamingos feed by both day and night, their circadian rhythms varying according to location and season. Breeding birds, for example, after being deprived of food during a long period of incubation, probably spend 15 hours or more per day feeding for two or three consecutive days before returning to the nest. In autumn, birds assembled for moulting in the Camargue spend much of the day resting in one of the extensive saltpans, where they are undisturbed. In the late afternoon they either walk or fly to smaller ponds, which presumably offer better feeding conditions and where night feeding allows the birds to exploit sites where they might feel insecure during the day (because of proximity to roads, human activities, etc.) (Britton *et al.* 1985). The same applies to flamingos which now feed in rice fields in the Camargue and in Spain: they converge on the paddies at nightfall, when they are undisturbed, and depart the following morning, often only when they are chased away. It may be that some invertebrate species are occasionally more accessible at night, as suggested by Britton *et al.* (1985), but there is little evidence of this. Observations in the saltpans of the Camargue indicate a tendency towards night feeding year-round, with periods of rest during the day. The opposite can be observed in Cyprus, where during migratory halts flamingos feed during the day, and in the evening either congregate to roost or depart on migration. In tidal areas in Sfax (Tunisia) and on the Banc d'Arguin (Mauritania), birds were observed to adjust their feeding according to the rhythms of the tide, foraging mainly at low tide and resting at high tide, the birds walking towards the shore as the tide rose and then back out to sea again as it ebbed, as earlier described by Trotignon (in Kear & Duplaix-Hall 1975). In short, flamingos are large birds feeding on small or minute organisms which, unless they occur in great profusion, oblige the birds to feed for periods exceeding either the light or the dark hours of the circadian period.

FORAGING AND DRINKING FLIGHTS

FORAGING FLIGHTS

Flamingos and other colonial waterbirds nest in areas where they are free from disturbance and/or predation, but they can breed successfully only if they are able to sustain themselves, and later their chicks, within reach of the colony. For conservation purposes, therefore, it is of prime importance to know the feeding range and locations of colonially nesting waterbirds, since food availability is just as vital to their survival as lack of disturbance during incubation.

The food resources necessary to sustain colonies of birds may be spread over a wide area and only a few individuals may actually feed in the vicinity of the breeding site. Areas which are suitable for the establishment of flamingo colonies are not necessarily wetlands of high biological productivity, and many dry out in spring and summer. As for many other colonial waterbird species, food availability, and the distance which birds must travel to feed, may influence colony size and affect breeding success. In Puffins, for example, mass failure during incubation and chick-raising may be attributable to a scarcity of food in the vicinity of the colonies (Harris 1984).

Long-distance foraging flights by flamingos were first reported from the southern Caribbean (Rooth 1965). Caribbean Flamingos breeding on Bonaire (Netherlands Antilles) formerly fed mostly in the saltpans on the island, principally on the chrysalids of brine fly and brine shrimp. In 1969, the flow of water through the saltpans was modified, changing the flamingo's food supply. Some birds responded by changing their diet to other aquatic invertebrates, but there was also an increase in the number of flamingos making foraging flights to the coastal wetlands of Venezuela, situated 140 km to the south (Rooth 1975). In the Old World, Demente'ev et al. (1951) reported that flamingos breeding in Kazakhstan undertook long daily flights in search of food, those at Lake Tengiz flying 30–40 km to the wetlands of the Transvolga steppes, and those at Kara Bogaz-Gol Bay on the Caspian Sea flying 50–60 km to forage.

We have investigated the feeding range of four of the major flamingo colonies established in the Mediterranean region, using data collected by various methods:

Camargue

PVC-banded birds identified as breeding at the Etang du Fangassier were sought throughout the Rhône delta, on the Languedoc lagoons to the west and on the wetlands of the Plan du Bourg and the Etang de Berre to the east. However, because band codes can be read only when birds are feeding in shallow water and at relatively close range, a sample of incubating birds was also dye-marked during three breeding seasons (1987–89; see Chapter 3). These individuals could be seen at a distance of one kilometre, and in the third year of this study practically all the dye-marked birds were located on their feeding grounds. They were foraging in all types of wetlands, from freshwater marshes to hypersaline waters. About one-fifth of the marked birds were found in the saltpans at Salins de Giraud and about one-third were feeding within 10 km of the colony. The study confirmed that flamingos which breed in the Camargue feed throughout the Rhône delta, some birds going to the Languedoc lagoons up to 70 km west of the colony.

Spain

Manuel Rendón and colleagues have studied the movements of flamingos between the breeding site in Andalusia and the foraging areas (Rendón et al. 1991; Rendón-Martos & Johnson 1996; Rendón-Martos et al. 2000; Rendón et al. 2001), some of which lie at a distance of around 200 km from the colony (Figure 32). Flamingos are able to breed at Fuente de Piedra (Malaga) only when the preceding year's autumn–winter rainfall has been sufficient to extensively flood the lagoon and the surrounding feeding areas. The lagoon often dries again in spring and it can be completely parched by early summer, several weeks before the chicks fledge. The surrounding wetlands, lying within 10–20 km of the colony, can support only relatively small numbers of flamingos and large numbers of birds must forage in the marshes and saltpans lying between the mouth of the River Guadalquivir and Cádiz, a distance of 140–200 km from Fuente de Piedra.

Foraging flights are nocturnal. Departing birds assemble in the evening near the shore of the lagoon closest to their planned direction. At sunset, or shortly afterwards, they take wing in one or two groups, circling to gain height before

122 *The Greater Flamingo*

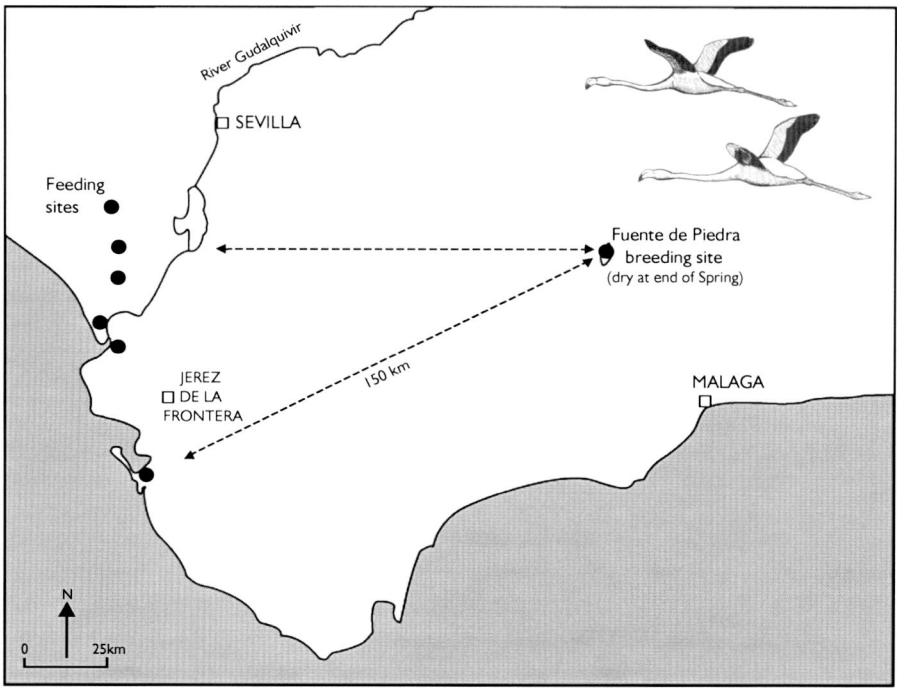

Figure 32. Many of the flamingos breeding at Fuente de Piedra in southern Spain forage at wetlands up to 200 km from the colony.

heading off at nightfall on a flight which may take them two hours or more. Birds returning to the colony seemingly also take wing in the evening from the foraging areas, since observers at Fuente de Piedra regularly hear the clamour of flocks arriving at the colony about two hours after nightfall (Rendón-Martos *et al.* 2000).

Many banded birds identified as breeding at Fuente de Piedra have been observed on the feeding grounds along the Atlantic coast of Spain, one or two days after being seen on the nest and shortly before returning to the colony (Appendix 13). The furthest of these was foraging 198 km from the colony. Some of the birds fed their chicks and left again shortly afterwards, suggesting that they sometimes travelled about 300 km in a single night (Rendón-Martos & Johnson 1996; Rendón-Martos *et al.* 2000).

Tunisia

During an aerial survey of Tunisian wetlands in June 1974, we observed a crèche of *c.* 8,000 flamingo chicks in a remote and rapidly drying part of the Chott Fedjaj, southern Tunisia (Kahl 1975c) (Figure 33). There were only *c.* 1,000 adults attending the crèche at the time, indicating that many of the parents were feeding elsewhere. Although large numbers of flamingos were feeding in this wetland and on others

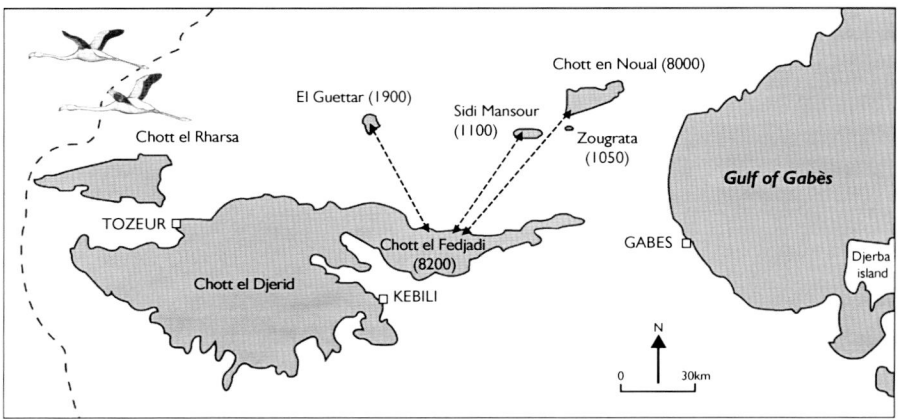

Figure 33. An aerial survey of Tunisian wetlands in May 1974, following abundant winter rains, revealed the presence of a crèche of c. 8,000 chicks in the remote Chott Fedjadj. Counts of adults, given here, indicated that many parents were feeding in wetlands up to 80 km from the crèche. The conservation of these areas is just as vital to the flamingos as that of the actual breeding site.

nearby, it was only by including birds feeding up to 80 km distant that the 16,000 parents could be accounted for, a large proportion of the adults being on the furthest and most extensive of these playas.

Turkey

The main breeding site of flamingos in Turkey is at Lake Tuz, a very extensive saltpan located in central Anatolia. In many places the shoreline is difficult to reach by land, and the lake is more conveniently surveyed by plane. Small colonies are also established in some years at Lake Seyfe, which lies 100 km to the south-east of Lake Tuz, but this wetland is used mainly for feeding. In June 1992, evening observations by G. Magnin and ARJ, made from hills conveniently lying to the south-west of Lake Seyfe, revealed large numbers of flamingos gathering in the south-western part of the lagoon (Figure 34). Just before nightfall all the birds took wing and circled many times in order to gain altitude before departing at nightfall over the hills in the direction of Lake Tuz, a situation reminiscent of our observations in Andalucia. Two days later, an aerial survey of Lake Tuz confirmed that flamingos were indeed breeding there and that the lake was almost dry. Only the later breeders, and their chicks, were still at the nest. The older chicks, aged about one month, were located where many adults were feeding, at the inflow of a small stream, 17 km west of the colony, and over 100 km from Lake Seyfe.

We suspect that a similar situation prevails at other colonies which have been less studied, and that over most parts of their range flamingos rarely find sufficient food resources in the immediate surroundings of the breeding site. The larger the colony,

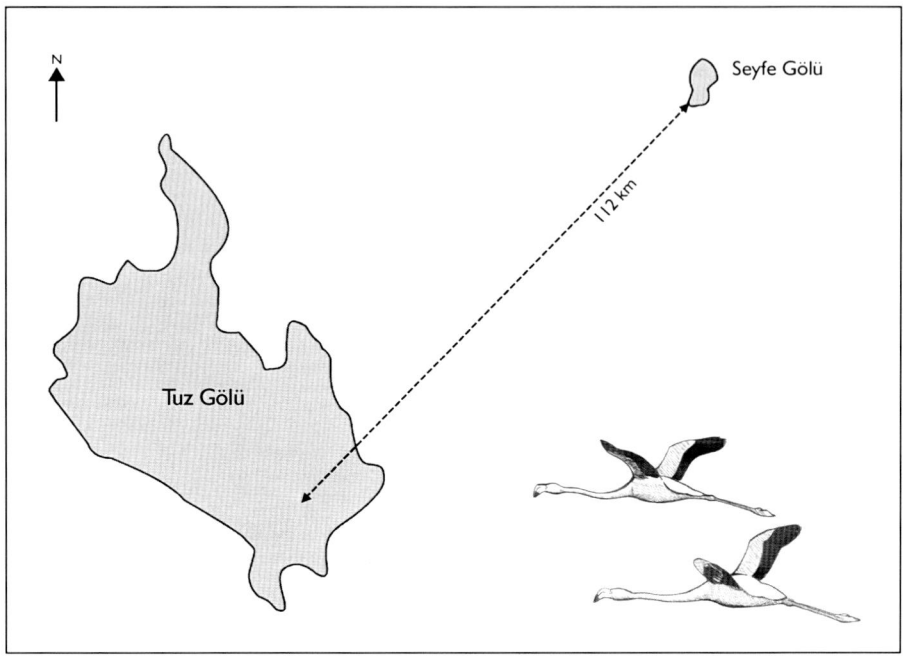

Figure 34. In Turkey, flamingos assemble in the late afternoon in June in the south of Lake Seyfe before undertaking evening flights over the hills to the south-west. This is an indication that they are breeding birds from Lake Tuz, 100 km away, which dries out in summer.

the greater will be its possible feeding range, depending on the area of wetlands around the colony, and on annual variations in flooding and food availability. During the breeding season, temporary sites tend to dry out, so parents at many colonies, for example Fuente de Piedra, will fly farther to forage than they did at the beginning of the season.

Among colonially nesting, non-soaring landbirds, the foraging flights of flamingos are among the longest, the furthest known being in Andalusia where some feeding areas lie at 198 km from the breeding site. Some of the birds involved fed their chicks at Fuente de Piedra and left again shortly afterwards, suggesting that they sometimes travelled around 300 km in a single night.

Drinking flights

Their salt glands enable flamingos to drink during feeding, but at times they may fly a short distance from foraging areas to fresh water, where they drink and perhaps bathe and preen. This behaviour seems to be characteristic of hypersaline situations, as observed, for example, at Sale Porcus (Sardinia) or at Fuente de Piedra

(Andalusia) when these lagoons are drying out. Such behaviour is not characteristic in the Camargue, although birds have been seen drinking on the beach adjoining the saltpans where they feed or, after incubation, flying to low-salinity areas to drink and bathe before moving on to foraging areas.

CHAPTER SEVEN

Mating system and mate choice

In this chapter we explore the factors behind mate selection and pair-bonding, as observed in both captive and wild birds. A long-lived serially monogamous species like the flamingo, which during its life may make several breeding attempts, is regularly faced with the task of choosing a reproductive partner. In this respect, colonial life may increase pairing options for flamingos, as large numbers of birds congregate and display at potential breeding sites long before breeding begins. This seems a favourable situation in which to study the pairing process between males and females. However, much of our current knowledge of the mating system and mate choice in flamingos has been based on studies of captive flocks, because of the various limitations inherent in their environments. Only recently have results from the Camargue long-term study begun to change our understanding of the flamingo's mating system.

There are many indications that animals do not pair randomly but select a partner on the basis of one or several characteristics (Bateson 1983). In this scenario, natural selection should favour behaviour which demonstrates good breeding condition to a prospective partner. Flamingos begin displaying several months before the start of breeding; groups of tens or hundreds of colourful adults stop feeding, raise their heads on fully stretched necks and begin flagging their heads from side to side. These are the first movements of the group displays which start in the Camargue as early as November and last for close to six months, usually ceasing completely by the second half of May. New birds may join in the display, while others return to feeding or comfort behaviour (see 'Group displays' below). Head-flagging usually leads to marching and the birds break into an impressive clamour. On a quiet day, the drone of these displaying groups can be heard more than a kilometre away. In southern France, displaying flamingos may be out in the middle of the national nature reserve, surrounded by thousands of wintering ducks, or on the Languedoc lagoons, 50 km or more from where they will later breed.

The displays consist of a series of easily recognised movements which the birds perform for long periods during the day, unless they are deterred by adverse weather. Males and females join in these ritualised performances, which increase in intensity as the days lengthen. The black of the remiges and the crimson wing-coverts, hidden on the closed wing, produce a flash of colour during the wing-salute. These are the flamingos which, three or four months later, will colonise the breeding island at the Etang du Fangassier.

GROUP DISPLAYS

Several studies have described the displays of flamingos (i.e. Allen 1956, Brown 1958, Rooth 1965, Studer-Thiersch 1974, and Kahl 1975a), and the scenario seems to be the same throughout the species' range. Various combinations of postures and movements (Plates 8–10) constitute the display repertoire. Our description has been largely inspired by that of Kahl (1975a), to whom we refer the reader for a more detailed account.

Alert posture or stretched attitude: Birds stand erect with neck stretched vertically upwards, the bill held high with the distal half nearly horizontal; they are seemingly looking nervously about. This posture is often a prelude to head-flagging.

Head-flagging: With their neck still stretched, but the upper part bent backwards and outwards, birds begin calling loudly and flagging the head jerkily from side to side. This movement is the first, and most prolonged, of the many which constitute a full display. Head-flagging is usually followed by a wing-salute or by marching.

Marching: Large groups of birds, sometimes up to 500 or more, stand in a tight mass, very erect and in a 'frenzied trance'. They move together, first rushing in one direction, then in another. Some birds, mostly males, may assume a *hooking posture* (Plate 10), whereby they incline their neck forwards 30°–60° from the vertical. The bill is 'hooked' downward with the tip almost touching the neck, whilst the scapulars and back feathers are often erect. After marching rapidly a few metres, the group slows down when some birds bend to dip their bills in the water (*false-feeding*). After a few seconds, all birds resume the erect posture and start walking. There is often aggression among members of these displaying groups.

Wing-salute: Birds cease marching and head-flagging but maintain the neck's stretched position, the bill slightly above the horizontal (Plates 8, 10). In a sudden, rather butterfly-like movement, they throw open their wings to their full extent, holding them as much as 40° behind the plane of the body. All birds are facing in the same direction, usually into the wind, and the salute seems not to be directed towards any particular individual. This position is held for 2–10 seconds, the time varying inversely with the intensity of the display (Kahl 1975a). During the wing-salute, the bird ceases the loud, disyllabic 'honking' given when head-flagging and utters a series of rather weak, short, low-pitched 'grinding' notes which can be heard only at close range. The wings are closed again in just as sharp a movement and this is almost invariably followed by a *twist-preen*.

Twist-preen: In a sudden movement lasting no more than 1–2 seconds, the bird twists its head and neck back to one side, drops down the wing on the same side and appears to preen behind the wing. An individual may repeat this movement several times in quite rapid succession.

Inverted wing-salute: This striking movement (Plate 9) is usually preceded by the *twist-preen* in displaying groups but may also be performed by solitary individuals. Birds bend forward from the erect position in a 'bowing stance' so that the cocked tail is held higher than the chest. The neck is extended straight forward and in line with the body. At the same time, the wings are flashed partially open, with the carpal joint pointing down and the black primaries pointing up. Seen from in front there is a sudden flash of red, and from the back a flash of black as the wings are held half-open for 1–2 s. Following this, the bird may continue *head-flagging* or it may terminate the display.

Scratching: Following one or more *twist-preens*, birds dip their bills into the water (*false-feeding*) then, still in a lowered position, shake their heads several times from side to side before scratching the neck just behind the chin.

Wing-leg stretch: One wing and the leg on the same side are stretched outward for 1–2 s, much as during comfort movements (Kahl 1975a). This is a posture which may be directed towards another bird since it is generally executed by individuals away from the groups which are *marching* and *head-flagging*.

Display flights: Small groups of between four and 15 birds, composed of both males and females, take wing after *marching* and begin circling as a tightly packed group, often flying over the breeding island. The manner of flying is distinctly different from the normal leisurely flight: the sternum is prominent and the wingbeats are rather rigid and shallower than usual. At times they appear to be synchronised briefly among several birds. The flock may circle only a few times, but usually the birds stay airborne for up to 30 minutes, some birds occasionally leaving the group and resettling. These flights, always centred over their display area, can attain considerable heights and extend to over a kilometre before the birds return to their point of departure. Allen (1956) and Rooth (1965) both refer to display flights by flamingos in the Caribbean, but Kahl (1975a) rather surprisingly does not mention these flights, perhaps because he observed displaying birds mainly in captivity. Display flights are common in the Camargue in spring.

The display movements are performed in the sequence in which they are presented here, but groups do not necessarily perform all of these movements every time they display. Birds may wing-salute without marching, just as they may march without assuming the hooking posture. Full displays, however, include all these movements with the exception of the wing-leg stretch. In the Mediterranean region, displaying usually starts in mid-December, but the numbers of birds involved and the intensity of displays seem to depend to a certain degree on the weather; in the Camargue, mild and wet autumns can induce displaying in November. The later in the season that birds display, the less likely it is that they will attempt to breed that year, and very few of the banded birds which have been observed still displaying once the colony is established have been observed later with egg or young.

Many of the postures adopted during displaying are similar to those of comfort activities outside the breeding season (Kahl 1975a). Head-flagging is possibly derived from the alert posture and from searching movements of the head. The two wing-salutes seem based on the wing-flapping and wing-stretching which flamingos often do after periods of inactivity. The twist-preen is clearly related to wing-preening, the inverted wing-salute to forward-stretch comfort movement with the wings stretched upwards over the back, and the wing-leg stretch to the normal stretching of a leg and wing. Each of these movements appears little changed from the 'ancestral' behaviour pattern, except that they are stiffly executed and are given within the context of a display bout (Kahl 1975a).

According to Rooth (1965), display is initiated, at least in the Caribbean Flamingo, by males, which display longer and more vigorously than females. Kahl (1975a) also refers to head-flagging usually being initiated by a male, while Brown (1958) refers to mass displays by males only (something we have not observed in the Camargue).

THE FUNCTION OF GROUP DISPLAYS

Throughout many parts of their world range flamingos are opportunists, breeding at irregular intervals, where and when natural climatic conditions permit, usually following periods of heavy rainfall. Flamingos occur over vast areas: in the Northern Hemisphere from Kazakhstan (50°N), where seasons are well defined, down to the tropics, where there is little variation in length of day between summer and winter. It is important, therefore, that birds are able to adjust spontaneously to favourable conditions when they occur. In the absence of seasonal factors for regulating physiological conditions, flamingos may possess a mechanism which allows a rapid synchronisation of hormonal levels so that they can establish colonies at the most opportune time. Darling (1938; see also Gochfeld 1980, Wittenberger & Hunt 1985) discussed the possible relationship between synchronised behaviour and physiological adjustment, and reproductive success in birds. Social facilitation (intensified display in response to the same behaviour by other individuals) is thought to enhance breeding synchrony through an acceleration of physiological cycles and ovulation (Darling 1938; Lott et al. 1967; Burger & Gochfeld 1991). For instance, Southern (1974) found that wing-flapping in Ring-billed Gulls was synchronising reproduction. Studies of captive flamingos have shown that displays increase in frequency and synchrony through social facilitation, also resulting in increased sexual activity (Stevens 1991; Pickering & Duverge 1992). Group displays could, therefore, stimulate synchronisation of laying (Ogilvie & Ogilvie 1986; Stevens 1991). However, this is speculative, and detailed field studies have not yet been carried out, either on the frequency and intensity of group displays or on breeding success in relation to the timing of breeding in flamingos.

A more likely function of group displays is mutual mate choice. Draulans (1988), in his study of Grey Herons, draws attention to the relationship between display and coloniality in the context of mate attraction. The attraction of a large number of potential mates to a colony site could substantially improve the selection of a suitable partner. In this case, conspicuous elaborate visual displays should be favoured in highly colonial species (Draulans 1988). Although flamingos, as opportunistic breeders, generally have a limited window of opportunity during which local conditions are favourable, they usually start displaying long before the onset of breeding. This time window has important implications for signal design and behaviour in relation to mate choice (Sullivan 1994). If mate choice is not random, but based on an evaluation of the potential partner (Bateson 1983), selection must optimise information-gathering within prevailing time constraints. When more time is available for mate assessment, facultative traits, the expression of which depends on the current or recent condition of the individual, should be favoured (Sullivan 1994).

To attract a partner, individuals must advertise their condition as efficiently as possible. Ornamentation coupled with vigorous display seems to function as an indication of condition (Zahavi 1975; Nur & Hasson 1984; Grafen 1990; Andersson 1994). Several studies have indicated that individuals showing an intense display effort (based on both display rate and duration) experience improved mating success (Johnson 1988; Verhencamp et al. 1989). Coloration due to dietary pigments, such as plumage carotenoids (Hill 1990, 1992; Olson & Owens 1998), is a further indication of the condition of the individual. Indeed, birds cannot directly synthetise carotenoids, which they must obtain from their food. Therefore, Endler (1980) suggested that, because carotenoids are scarce in nature, carotenoid-dependent ornaments indicate foraging ability, and hence can be used by individuals to gauge the quality of potential partners. However, carotenoids are also known to stimulate the immune system and to act as free-radicals scavengers, and Lozano (1994, 2001) suggested that carotenoid-dependent ornaments may actually indicate the bearer's immune condition and health status. On this basis, it is interesting to note that some display movements, such as the wing-salutes, appear to be both energetically costly and make conspicuous the red parts of the wing, whose bright coloration must be linked to the richness of carotenoids in the diet and the efficiency of the metabolic process (Fox 1975).

PAIR FORMATION AND COPULATION

Flamingos form pairs from a few days to a few months prior to breeding. In southern Europe, pairs are usually noticeable from the end of December onwards, either in close proximity within groups, or flying together as a pair, the smaller female generally preceding the male. Paired birds are recognisable in flocks because they assume the alert posture and false-feed together (Studer-Thiersch 1975a). In extreme cases, the male may be almost twice as tall as the female. It is unlikely that pairs form prior to a long migratory flight, but rather where and when conditions for breeding are suitable, and therefore not far from where they will nest. Early pairing in waterfowl indicates that some individuals are dominant at the end of winter (Ankey et al. 1991). Studies of waterfowl have shown strong correlations between body condition at the end of winter and breeding performance, individuals in good physical condition being more successful (Gloutney & Clark 1991; Bêty et al. 2003). No such data have yet been obtained on flamingos, although preliminary analyses of incubation shifts (Chapter 8) showed that pairs which changed shifts every 1.5–2 days tended to be more successful than those who changed at intervals of only 3–4 days (Johnson 1989b), perhaps suggesting greater foraging efficiency in the former.

Paired flamingos remain together both away from the colony and at the breeding site until shortly after egg-laying. The male and female will subsequently be together only briefly, during incubation changeovers. They again spend more time together just before the egg hatches, and while the chick is still in the nest. They may as a pair lead the chick into the crèche. By sometime in June in the Camargue, pairs are usually no longer apparent to the observer in the field, although both parents will feed the chick until it fledges.

Copulations take place away from the displaying groups, either in the vicinity of the breeding island, as observed also in East Africa by Brown et al. (1973), or elsewhere, for example during foraging, as observed by Rooth (1965). In the Camargue, copulations can be seen as early as the first days of January but are more frequent in spring, terminating by the end of May. Copulation in flamingos (Plate 11) has been described by Suchantke (1959), Brown et al. (1973), Poulsen (1975) and Studer-Thiersch (1975a), to which studies we can add our own extensive observations. Strolling behind the female, the male signals his readiness to copulate by touching the feathers of her lower back with his bill. If willing, she dips her head in the water false-feeding, slows down and consents to copulation, standing still and partially opening her wings. His breast rubs against her tail and he then mounts with beating wings, placing his feet on her scapulars. As he lowers his body for cloacal contact, his legs, but not his feet, are held below her wings (Brown 1973). This position gave rise to the erroneous belief that his legs were placed under her wings (Allen 1956; Brown 1958; Rooth 1965), which may seem to be the case from a distance. In addition, because pairs usually copulate in water where parts of the female body remain submerged, it is somehow difficult to have a clear picture of how pair members position themselves during copulation. However, the observation of some occasional copulations on land reveals that both birds hold their necks pointing forwards, with the female eventually touching the ground with the tip of her bill, perhaps to help her to keep her balance while supporting the superior weight of her partner. Cloacal contact is achieved by his bending his tail under hers, which is pushed to one side, and this position is maintained for 1–2 s. The whole act lasts 6–10 s, after which the male always climbs forwards off the female's back. He may assume the hooking posture momentarily, and false-feed or preen, whilst she either false-feeds or preens.

There is no information available concerning the timing of copulations in relation to egg fertilisation and laying. Duplaix-Hall & Kear (1975) doubted that flamingos could store sperm because they lay only a single egg. However, various studies of several seabird species which also lay a single egg (Birkhead & Møller 1992) have demonstrated a capacity for sperm storage, with the average interval between last copulation and egg-laying reaching up to 60 days in the Grey-faced Petrel (Imber 1976). Only scant anecdotal data are available on the frequency of copulations between partners in flamingos. In the Camargue, we have observed a pair copulating three times within an hour. Given the lack of synchrony between copulation and fertilisation in birds (Birkhead & Møller 1992), frequent copulation may be a suitable strategy when breeding opportunistically in an unpredictable environment.

EXTRA-PAIR COPULATIONS

Extra-pair copulations regularly occur in socially monogamous avian species (Birkhead & Møller 1992). They have been observed in several colonial waterbird species (Frederick 1987; Hatch 1987; Aguilera 1989; Graves et al. 1991; Hunter et al. 1992), and some studies have provided evidence for extra-pair fertilisations (Graves et al. 1991; Hunter et al. 1992; Austin et al. 1993). Although the occurrence of extra-pair copulations has been recorded in captive flocks of flamingos (Studer-Thiersch 1975a; Pickering 1992), the frequency of extra-pair copulations, and the possible occurrence of extra-pair paternity in Greater

Flamingos in the wild, remains unknown. Observations in captivity have shown that unpaired males, or males whose partners are not ready to mate, try to copulate with other individuals, and sometimes do achieve contact with another female (Studer-Thiersch 1975a). In the wild, unpaired males have been observed to seek copulations with paired females, and paired males react aggressively to all males approaching their female. The male maintains close proximity to the female, a behaviour similar to mate-guarding in other monogamous bird species (Birkhead & Møller 1992). A female flamingo is particularly vulnerable to rape attempts when she copulates with her mate, because once she is in a copulating position, she remains motionless as long as she feels pressure on her back. In this situation, joint assaults by several unpaired birds may be successful if the male is unable to ward off the intruders (pers. obs.). Although benefits of extra-pair copulations are obvious for both paired and unpaired males (Trivers 1972), females may also be interested in obtaining extra-pair fertilisations in order to increase their probability of fertilisation, or to acquire better genes for their offspring (Birkhead & Møller 1992). Although there is no evidence of a paired female flamingo actively soliciting extra-pair copulation, as has sometimes been observed in other colonial waterbirds (Birkhead & Møller 1992), we cannot rule out the possibility that active solicitations of extra-pair copulations by females are secretive or occur far from the colony, and are thus difficult to observe.

SEASONAL MONOGAMY IN GREATER FLAMINGOS

Monogamy is generally thought to be the predominant mating system in birds, involving about 90% of all species (Lack 1968b). This predominance is attributed to the relative success, in terms of the number of descendants, of chick-rearing shared by male and female. Although this argument does not explain monogamy in all bird species (see Davies 1991), it certainly applies to most colonial waterbirds (Dubois *et al.* 1998). Among storks, herons and seabirds, male and female share incubation and chick-feeding, and the loss of one partner usually leads to breeding failure (Oring 1982; Salathé 1983; Cézilly 1993). In the case of flamingos, the slow development of the crèche (see Chapter 8) and the need for synchronised breeding within the colony effectively prevent individuals from making two successful breeding attempts within the same season. However, bi-parental care is only one aspect of avian monogamy, and this mating system has recently turned out to be more complex than previously supposed (Wickler & Seibt 1983; Birkhead & Møller 1992; Cézilly & Nager 1995; Black 1996). For instance, recent studies using genetic markers have led to a distinction between social bonding and sexual exclusivity, as long-term pair-bonding does not always ensure sexual fidelity (Birkhead & Møller 1992). In addition, there is much variation among monogamous species in the extent of social mate fidelity, both during and between breeding seasons (Choudhury 1995; Ens *et al.* 1996).

Mate fidelity between consecutive breeding seasons

Prior to analysis of the Camargue data, the existing knowledge of pair-bonding and pair-bond duration in flamingos was essentially based on observations of captive populations (Studer-Thiersch 1975a; Pickering 1992; Stevens *et al.* 1992; King 1994). Studer-Thiersch (1975a) reported mate fidelity to be predominant in a captive flock of *c.* 50 Greater Flamingos at the Basel Zoo. She emphasised the close proximity of mates, both during and outside the breeding season, and noted the rarity of mate-changing, although no quantitative account was provided. Studying the flocks of flamingos at the Wildfowl and Wetlands Trust in Slimbridge, UK, Pickering (1992) estimated between-season mate fidelity to reach 94% in a flock of 30 individuals. From such accounts, flamingos were considered to be highly faithful to their mates over consecutive breeding seasons (Cramp & Simmons 1987; del Hoyo *et al.* 1992). Such a high rate of fidelity was not considered aberrant given the longevity of the species (Lebreton *et al.* 1992; Johnson 1997b) and the general belief that mate fidelity should be favoured in long-lived bird species (Rowley 1983; Ens *et al.* 1993).

Analysis of the Camargue data (Cézilly & Johnson 1995) radically changed our perception of the Greater Flamingo's mating system. In 1983, for the first time since PVC-banding started, two pairs with both partners banded were observed at the Fangassier colony. Since then, pairs of banded birds have been recorded each year (Table 10). Although close proximity between the birds is by no means an indication that the pair will successfully reproduce, or even attempt to do so, among the 175 PVC-banded pairs observed during the period 1983–98 (Table 11), 90 are known to have attempted breeding. In striking contrast to the observations of the captive flamingos, mate fidelity seems to be the exception, with only one pair maintaining a bond over two consecutive breeding seasons, in 1993 and 1994. The following year, both birds attempted breeding with different partners. These observations indicate a divorce rate of 98% between consecutive seasons for the species in the wild (Table 11; see also Cézilly & Johnson 1995).

Table 10. The number of PVC-banded pairs of flamingos observed at the Fangassier colony, Camargue, 1983–98 (n = 175).

year	number of pairs	year	number of pairs	year	number of pairs	year	number of pairs
1983	2	1987	7	1991	22	1995	20
1984	5	1988	10	1992	19	1996	7
1985	12	1989	6	1993	13	1997	5
1986	12	1990	17	1994	10	1998	8

Table 11. Mate fidelity between two consecutive seasons in the Greater Flamingo at the Fangassier colony.

Number of pairs with both partners banded observed 1983–98	175
Number of the above pairs for which both partners were observed at the Fangassier the following year (1984–99)	97
Both partners unpaired.	17
One member paired to a new partner, the other not seen paired (29 males, 20 females).	49
Both partners acquire new mates.	30
Partners reunite.	1

Mate fidelity within a breeding season

Some bonds appear to be only transitory even *within* a breeding season. In seven cases, pairs observed early in the breeding season were known to have separated a few weeks later. These early mate changes may correspond to 'trial liaisons' (Choudhury & Black 1993), a two-step mating 'partner-hold' strategy (Sullivan 1994), by which individuals maintain proximity while still assessing the prospective mate, eventually switching to a better mate if there is sufficient time for re-pairing. Such a strategy has been observed in the Barnacle Goose (Choudhury & Black 1993) and the Northern Mockingbird (Logan 1991). Flamingos may also change mates within a single breeding season when a first breeding attempt ends in failure (Cézilly & Johnson 1995). Many colonial waterbirds are able to lay a replacement clutch if the first is lost early in the season, and flamingos do so both in captivity (Studer-Thiersch 1975a; Pickering 1992) and in the wild. However, it appears that in flamingos a failed breeding attempt may lead males to divorce and re-pair with a different partner. Indeed, out of 513 second-breeding attempts recorded at the Fangassier colony from 1983 to 1998 (Figure 35), 345 (67.3%) were by males, probably with different partners. It seems that more males than females are able to re-pair because a surplus of unpaired birds at colonies enables males to select another mate, whereas females seem to be more often unable to recover from the energy expenditure associated with producing an egg. Of the 17 banded birds recorded making their third breeding attempt during one season over the same period, 16 were males.

Why do flamingos have high divorce rates?

In the light of our results, the mating system of the Greater Flamingo should now be considered to be seasonally or serially monogamous (Wickler & Seibt 1983). This mating system contrasts strongly with that of other long-lived avian species that tend to maintain pair bonds until the death of one partner. The adaptive significance of pair-bonding in birds has been the subject of much discussion (Coulson 1966; Rowley 1983; Johnston & Ryder 1987; Cézilly & Nager 1995;

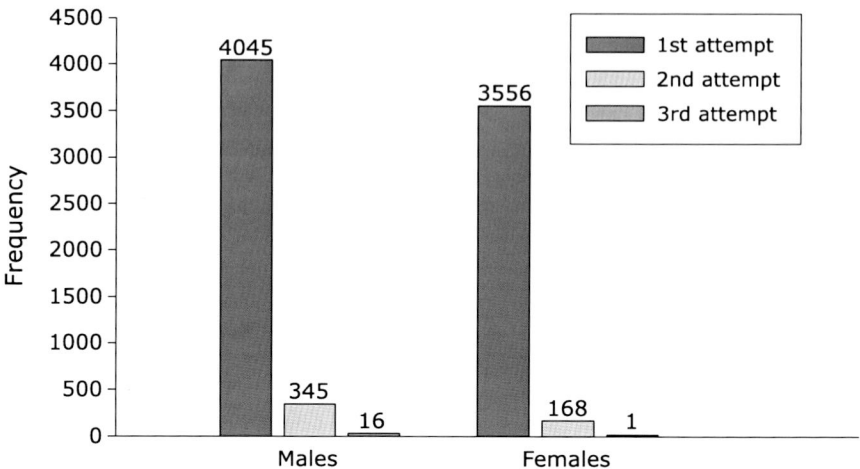

Figure 35. The proportion of male and female flamingos (banded birds) making a first, second and third breeding attempt in the same season at the Fangassier colony 1983–98.

Choudhury 1995; Ens *et al.* 1996; Cézilly *et al.* 2000), and various hypotheses have been offered to explain variation in divorce rate, both between individuals and between populations.

Coulson's incompatibility hypothesis suggests that the maintenance of pair bonds is not related directly to intrinsic qualities but to the compatibility of partners, this compatibility being dependent on factors such as behaviour, immunity, etc. This hypothesis is based principally on observations of long-lived seabirds where it is primarily the younger breeders which divorce, usually following a failed breeding attempt (Coulson 1966; Brooke 1978), but it cannot explain the absence of long-term pair bonds in flamingos where changes of partner are independent of age and breeding success. According to the 'best option' hypothesis (Ens *et al.* 1993), the probability of partners changing is dependent on the availability of other potential partners, and in a monogamous situation where sex ratios are balanced, adult mortality is an important factor regulating the availability of partners, such that high divorce rates are expected to occur mainly in short-lived species. On the other hand, Rowley (1983) suggested that lasting pair bonds should occur only in long-lived species, because of the benefit that was to be gained through accumulative experience during several breeding seasons. Obviously, both hypotheses fail to explain why flamingos do not have lasting pair bonds in the wild. Rowley (1983) emphasised the importance of lifestyle in determining the frequency of reuniting among bird species. Although some individuals migrate between a breeding area and a wintering area, flamingos are usually nomadic, wandering over a wide area in response to unpredictable climatic or seasonal conditions (see Chapter 5). In addition, most traditional breeding sites are very unstable and cannot be used on a regular basis. These factors have certainly favoured opportunisitic breeding behaviour in the species. Flamingos must be ready to exploit favourable conditions as soon as they occur. Such a situation, together with the usual crowding of colony sites, might be sufficient to prevent the development of long-term mate fidelity. Other species of colonial waterbirds show a correlation between low mate fidelity and low site fidelity (Cézilly *et al.* 2000). Simpson *et al.* (1987) found that all the birds in a Great Blue Heron population changed mates between consecutive breeding seasons when site fidelity was lower than 10%. Cuthbert (1985) also found that habitat instability was a major determinant of mate change between consecutive seasons in the Caspian Tern.

Flamingos are also unusual among colonial waterbirds in their high rate of divorce within a breeding season. Other species for which mate fidelity within a breeding season has been documented, for example Sooty Tern (Ashmole 1963), Caspian Tern (Cuthbert 1985) and Little Penguin (Reilly & Balmford 1975), do not show this behaviour. One reason why flamingos may change partners during one breeding season may be linked to asymmetric costs of reproduction between males and females. The energetic requirements of breeding are borne mainly by the female during egg production, whereas the male's only investment is to help to guard the nest a few hours or days prior to egg-laying. Observations at the Camargue colony strongly suggest that females have a limited capacity to produce a second egg in less than 15 to 35 days (Cézilly & Johnson 1995). A critical point is that most colonial species require synchrony in laying in order to reduce predation of the smaller chicks. The limited time available for breeding may explain why males choose a new female who can produce a new egg more rapidly. Such behaviour is probably facilitated by the fact that there are still many flamingos displaying around the colony in May and many pairs which have been prevented from breeding by lack of space on the breeding island, which is filled to capacity every year. To divorce within a breeding season may thus have evolved as a strategy by which males maximise their reproductive efficiency.

Finally, the discrepancy between social bonding in the wild and in captivity deserves further comment. It is obvious that the captive situation differs markedly from the natural environment of the birds. Captive flocks are relatively small (commonly 30–50 individuals), whereas several thousand flamingos normally gather around the main breeding sites during the breeding season. In addition, captive birds remain in close proximity throughout the year, whereas in the wild they disperse over a large area outside the breeding season. The influence of captivity on pair bonds has been noted in other species. Although penguins have a naturally well-developed pair bond, Bowles *et al.* (1988) still observed lower divorce rates among these birds in captivity than among those in the wild.

AGE-ASSORTATIVE PAIRING

Among bird species, assortative pairing can be observed for a large variety of traits, such as ornament size, body weight, body size, or colour morph. Assortative pairing occurs when pair members in a population are more similar (positive assortative pairing) or more dissimilar (negative assortative pairing) than if they had paired at random (Cézilly 2004).

In several bird species, partners tend to be of similar age or breeding experience (Reid 1988). Age-assortative pairing in birds is, however, somewhat ambiguous, as it refers to both a pattern of mating and to the underlying process of mate acquisition (Burley 1983; Reid 1988). Age-assortative pairing can occur passively, for instance in species where individuals start breeding at a similar age and show both high mate fidelity and adult survival, as regularly observed in colonial waterbirds (Nisbet *et al.* 1984; Bradley *et al.* 1995). Since individuals of such species rarely change mates, the initial similarity in age between pair members is preserved over time and a strong correlation between ages is observed (Reid 1988). Alternatively, age-assortative pairing can occur through a particular preference linked to age or some attribute closely associated with age. Active choice underlying age correlations within pairs of breeding birds can correspond firstly to a preference for a mate of similar age ('homotypic' preference) or secondly to a preference for either an older or a younger mate ('directional' preference). The case of a homotypic preference has been identified in the Barnacle Goose (Black & Owen 1995), for which mate fidelity is high (Forslund & Larsson 1991); homotypic preference for age appeared to be a near-optimal reproductive strategy given the cost of forming a new pair (Black & Owen 1995). Several authors (Nisbet *et al.* 1984; Perrins & McCleery 1985; Shaw 1985; see also Manning 1985) have considered that in monogamous species where reproductive success increases with age and/or experience, selection should exist for a directional preference for older/more experienced individuals, hence resulting in age-assortative pairing. However, empirical evidence for a directional preference underlying age-assortative pairing remains scarce.

From 1983 to 1998, 169 PVC-banded pairs of known age were observed at the Camargue colony, the oldest birds being up to 21 years of age. In 30 of these pairs the partners were of the same age, in 75 pairs the female was older and in the remaining 64 pairs the male was older, indicating that the age differentials are more or less random. The relationship between the age of partners is shown in Figure 36. There is a significant correlation between the age of pair members, suggesting age-assortative pairing. Since flamingos do not maintain pair bonds beyond one reproductive season, the observed pattern cannot be regarded as the mere consequence of birds starting their reproductive career at similar age. However, the observed correlation between the age of pair members remains biased because fewer birds were banded in earlier years, thus the existing pairs are more likely to have been banded birds of younger age classes, and more time is required to obtain

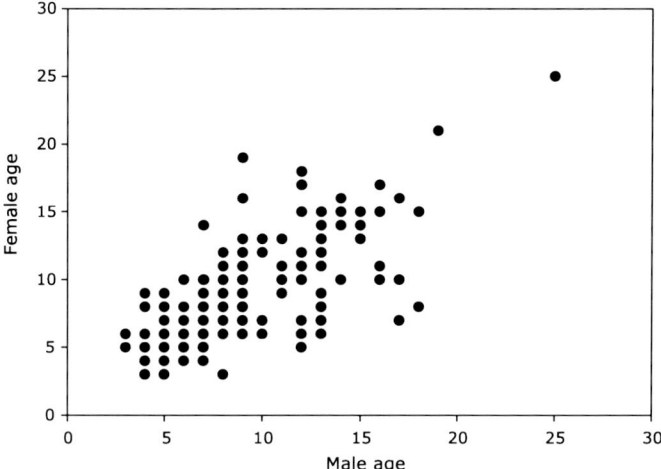

Figure 36. The correlation between the ages of the two members of a pair of banded flamingos.

samples of older birds. In order to reduce this bias, we considered pairing patterns only for the years 1990–92, when a large number of banded cohorts were available and several pairs with both partners banded were observed in each year (Cézilly et al. 1997). Using correlation coefficients and comparing observed data with expectations based on the assumption of random pairing between individuals, we showed that for each single year, age-assortative pairing was statistically significant, even after controlling for potential confounding factors (Cézilly et al. 1997). In particular, the observed pattern of age-assortment could have been due to birds of different ages returning to the colony site at different times of the season, as has been observed in some other colonial waterbird species (Fisher & Fisher 1969; Ainley et al. 1983; Nisbet et al. 1984; Slater 1990). In such a case, early arrival and early pairing of older individuals would increase the chance of significant age correlations in the population. However, in the Camargue, flamingos of all ages gather in increasing numbers around the breeding site from December to March and, therefore, the observed patterns of age-assortative pairing did not appear to be an artefact of differential date of return to the colony with age (Cézilly et al. 1997). Furthermore, a finer analysis of display patterns for the year 1991 suggested that the observed pattern of age-assortative pairing was linked to a directional preference for older individuals, rather than birds attempting to breed with a bird of the same age (see Cézilly et al. 1997).

Age-assortative pairing in flamingos can occur in the absence of mate fidelity between years, a situation that has never been observed in other bird species. In addition, this pairing pattern could result from a directional preference for older individuals. In species with strong mate fidelity, the cumulative and common experience of pair members may be more determinant in terms of breeding success than the sum of the individual experiences of the two pair members (Reid 1988; Black & Owen 1995; Cézilly & Nager 1996) and it has been argued that such a situation should favour a homotypic preference (Burley & Moran 1979; Black & Owen 1995). On the other hand, in species that maintain weak or no pair bonds between consecutive breeding attempts (Shaw 1985; this study), birds may select older breeding partners because age indicates survival ability, presumably an inheritable character (Manning 1985; but see Hansen & Price 1995). An older individual would then represent a fit breeding partner. Finally, it is not clear whether birds select their mates on the basis of reproductive experience or on the basis of true age since the two variables are most often closely associated. Burley & Moran (1979) demonstrated experimentally that Feral Pigeons showed a directional preference favouring more experienced individuals among birds of a similar age. In this study, we were not able to assess reproductive experience precisely since some birds may have bred undetected elsewhere in the Mediterranean or West Africa, or even in the Camargue. In addition, the proximate cues used by flamingos in determining age or experience are not straightforward, since once adult, plumage does not change with age (Johnson *et al.* 1993). Birds may, however, select their mate on the basis of display behaviour. Presumably, display behaviour is costly in terms of time and energy (see Vehrencamp *et al.* 1989) and may serve as an honest indicator of individual quality (Simmons 1988; Hidalgo de Trucios & Carranza 1991). Future studies will examine the relationship between the age of individuals and the qualitative and quantitative aspects of display behaviour.

CHAPTER EIGHT

Breeding biology

Since 1969, many thousands of flamingos have gathered at the Etang du Fangassier in the Camargue each spring. The breeding birds are tightly packed onto one or the other of the muddy islands and they are surrounded by equally impressive numbers of hopeful breeders who have still to seek out a place. Overcrowding and a loud clamour of calls are typical of large flamingo colonies, which constitute one of the most extraordinary sights that wildlife can offer. But there is more to the scene than meets the eye.

   The initiation of breeding is actually a complex phenomenon that shows variation in space and time, and we are only just beginning to understand which environmental factors play a key role in colony formation and trigger breeding activity. Even once the eggs have been laid, there is still a long way to go before the young will join the crèche and eventually fledge. At each stage of reproduction, success heavily depends on the ability of pair members to efficiently coordinate their parental duties, particularly when facing harsh climatic conditions. The present chapter summarises our current knowledge of the breeding biology of flamingos. We consider the different factors affecting the timing of breeding, and recapitulate the successive phases of reproduction, from colony establishment, through egg-laying to the fledging of the young, with a particular emphasis on the causes of variation in breeding success at each stage.

TIMING OF BREEDING

Geographical variation in egg-laying dates

Gonadal development and circannual rhythms in birds are dictated largely by the photoperiod, or increasing length of daylight hours (see Gwinner 1996), which corresponds to a period of increased food availability during egg production and during the chick-raising period. The final expression of reproduction, however, is dependent upon certain ecological parameters (Murton & Westwood 1977). The dates of egg-laying at flamingo colonies in north-west Africa, the Mediterranean region and south-west Asia are shown in Figure 37. Baker (1938 in Campbell & Lack 1985) stated that in North America spring arrives 4–5 days later for each 1° increase in latitude, and for each 125 m rise in altitude, and it arrives about 4 days later for each 5° longitude east. Similarly, Kear & Duplaix-Hall (1975) reported that flamingo eggs are laid one month later for every 10°N latitude (and also that flamingos lay eggs one month earlier in captivity than in the wild). We computed an 'effective' geographic index for the Palearctic using the relationships suggested by Baker (1938) as the sum of latitude (in decimal degrees), altitude (derived as altitude/125), and longitude (as decimal degrees/5) (Figure 38). We then tested whether mean laying date could be explained by this relationship. However, it should be noted from the outset that our estimate of mean laying date is admittedly crude, often based on general descriptions (see Chapter 12). Thus, if a laying date was reported as early April, we used 5 April as our best approximation, and for mid-April, we used 15 April, etc. (see Appendix 14). Despite this lack of accuracy, we found a significant correlation between this geographic index and laying date ($r = 0.67$, $P = 0.005$). When the individual components were evaluated separately, latitude and longitude were both correlated with laying date ($r = 0.63$, $P = 0.0008$, and $r = 0.60$, $P = 0.014$, respectively), but altitude was not ($r = 0.34$, $P = 0.205$). The absence of this latter effect may be due to the fact that 60% of the sites included in our analysis were at or very near sea level, and so had little variability. Indeed, dropping altitude in the computation of the index suggested by Baker (1938) had no effect on the strength of the correlation ($r = 0.68$, $P = 0.004$).

An interesting aspect of this evaluation, however, is not only the sites that seem to fit Baker's theory, but rather those that do not. For example, two of the three sites in Mauritania, both of which are in the Parc National du Banc d'Arguin, are quite different from the general pattern. In fact, if these two sites are removed from the data, the association of the other sites becomes dramatically stronger ($r = 0.90$, $P < 0.001$; Figure 38 and Appendix 14) when using the index including only

Breeding biology 143

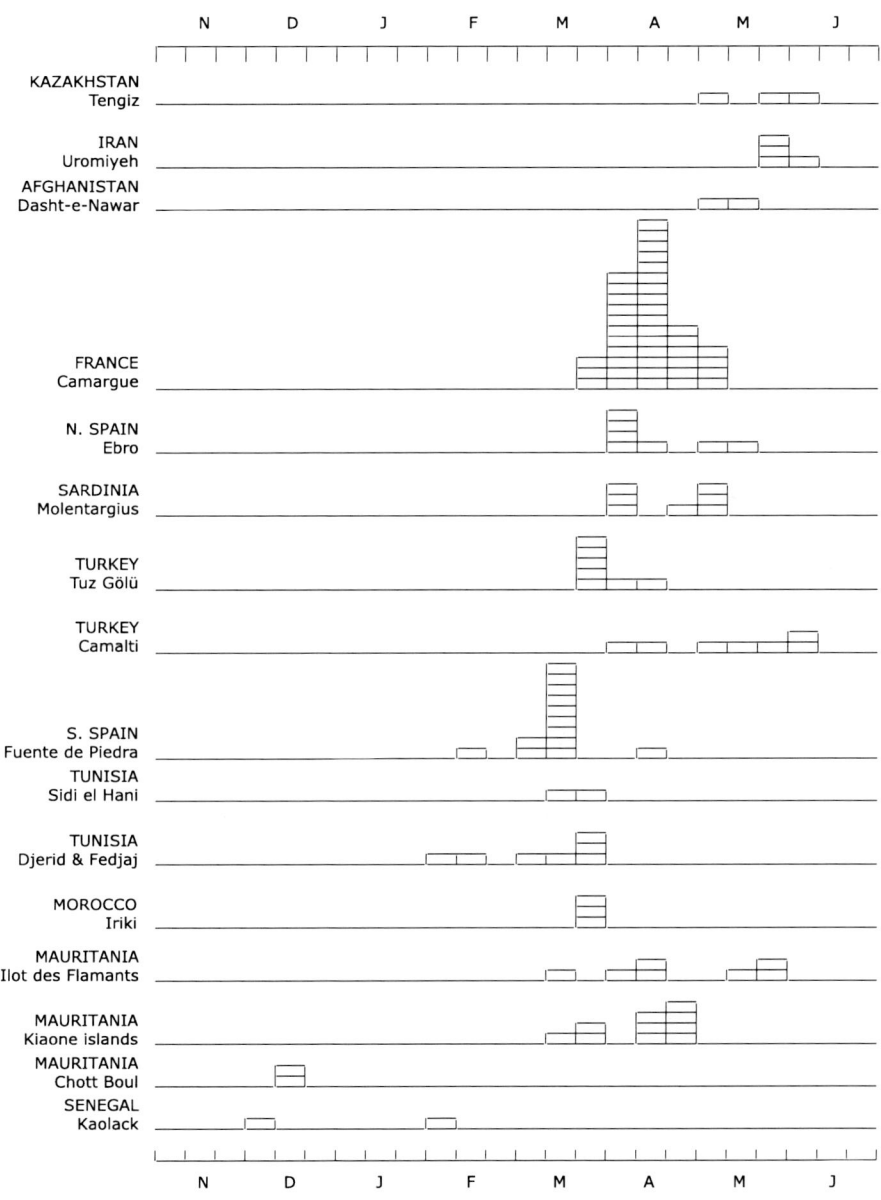

Figure 37. The start of egg-laying at flamingo colonies in north-west Africa, the Mediterranean and south-west Asia.

144   *The Greater Flamingo*

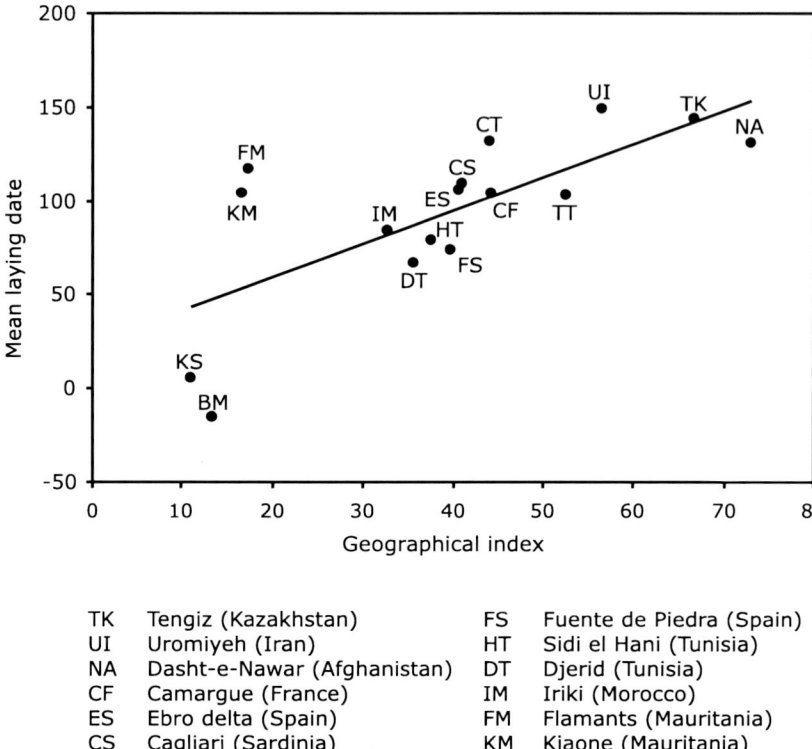

TK	Tengiz (Kazakhstan)	FS	Fuente de Piedra (Spain)
UI	Uromiyeh (Iran)	HT	Sidi el Hani (Tunisia)
NA	Dasht-e-Nawar (Afghanistan)	DT	Djerid (Tunisia)
CF	Camargue (France)	IM	Iriki (Morocco)
ES	Ebro delta (Spain)	FM	Flamants (Mauritania)
CS	Cagliari (Sardinia)	KM	Kiaone (Mauritania)
CT	Camalti Tuzlasi (Turkey)	BM	Chott Boul (Mauritania)
TT	Tuz (Turkey)	KS	Kaolack (Senegal)

Figure 38. The mean egg-laying dates at flamingo colonies in north-west Africa, the Mediterranean and south-west Asia corrected with geographic index for altitude, latitude and longitude. See Appendix 14.

latitude and longitude. Why is the Banc d'Arguin so different? Naurois (1969a) suggested that nesting on the Ilot des Flamants may be retarded in some years by high spring tides, which can submerge the island in March. However, this would not explain why nesting on the Kiaone Islands, which are rocky islands several metres above sea level, would also be delayed. The delay could also reflect some aspect of food availability. We are not certain of the primary food source of flamingos in the Banc d'Arguin, although Wolff & Smit (1990) and Gowthorpe *et al.* (1996) suggested that flamingos probably feed to a large degree on gastropod snails, which are also consumed by the two million small waders which habitually winter on the Banc d'Arguin. If these authors are correct, then delayed breeding could be in response to some aspect of the life cycle of the snails, or related to avoiding competition with the large numbers of migratory waders. The Banc d'Arguin is an extremely important site for many species of many taxa, and was declared a National Park in 1976. It is hoped, therefore, that research currently being initiated into the functioning of this environment might shed more light onto the flamingos' habit of later breeding in Mauritania.

Egg-laying at Camalti Tuzlasi in Turkey has also occurred rather later than might be expected for a coastal site. This may be linked to the flooding regime of the saltpans and/or the fact that this colony is still quite young, having begun in the 1980s. In contrast, egg-laying at Lake Tuz, which is on the same latitude as Camalti Tuzlasi, occurs quite early for a site which lies at almost 900 m altitude, probably because of early drying out (in June) owing to its shallow water.

THE INFLUENCE OF RAINFALL AND WATER LEVELS ON TIME OF BREEDING

Some of the variability in the time of nesting may be attributable to environmental factors, particularly rainfall or water levels. Such factors can strongly affect whether or not flamingos breed, and when, in a given year. In East Africa, flamingos can breed at any time of the year (Brown & Britton 1980). Similarly, in the tropics, where there are no seasonal temperature constraints, Caribbean Flamingos bred continuously from August 1966 to November 1967, during exceptionally heavy and prolonged rains (Gerharts & Voous 1968). Flamingos often breed in unstable environments where water levels, and hence food availability, fluctuate widely. In natural wetland habitats, lying in arid and semi-arid areas (e.g. playas), breeding by flamingos is strongly dependent on precipitation in the catchment area. Abundant rainfall is infrequent and unpredictable in such areas, and flamingos can consequently establish colonies there only sporadically. For example, Berry (1975) noted that flamingos breed at Etosha Pan in Namibia only when precipitation over the Pan exceeds 500 mm, the mean annual precipitation being only 430 mm. However, Simmons (1996) more recently revised this, suggesting that the mean during the rainy season (August to May) is 389 mm, with the threshold for flamingos attempting to breed being 400 mm (Figure 39a). Similarly, rainfall has a major effect on breeding success at the Makgadikgadi Pans in Botswana (McCulloch & Irvine 2004).

In Spain, M. Rendón-Martos showed that breeding at Fuente de Piedra takes place only if winter rainfall in the catchment area during the preceding months (October–February) exceeds 289 mm (Figure 39b). The only exception to this occurred in 1991, when the preceding year had been outstandingly wet (Rendón-Martos & Johnson 1996) so that residual water levels persisted until the breeding season of 1991.

Another example of the potential effects of rainfall recently occurred in Italy. Throughout the Mediterranean region, egg-laying normally occurs only from February to May. However, in autumn of 1996, nine pairs of flamingos unexpectedly began breeding in Apulia (Margerita di Savoia) in September. They were all successful and the nine chicks fledged at the end of December. This unusual event, at a site newly colonised by flamingos, was preceded by a period of exceptionally heavy rainfall during August–September (Albanese et al. 1997).

We have also monitored the periodic breeding by flamingos in Tunisia since a large active colony was discovered in the Sebkret Sidi el Hani in 1972. Nesting has since been observed in three regions: at Sidi el Hani (Tunisian Sahel), in the Sebkhas Sidi Mansour and El Guëttar (semi-desert zone), and at the Chotts Djerid and Fedjaj lying on the edge of the Sahara Desert (Figure 39c). Additionally, flamingos attempted to breed at Sebkhet El Djem in 1976 (Morgan 1982). Monthly precipitation levels from the meteorological stations closest to the nesting sites, namely Kairouan, El Djem/Souassi and Tozeur were obtained from the National Meteorological Office in Tunis. Wet years occurred irregularly during the period of study, and are grouped towards the beginning and the end, with a long intervening series of relatively or exceptionally dry years. As in Fuente de Piedra, periods of exceptionally high rainfall may replenish the underground water table to the extent that even average rainfall the following year can result in water levels sufficiently deep to allow flamingos to breed. We calculated totals for the wettest months, i.e. from September to May (Figure 39c). This is the period when >90% of the annual precipitation occurs at all stations, and it extends from the pre-breeding season to incubation. Rainfall is greater at the two more northerly stations, but all of these show important annual variations. The data indicated that the occurrence of breeding in any of the three regions was positively associated with rainfall at the two northernmost gauging stations ($\Pi^2 = 4.58$, 1 df, $P = 0.03$ and $\Pi^2 = 4.32$, 1 df, $P = 0.04$ for Kairouan and El Djem/Souassi, respectively), but that there was no association between breeding and rainfall recorded at the Tozeur gauging station ($\Pi^2 = 1.32$, 1 df, $P = 0.25$). The association was slightly stronger if an average of the two northern stations was used ($\Pi^2 = 5.11$, 1 df, $P = 0.02$).

In contrast to the situation in naturally unstable environments, in habitats where water levels are more predictable (e.g. coastal habitats and saltpans) rainfall appears to have less effect on whether birds initiate breeding within a given year, although in some cases it may influence other aspects of breeding, such as the date of initiation, the number of breeders, or success of breeding (see below). Management of the saltpans in the Camargue by the salt industry results in the maintenance of stable water levels in this habitat from year to year. Although this spring flooding regime is favourable to flamingos, it is the opposite of a natural situation in that the water surface areas are expanding (throughout the saltpans) at a time when water is normally receding in the surrounding wetlands, these being subject to a more natural hydrology. However, flamingos do not forage exclusively in managed saltpans; they exploit the large Vaccarès-Impériaux complex of brackish lagoons and the surrounding salt-steppe. This low-lying steppe, with sparse halophytic vegetation, covers about 4,500 ha and is exploited by flamingos when flooded. Water flows through this mosaic of marshes according to the strength and direction of the wind, and water levels may fluctuate widely from one year to another. Varying water levels promote major changes in the abundance and availability of invertebrates in temporary, seasonal and semi-permanent wetlands. The natural flooding of the brackish lagoons and salt-steppe induces the rapid growth of chironomid and amphipod populations, two important food sources for flamingos.

Breeding biology 147

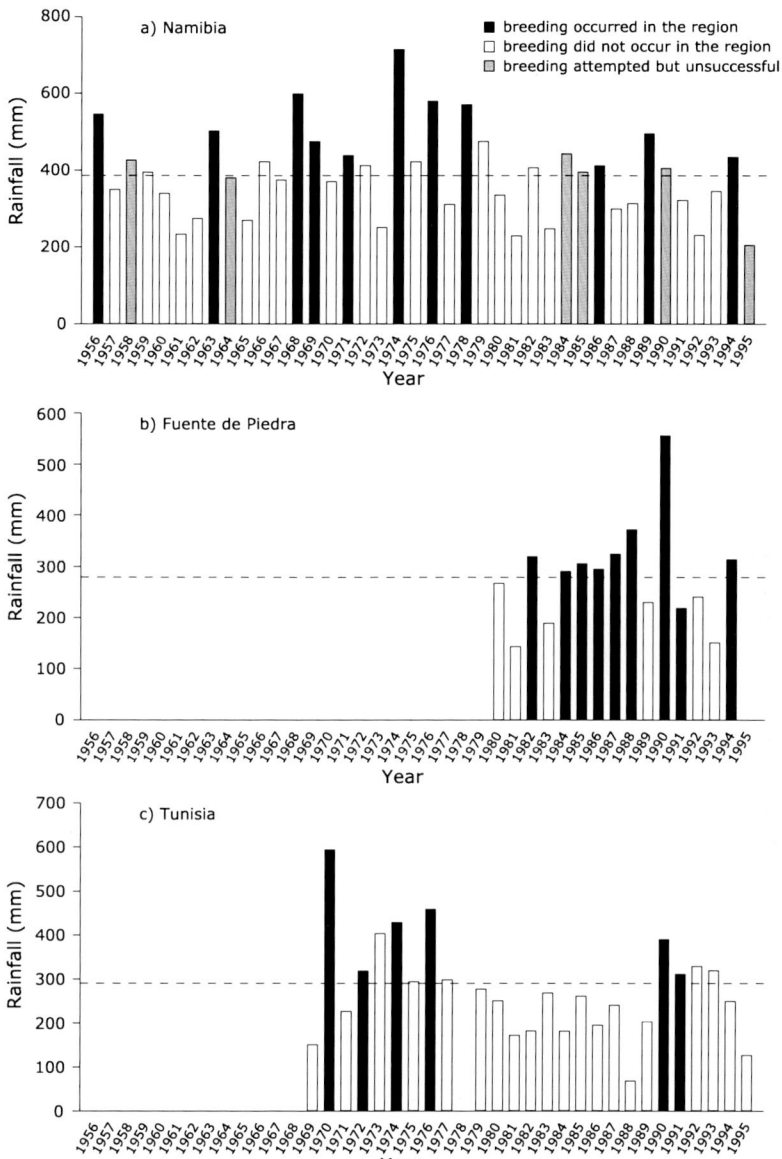

Figure 39. Annual rainfall and breeding occurrence a) in Etosha National Park, Namibia (after Simmons 1996), b) at Fuente de Piedra, Spain (after Rendón Martos 1996) and c) in Tunisia. Also shown (dashed lines) are the mean rainfall reported by Simmons (1996) and the threshold levels reported by Rendón Martos (1996) and that proposed for Tunisia for initiation of breeding at these sites. In 1970 breeding was suspected, but not confirmed, in Tunisia, based on the arrival in the Camargue of post-fledging juveniles that had not originated from alternative known colonies.

The abundance of invertebrates may be particularly important for females in providing certain nutrients required for egg-production. Cézilly *et al.* (1995) showed that the number of breeding birds at the Fangassier colony was highly correlated with water levels in the Vaccarès system in March (Figure 40), when breeding pairs start visiting the breeding island. However this result was limited to the period 1984–1991. We extended the analysis to the period 1984–1997, and found that the number of breeding pairs actually shows an exponential increase with water levels ($r = 0.855$, $n = 14$, $p = < 0.0001$). However, although rainfall appears to have less influence on the timing of nesting in more stable habitats, it can still have an effect. Since 1947, egg-laying at colonies in the Camargue has begun between 27 March and 5 May (Appendix 16), with an average laying date of 15 April. Each spring, the salt company floods the Etang du Fangassier, where the flamingos breed, as part of the annual salt production process. In addition to technical considerations, the date of flooding is determined by the presence of winter rainwater in evaporators and low-salinity seawater at the intake (diluted by river water owing to prolonged south-east winds). This variability in flooding date has a significant effect on the date on which flamingos initiate egg-laying ($r = 0.707$, $n = 22$, $P < 0.001$; Figure 41). Since 1975, egg-laying has started 2–33 days after the flooding of the Etang du Fangassier (average = 20.9 days ± 7.6 sd). The one instance where initiation occurred two days after flooding was in a year when an extensive area of winter rainwater remained around the breeding island,

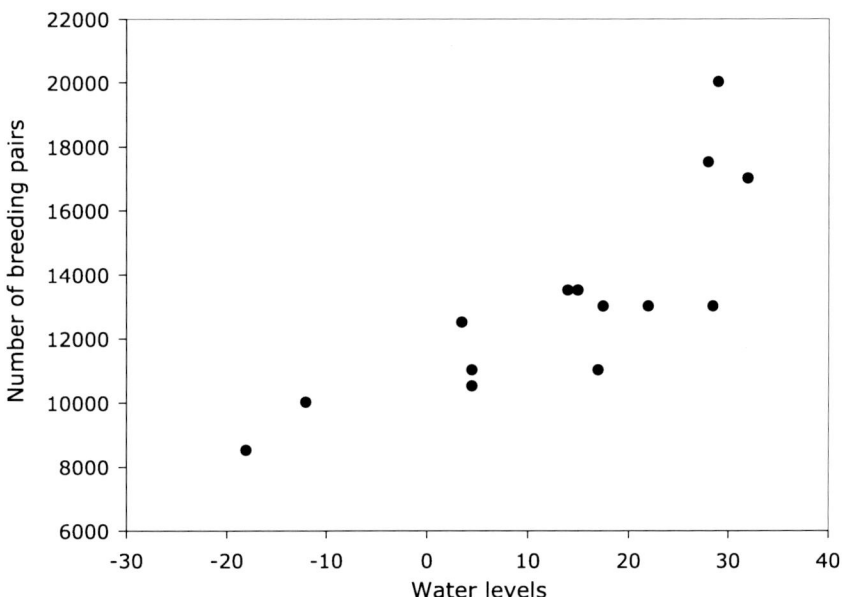

Figure 40. The number of breeding pairs of flamingos in the Camargue as a function of mean water levels in March 1984–97 (expressed as montly averages of relative values to the general level in France).

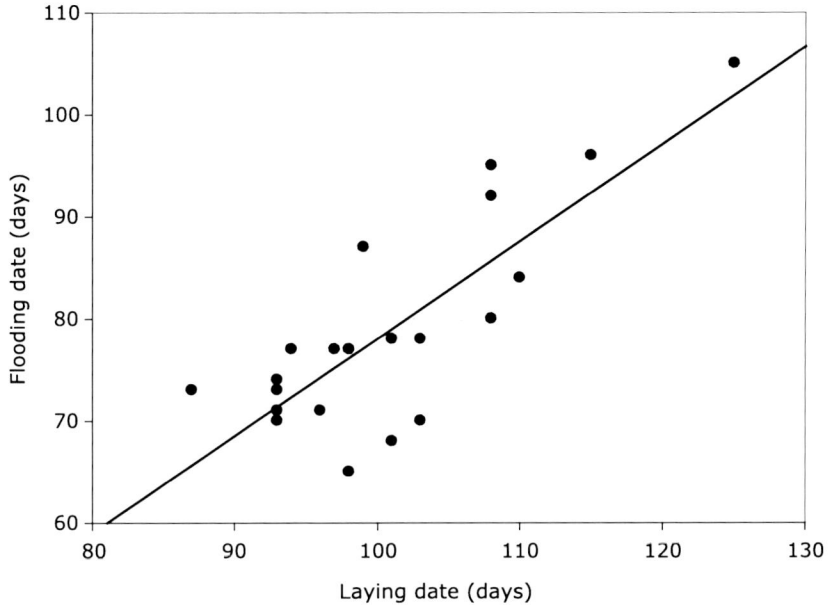

Figure 41. The start of egg-laying in relation to the date of flooding of the Etang du Fangassier, Camargue.

which is usually dry prior to flooding by the salt company. The latest recorded date of flooding in the Camargue (15 April) was in 1989, owing to a strike by the salt-workers. This also happens to be the year with the latest recorded start of egg-laying (5 May).

In conclusion, the total number of birds that attempt to breed each year might be determined by water levels, and hence food availability, in that part of the foraging range of the colony that shows little stability, whereas the precise timing of breeding may be related to local conditions in the immediate vicinity of the colony. Periods of reduced food abundance are indeed expected to lead to reduced investment in reproduction, decreased breeding success, or even abstinence from breeding in long-lived species that lay small clutches and show deferred maturity, such as flamingos (Cézilly et al. 1995). Because flamingos lay a single egg, they are also unable to trade off the number (i.e. by adjusting clutch size) against the quality of young they produce, as has been observed in colonial waterbird species laying several eggs in a clutch. When food is in short supply at the onset of the breeding season, as is the case when water levels are low, some adults may refrain from breeding if the probability of successfully rearing young is too low. On the other hand, once birds are ready to breed, flooding around the breeding island, presumably signalling safety from predators, might be of crucial importance in triggering nest-building.

TIMING OF 'NEW' COLONIES

The timing of breeding may also be affected by whether or not the colony is 'newly established'. There appears to be a tendency for late breeding during the first year(s) of occupation of a new site, as was the case in several recently established colonies in Spain, Italy and Turkey. At El Hondo in Spain, egg-laying in the first year (1996) was initiated in May, and the following year in late April to early May (Aragoneses & Echevarrias 1998). In the Ebro Delta in Spain, egg-laying in the first year (1992) started in June or July (see Luke 1992), and the breeding attempt failed, whereas in the six following years (see site sheet) the birds began egg-laying in April or May and breeding was successful. Similarly, at Margherita di Savoia, in Italy, the first breeding attempt was made at the end of the summer and was unsuccessful (Albanese et al. 1997). This colony has since had extremely variable and extended laying dates, including successful breeding attempts in two autumns (1996, 2000). Finally, at Camalti Tuzlasi in Turkey, during the first four years of breeding (1982–86), egg-laying did not begin until mid-May or early June, while during 1990–94 colonies were established in April or early May.

It is not clear whether the unusual timing of breeding in newly established colonies is a result of new surroundings or of the age of its occupants, or both. It may be that the individuals have simply not yet established a routine, and have taken more time to establish pair bonds, select the site, etc. Alternatively, the timing may be influenced by the age of the breeding birds. At several newly established colonies (e.g. the Italian colony described above and two more recently established colonies, at Comacchio in Italy and at Laguna Petrola in Spain) a larger proportion of the birds are often relatively young, perhaps even breeding for the first time, as indicated by residual immature plumage or identification of banded individuals. Birds in immature plumage are rarely observed actually breeding at well-established colonies, so the erratic timing at these new colonies may simply reflect the inexperience of their occupants.

COLONY ESTABLISHMENT

The number of individuals in the immediate surrounding of traditional breeding sites often starts increasing several weeks prior to egg-laying. In the Camargue, groups of flamingos start visiting the breeding island from up to two weeks before the start of egg-laying. The birds that arrive early in the season, however, seldom stay on the island overnight, and are not necessarily those which will later attempt to breed. Synchrony in egg-laying amongst a certain number of pairs is important at the onset of the breeding season. Birds which lay before the first main wave of breeders are likely to abandon their eggs when the others, which are not ready to lay, fly off in the evening to forage. In the Camargue, several thousand birds in breeding condition generally take possession of the breeding island from nine days down to a few hours (average 3.4 days, n = 24) prior to egg-laying (Table 12). From this moment on, the colony is permanently occupied, day and night, until the end of the breeding season.

DESPOTIC ESTABLISHMENT OF BREEDING BIRDS

In most areas, the number of 'hopeful' breeders at the onset of colony establishment by far exceeds the number of nesting pairs that the breeding site can accommodate. In such a situation, intraspecific competition for breeding space is intense, and may force some individuals to attempt breeding in sub-optimal habitats. This has been observed at the Fangassier colony, with pairs regularly attempting to breed on a nearby dyke after the breeding island has become saturated. Most of the time, breeding in sub-optimal habitats is doomed to failure. However, it is difficult to determine whether breeding failure is the direct consequence of the lower quality of the site (due for instance to increased exposure to terrestrial predators) or is linked to the intrinsic quality of breeding pairs. Indeed, in the Camargue, we have regularly observed that breeding attempts on the dyke are mostly made by young individuals, whose breeding success is typically low (see below). In addition, owing to the limited number of birds attempting to breed in a sub-optimal habitat, social stimulation may be too low to maintain birds on their nest in the face of adverse conditions, although direct evidence for this mechanism is still lacking.

More recently, the 'despotic' establishment of flamingo colonies has been studied in southern Spain (Rendón et al. 2001). The spatial distribution of individuals conforms to a despotic distribution when some individuals are prevented, through interference or exploitation competition from other flamingos, from settling in the

Table 12. The start of egg-laying in relation to the date of the first flamingos visiting the breeding island at the Etang du Fangassier.

Year	Site occupied	Start of laying	Year	Site occupied	Start of laying
1976	Apr 10	Apr 15	1988	Mar 31	Apr 4
1977	Apr 17	Apr 20	1989	May 3	May 5
1978	Apr 20	Apr 22	1990	Apr 2	Apr 6
1979	Apr 22	Apr 25	1991	Mar 25	Mar 27
1980	Apr 17	Apr 18	1992	Apr 10	Apr 13
1981	Apr 14	Apr 14	1993	Apr 5	Apr 8
1982	Apr 2	Apr 3	1994	Mar 23	Mar 28
1983	Apr 8	Apr 11	1995	Apr 4	Apr 9
1984	Apr 5	Apr 11	1996	Apr 14	Apr 18
1985	Apr 4	Apr 7	1997	Apr 14	Apr 18
1986	Apr 2	Apr 3	1998	Mar 30	Apr 8
1987	Mar 29	Apr 3	1999	Apr 02	Apr 07

The breeding island is colonised on average 3.4 days prior to the start of laying (range 0–9 days, n = 24).

best-quality site and are then forced to use lower-quality habitats. One important criterion to demonstrate the existence of despotic spatial distribution in the wild is to show that individuals ending up breeding in the lower-quality habitats have indeed attempted to settle first in the best-quality habitat. This is exactly what Rendón et al. (2001) have observed in southern Spain. There, flamingos use both the Fuente de Piedra Lagoon and the Guadalquivir Marshes as breeding sites. However, the Fuente de Piedra site offers a much better protection against terrestrial predators, and birds breeding there experience a higher breeding success than birds breeding in the Guadalquivir marshes. Accordingly, flamingos settle first at Fuente de Piedra, independently of differences in the availability of trophic resources between the two sites (Rendón et al. 2001). Direct observations of settlement patterns at the two sites revealed that 18 marked individuals that prospected for a breeding place in Fuente de Piedra were later recorded breeding in the Guadalquivir marshes in the same year. As in the Camargue, breeding in the sub-optimal site following unsuccessful settlement in the high-quality site involved mainly young individuals.

NUMBER OF ATTEMPTS PER YEAR

As mentioned in Chapter 7, both male and female flamingos may make a second, or even a third, breeding attempt during the same season, but only if the previous attempt(s) failed. Successful breeding, from pairing to the chicks fledging, requires about four months. This corresponds to the period when environmental conditions are most suitable throughout the temperate areas where flamingos occur. In the tropics there are a few records of flamingo colonies being established twice in the same year, for example at Lake Elmenteita, Kenya, in 1968 (Brown 1973) and at Aftout es Saheli, Mauritania, in 1987 (Wetten et al. 1990–91; J.-L. Lucchesi pers. comm.), but it is not known how many, or if any birds made a second attempt following a first successful breeding within one year.

'FALSE' BREEDING ATTEMPTS OR 'PRACTICE' NESTING

Groups of flamingos, generally small but sometimes numbering thousands of birds, may engage in nest-building at a new site quite late in the season. The nests may be very elaborate, but these 'breeding attempts' seldom progress to egg-laying. Such nest-building activities have been reported from all parts of the species' range, for example in the Mediterranean (Johnson 1970, 1976; Amat & Garcia 1975; Garcia Rodriguez et al. 1982; Kirwan 1992; Handrinos & Akriotis 1997), in Asia (Montfort in Roberts 1991; Wickramasinghe 1997) and in Africa (Middlemiss 1961; Anderson 1994). In some localities this activity has, however, led to the establishment of true colonies a year or more later. 'False-nesting' has been reported at almost all of the recently established colonies in the western Mediterranean: in

Italy at Molentargius (Schenk *et al.* 1995; Grussu 1999), Orbetello (Baccetti *et al.* 1994) and Margherita di Savoia (Albanese *et al.* 1997), and in Spain at the Ebro delta (Luke 1992) and El Hondo (de Juana 1996; Aragoneses & Echevarrias 1998). Several of the birds seen incubating late in the season at Molentargius, Sardinia, in 1993 were in immature plumage, which leads us to suspect that many of these efforts are made by young flamingos. The sites thus colonised are perhaps marginal for breeding when compared to the traditional locations, which, partly due simply to their perennity, generally offer greater security from predators and more extensive feeding areas. We still do not know whether engaging in 'false-nesting' is of benefit for the birds. Possibly, young individuals may gain some experience through 'practice-nesting' but the direct consequence of it for future reproductive success remains undocumented at the moment.

BREEDING STAGES

NEST-BUILDING

Nest characteristics

Flamingos generally build a truncated, cone-shaped mound of mud or sand and lay their egg in the depression on the top. The mound, which has a diameter of 25–30 cm on top, offers protection against a rise in water level, and the depression prevents the egg from rolling out of the nest. In very hot climates, Brown & Root (1971) noticed that the top of the mound is considerably cooler than the surrounding mud, and the nest therefore also plays a role in preventing the embryo from overheating. The mound is built by both partners, who scrape up the earth surrounding the spot chosen for breeding. The sitting flamingo uses its bill to scrape and draw towards its body whatever material is within reach of the nest: pebbles, mollusc shells, small sticks or the roots of plants, old eggshells and feathers, the last presumably from birds which are moulting their contour feathers during incubation. All of these items can be found encrusted in the mud or sand mound, which is compressed by trampling. Flamingos do not usually carry material to the nest.

The height of flamingo nest mounds varies considerably. When breeding on stony ground, as on the Kiaone Islands in Mauritania, at Lake Uromiyeh in Iran or at Lake Elmenteita in Kenya, the birds cannot construct mounds at all and they lay their eggs on bare earth. In the Camargue, flamingos attempting to breed on the dyke in the vicinity of the main colony even laid eggs on top of some hay bales one year. The closer a mound is to water, the taller it is likely to be, and those mounds either standing in water or in a spray zone are the tallest. After years of occupation, and as islands suffer from erosion, some nests may be almost a metre in height and consist of 50 kg or more of mud (see Figure 42).

Figure 42. Foraging immature Yellow-legged Gull. In the Camargue, the chicks of later breeders are particularly vulnerable to predation by Yellow-legged Gulls. Note the size of some of these nests.

The nest mound is the pair's territory and one of the two parents will always be present, from a few days prior to egg-laying until the chick moves into the nursery. On sandy substrates, nests, once abandoned, quickly disintegrate, being trampled flat by the chicks or eroded by the weather. In Kazakhstan they may be sheared away over winter by ice floes. Clay mounds, on the other hand, withstand time much better and even after several years' disuse may appear to have been recently occupied and may be reused.

Some nests may be used by two successive breeding pairs during the same season, as observed in some years in the Camargue, at Fuente de Piedra in Spain (Rendón et al. 2001) and in Africa (Brown et al. 1982). This happens either following failure by the first occupants, or when breeding starts early and the first chick moves into the crèche while later breeders are still searching for a place to nest. In tropical climates, where the breeding season may extend to over a year, Rooth (1965) reported some nests being occupied by Caribbean Flamingos at least four times in 18 months. Where breeding is an annual event, pairs will often use mounds constructed in previous years. However, given the absence of long-term pair-bonding (see Chapter 7), there is no evidence that the same individuals reuse exactly the same mound from year to year. Nest-building usually begins a few days prior to egg-laying and is often continued throughout incubation. Studer-Thiersch (1975a) observed that in captivity it is the female who chooses the location of the nest. Whether this also applies in the wild is still unknown.

Nest density

As mounds become increasingly voluminous, nest density throughout the colony decreases (see Figure 42). However, when there is strong competition for a place to breed, birds will occupy the depressions separating the mounds, where they may breed successfully only if the depression is not flooded by heavy rains or spray. If the depression becomes submerged, they may abandon the nest; in some cases the incubating bird might even become trapped in the mud and, unable to stand, eventually succumbs (see Chapter 9). Nest density is the result, on one hand, of the social behaviour and gregariousness of flamingos, whereby some birds at a similar reproductive stage tend to group nests within the colony, and on the other hand, of colony structure depending on the topography of the site chosen for breeding.

In the Camargue, the breeding island is fully occupied most years, and a secondary colony often forms on the closest dyke. Being several kilometres long, space there is plentiful, yet nest density is always similar to that observed on the island, at least within groups. Maximum nest density in the Camargue has been measured at 2.7 mounds per m^2, but globally it is lower, around 1.3 mounds per m^2, because nests are generally grouped (Swift 1960). In Mauritania, Naurois (1969a) reported 2–3 eggs per m^2 on the Kiaone Islands, while Brown *et al.* (1982) gave *c.* 1.5 nests per m^2 as the maximum density. A most unusual colony was established in Sardinia in 1993, at the Molentargius Lagoon. Nests were built on the remains of narrow dykes separating saltpans, with mounds arranged in long lines, the colony being the width of only one nest in places. Similarly, in Kazakhstan, for a colony established on Tababa Island on the south-eastern shore of the Caspian Sea, Demente'ev *et al.* (1951) gave an average density of one nest per 1.14 m^2 (= 0.88 nests per m^2).

EGG-LAYING PERIOD

Once a pair of flamingos has chosen a site, both partners remain there, nest-building and guarding their territory for some hours or days before the egg is laid. During this time birds become familiar with the site and the surroundings. We consider colonies definitely established with the laying of the first viable clutches, and we define the spread of egg-laying as the period during which viable clutches, from which chicks survive to creching, are laid. Throughout this period, new breeders may regularly join the colony, one pair at a time or in groups, and replacement clutches may be laid. In the Camargue the spread of egg-laying has varied from 8 to 74 days (Appendix 16) with an average of 38 days (n = 37).

Egg characteristics

Eggs are laid at any time during the day or night. During laying, the female tilts her body forwards and compresses her tail downwards to prevent the egg from rolling out of the nest. If it does so, it will be abandoned. This is rather surprising since many birds lay their eggs and incubate in depressions between mounds. Flamingo eggs vary considerably in shape and size. They are generally elliptical-ovate (Figure 43) with one pole slightly more pointed than the other but some are almost biconical (see Campbell & Lack 1985). The mean dimensions of 120 eggs laid in the Camargue from 1969–83, all collected abandoned, are 89.6 mm (range 81.1–105.0 mm) × 55.2 (range 49.6–61.6 mm). Ogilvie & Ogilvie (1986) referred to two types of eggs, the smaller averaging 78.0 mm × 49.0 mm and the larger 90.0 mm × 55.0 mm. We have no evidence of there being any geographical variation in egg size and Ashtiani (1977) gave 88 mm × 52 mm as the average size of 38 eggs from Lake Uromiyeh, Iran in 1976 (see also Brown 1958, Ali & Ripley 1978, Anderson 1994). The minimum length reported in captivity and from southern Africa (Uys *et al.* 1963, and in Kear & Duplaix-Hall 1975) is 77.0 mm and minimum width 47.7 mm. Runt eggs are rare. One recovered in the Camargue in 1974 measured 51.3 mm × 39.3 mm and another was unusually narrow (86.0 mm × 46.0 mm). Shivrajkumar *et al.* (1960) found a runt egg in the Rann of Kutch (India) which measured 41.0 mm × 32.0 mm. It weighed only 17 g, about one-tenth the weight of a normal egg.

The mean weight of 16 freshly abandoned eggs collected in the Camargue (Appendix 17) was 172.9 g (range 153–195 g). If the body mass of a female flamingo is *c.* 2,530 g (Dunning 1993), then the egg represents 6.8% of her body weight. Perrins (1996) indicates that in birds, on average, one egg amounts to about 10% of female weight.

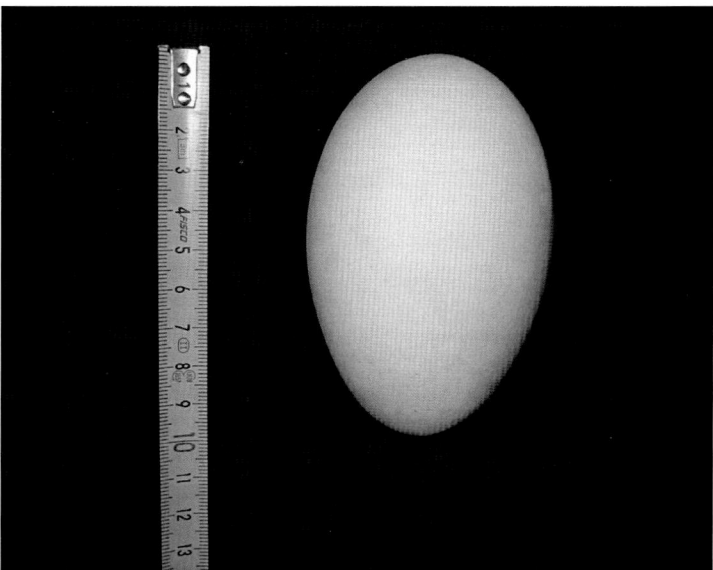

Figure 43. Flamingos lay a single egg, which is elliptical-ovate in shape and pure white, being coated with a calcareous substance.

The yolk is a deep orange to blood red, and the shell is off-white and coated with a pure white calcareous substance. However, unless the nest remains completely dry, the egg will quickly become discoloured by mud. In the Camargue, where incubating flamingos may be frequently exposed to rain or spray, the eggs rarely stay white for long, and some become so coated with mud that they resemble stones. We do not know whether the pore system in such cases remains unaltered, or whether the mud restricts gaseous diffusion (Board 1982). After hatching, flamingo chicks consume pieces of their eggshell (Brown 1958; Berry 1972), presumably, as noted by the latter author, to supplement their reserves of calcium for rapid bone growth.

Clutch size

Flamingos are uniparous but there are many records of nests holding two eggs, and very occasionally even three. Although females may sometimes lay two eggs (Brown 1958) and raise two chicks, it is exceptional. Most two-egg clutches are probably mixed, being the result of two females having laid in the same nest. This may correspond to intraspecific brood parasitism as has been observed in some waterfowl species (Lyon 1993; Power 1998). In captivity, Studer-Thiersch (1975a) observed that two eggs were laid in the same nest when two females were paired to the same male. However, cases of bigamy have not been documented in the wild. Where there is a high level of competition for a nest mound, as is now the case in most years at the Fangassier, an abandoned nest, along with its original egg, may be quickly taken over by another pair of flamingos in search of a place to breed, and this nest will soon hold a second egg.

THE INCUBATION PERIOD

Incubation

Incubation begins at laying and lasts c. 29 days, during which time the egg is never left unattended. In captive birds, Pickering (1992) recorded a mean of 29.7 days (range 26–32), with males taking a slightly greater share of duties at the nest (mean ratio 54:46). In the Basle Zoo, Switzerland, Studer-Thiersch (1975a) observed that at the beginning of incubation males sit more than females. The incubating flamingo does not leave the nest and during this time sits tight, occasionally standing to turn the egg and/or preen (Plate 13). Flamingos sit facing the wind when it is strong, and in well-synchronised colonies of several thousand pairs there may not be a single bird standing during a storm. During incubation birds cannot feed, nor are they fed by their partner, and in dry weather they cannot drink. At the onset of rain, following a dry period, many birds will drink the first drops of water that fall on their back and scapulars; to see the entire colony of 8,000 birds behaving in such a manner, as we have observed in the Camargue, is a most impressive and memorable sight. When at the nest, sitting or standing, birds occasionally fluff, or erect their back feathers in a threat or 'chrysanthemum posture' (Uys et al. 1963). Bill squabbles between neighbours sometimes occur, and passing birds moving into and out of the colony are greeted with a series of blows from the breeders they stumble over.

On several occasions flamingos have been recorded incubating obviously infertile eggs for unusually long periods. In 1986, three birds were monitored in the Camargue incubating for 59, 61 and 68 days respectively before abandoning the nest. Birds in Basel Zoo often sit for long periods, in extreme cases for up to 10 weeks (Studer-Thiersch 1975a). Occasionally, some birds make a second or even third breeding attempt in a season (Figure 35) if early attempts fail. Pickering (1992) recorded re-laying in captivity only six days after the loss of the first egg. In the wild, however, the time interval between the loss of a clutch and the laying of another egg by the same female appears to be in the range 10–15 days. In Barnacle Geese, long delays before re-laying are normally associated with birds having lost their egg after a long incubation (Mitchell et al. 1988). Whether this is also true of flamingos remains to be documented.

Nest relief and attentive periods

One or two days after the egg is laid one of the partners will leave to forage, and will be absent for a period lasting, in the Camargue, from one to four days (see below) (Johnson 1989b, 2000a). The time that a partner spends feeding will depend on food availability, an individual's foraging efficiency and perhaps also the distance it has to fly to feed. Meteorological conditions are also important. In the Camargue, strong winds are common in early spring; they hamper birds feeding in the open saltpans and prevent them from exploiting the exposed lagoons, in particular those which have a slippery substrate (and in which brine shrimp abound).

During the change of partners there is no ceremony. The returning bird approaches the nest, waits for the incubating bird to stand and leave the nest, then generally foot-shakes before taking over incubation. As the bird sits, it makes a side-to-side movement of the body in order to position the egg under the brood patch. This patch is generally only poorly developed in flamingos, and it is rare to see birds with completely unfeathered patches (Plate 8). The bird which has been relieved walks to the water and usually bathes or preens before flying off to feed. Attentive periods of up to two days are common in the Camargue, with a maximum of about four days for successful pairs (Johnson 1989b, 2000a; Cézilly 1993). Excessively long periods of attentiveness at the nest have been recorded in the Camargue, on two occasions 9.5 days and 9 days, before the incubating bird finally abandoned the nest. Indeed, there are indications that the longer a partner is left unrelieved on the nest, the more likely a pair is to fail in their breeding attempt (Cézilly 1993; see below).

The nestling period

Chick development

Flamingo chicks take 24–36 hours to hatch after the first crack is made in the shell (Brown 1958; Ogilvie & Ogilvie 1986). They are precocial and at birth (Plates 14, 15) are clad in white down and have rather swollen, orange-red legs, which are not particularly long, and a red bill. They average 225 mm in total length (Fjeldså 1977), weigh 73–98 g (Berry 1975) and are semi-precocial, remaining in the nest for the first week of life. They are brooded and often also fed under the parent's wing. Some stages in the growth of flamingo chicks are summarised in Appendix 18. They stand uneasily at the age of 3–4 days, but at one week are on their feet, at times trampling, a movement which later in life will allow them to obtain prey items hidden in the substrate. They soon begin to wander away from the nest, usually under the watchful eye of one or both parents and they are no longer brooded. In the Palearctic region, the colour of the soft parts gradually darkens; both the bill and the legs are dark grey or black and the down is uniformly grey when they leave the nest at 7–10 days. In hand-reared chicks in Namibia, a secondary down began replacing the natal down at 3–5 weeks and the change was completed at 7 weeks (Berry & Berry 1976). These authors also record chicks as having a coral-red bill until the age of 11 weeks. The egg tooth is normally barely visible after one week, but in hand-reared chicks (Berry & Berry 1976) some individuals have kept this for as long as 20–40 days. The bill, which is straight at hatching, gradually assumes its characteristic downward curve. Adults continually call to their chick, standing over it with head lowered (Brown 1959; Studer-Thiersch 1975a) (Plates 1, 14, 15).

Creching

As the chicks wander through the colony, adults still incubating give them warning blows with their bills. Social behaviour of creching chicks is difficult to study in the wild, and from an ethological view might merit further research if the appropriate tools for this could be developed. Observations in the Camargue (Tourenq et al. 1995) have shown that aggressiveness of parents towards alien chicks and adults varies with the development stage of their own chick. In Kenya, Brown (1973) observed an excited and aggressive male at Lake Elmenteita, who was brooding a newly hatched chick, 'kidnapping' a chick from a neighbouring nest, probably to brood it as well as its own chick. No such behaviour has been reported at the Etang du Fangassier, although observers have occasionally seen a nest containing two chicks. Parents are slightly but not significantly more aggressive when their chick is 4–6 days old and still at the nest, than they are shortly after it hatches. When their chick starts to wander away from the nest, and is aggressed by other adults, its parents become significantly more aggressive, delivering blows to other chicks and adults which they encounter. The level of aggression decreases once their chick is in the crèche at the age of 10–12 days. Nurseries of chicks, often composed of birds of similar age, assemble on the island, where they may remain for several days, occasionally longer in the Camargue, but sooner or later they take to the water. At Lake Uromiyeh (Iran) Ashtiani (1977) states that chicks enter the water at the age of 16 days. They can, however, swim well (Figure 44) shortly after hatching. Chicks of all ages regularly exercise their wings, and when they are about 10 days old their plaintive calls can be heard by an observer at 70 m. Up to the age of three weeks, chicks are very vulnerable to predators. At some of the Mediterranean colonies many fall victim to Yellow-legged Gulls (Salathé 1983), and in East Africa to Marabou Storks (Brown 1975).

Crèches may remain close to the island, as in the Camargue and at other sites which do not dry out, but in extensive temporary wetlands chicks may trek considerable distances to reach water as the lagoon dries (Plate 18), guided by the parents departing to and returning from the foraging sites. All chicks may unite in a single crèche, or there may be several crèches in which they remain until fledging. Crèches can be very impressive and contain many thousand chicks, for example 14,500 in the Camargue in 2000, 15,300 at Fuente de Piedra (1998) and 20,000 at Lake Uromiyeh (Iran) in 1973 (and reportedly half a million in the Rann of Kutch (1960), although this figure is astonishing, to say the least!).

Figure 44. Swimming chicks. Chicks can swim well soon after hatching; these are six weeks old and the characteristic downcurved bull is now well developed.

Feeding and fledging

During the day, the chicks are largely inactive and spend their time resting in a rather tightly packed group which is often partly in water and partly on land. On land, chicks often rest on their tarsi. Feedings in the crèche usually take place at dusk and continue into the night. Incoming parents arrive in the late afternoon in skeins of tens or hundreds, and with their arrival chicks begin to disperse from their diurnal resting place, which in the Camargue may involve a walk of almost 1 km. Parents alight in the crèche or nearby, and call as they walk through the crèche in search of their chick. In some cases they locate their chick almost immediately, in others, only after much searching. The chick runs up to the parent in a crouched posture and begs for food. Chicks supposedly recognise their parents by voice. However, visual cues such as the parents' bill pattern may also help chicks to recognise their parents, while parents may relocate their chick more rapidly if it repeatedly goes to the same place in the crèche, as observations in the Camargue lead us to suspect, and by its size, but these are cues which have so far been little studied in the wild.

Hundreds or even thousands of feedings take place simultaneously each evening, but not all chicks are fed daily. Some of the earlier literature (Ali 1945; Gallet 1949; Allen 1956) states that hungry chicks are fed by any adult. Brown (1958) was not of this opinion, however, and our own observations at the nurseries in the Camargue (over 1,600 hours during the period 1981–2000) tend to agree with his findings. Further, more than 50 banded parents with banded chicks have been seen feeding them on at least two different evenings, some of them on up to four evenings (1985–2000), and only one of these parents was seen feeding an unbanded chick in the same season. Banded parents observed feeding unbanded chicks in the crèche have always fed what observers believe, from knowledge of the chick's age/size, to be their own chick, and no other individual.

After a few moments of preparation (swallowing movements and head-shaking) the parent is able to secrete the crop liquid to its chick. During feeding, described in detail by Lomont (1953) and Uys *et al.* (1963), the chick stands in front of the parent and faces in the same direction (Figure 45), all birds in the crèche facing the wind if it is blowing. The chick adopts a submissive attitude as it holds its bill up to the parent's. When feeding a small chick, the parent leans forward with kinked neck, but when feeding a large one almost her size, a small female will be fully upright. Prior to and during feeding, the two birds may walk slowly forwards, or backwards, whilst their bill contact is maintained, the tip of the parent's bill resting in the trough at the tip of the chick's lower mandible (Figure 45). The chick's crop gradually distends as it receives food and at the end of a long feeding forms a substantial bulge at the base of the neck (Figure 45). Lang *et al.* (1962) and Lang (1963) were the first to examine the liquid which the adult feeds to its chick, which is a holocrine secretion from the parent's upper digestive tract and not regurgitation of predigested food. The oesophageal fluid fed to small chicks is deep red, but as the chicks grow older, it is not so intensely coloured. The coloration is due not only

Figure 45. Feeding a chick. This banded male aged 12 years is feeding its 55-day old chick. Both parents feed the chick, which is located in the crèche by vocal and visual signals.

to carotenoids but also to blood, as suggested a century earlier by Bartlett (in Studer-Thiersch 1966). It is produced by an unidentified tissue, has a nutritional value comparable to that of the milk of mammals (it has more fat and less protein than pigeon's milk), and is the chick's sole source of nutrition practically until fledging. This manner of feeding is unique in birds.

When small, chicks are fed several times a day by the attendant parent (Plate 16). Feeding bouts are brief, from 5–15 seconds in newly hatched young (Brown 1958) to 1–6 minutes, three or four times a day, when the chick is one week old (Ashtiani 1977). It is the parent who interrupts the feeding and, when the chicks are in the crèche, generally has to run or even fly away to escape the chick, which invariably begs for more, pursuing the parent with lowered neck and still in a submissive stance.

Holocrine secretion composition has been studied in the Greater Flamingo. It consists of about 8% protein, 0.2% glucose and 18% fat (Fisher 1972) and is particularly rich in carotenoids (Lang 1963). Recent data on nutrient hydrolysis in nestlings of Andean and James's Flamingos (Sabat *et al.* 2001) suggest that lipids might be more important than proteins in the nutrition of flamingo fledglings. However, it is highly probable that chicks of different species of flamingos have different digestive capacities of degrading and assimilating different dietary substrates. Such differences probably reflect the chemical composition of parental holocrine secretions, and, to a certain extent, interspecific variation in adult diet and nutritional physiology.

As the chick grows older, it is fed less frequently, perhaps not even daily once it is in the crèche. Feeding bouts tend to increase in duration with the age of the chick, and last on average 15 minutes (n = 101) (Figure 46, Appendix 19). Feeding bouts exceeding 16 minutes are by male parents feeding chicks of 60 days or more (Appendix 19) (Cézilly et al. 1994a). Feedings rarely exceed 20 minutes, but two chicks in the Camargue were fed for 42 and 53 minutes respectively. By the age of 30 days, as their bill structure develops, chicks may be able to find some food for themselves (Ogilvie & Ogilvie 1986), provided there is food where they are creching. In Namibia, Berry & Berry (1976) recorded independent feeding by hand-reared chicks at 7–14 weeks of age. Zweers et al. (1995) state that chicks at six weeks of age still have no lamellae and therefore have an extremely low filter capacity, even though they show filter behaviour. It is uncommon to see chicks being fed after their departure from the crèche, although there is one exceptional observation in the Camargue of a chick being fed until February, at least five months after fledging.

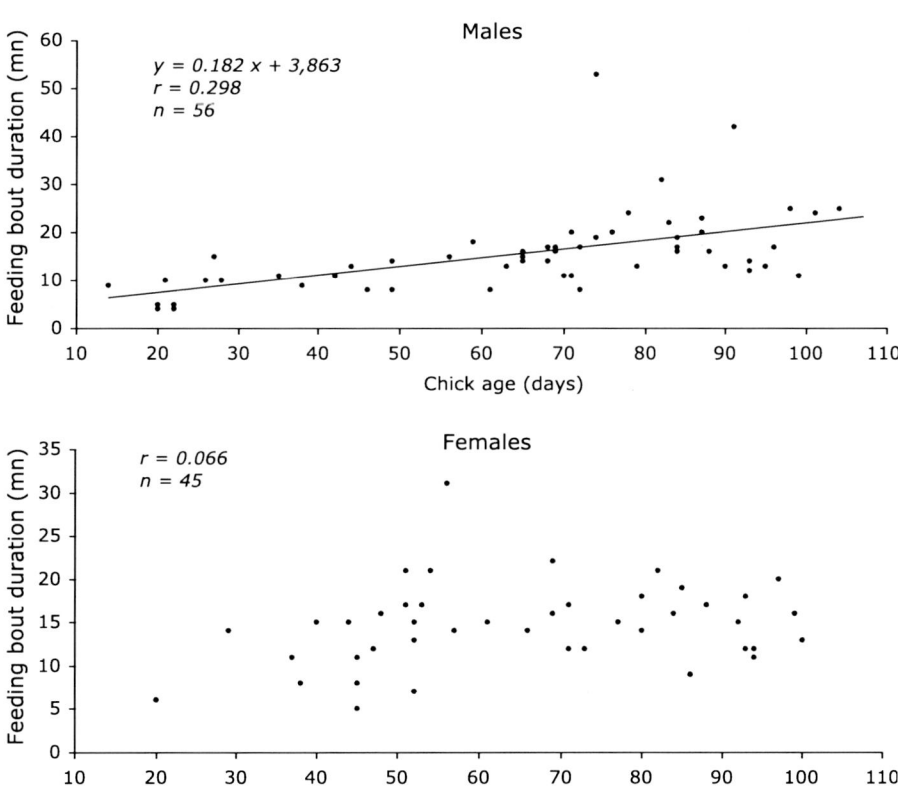

Figure 46. Feeding-bout duration in relation to chick age: a) males b) females. See Appendix 19.

The daily food intake of hand-reared chicks in Namibia was equivalent to 9.6% of their body mass (Berry & Berry 1976). At the age of 28 weeks they showed no further significant gain in body mass or external measurements. In Basle Zoo, however, Studer-Thiersch (1986) noted that the tarsus did not reach its full length until birds were about 18 months of age, and measurements of dead birds in the Camargue (Chapter 2) indicate an even longer period of growth.

The first chicks to actually become airborne are usually helped by windy weather. Large groups of young run together, upwind, and they begin flying just a few metres, alighting in a rather clumsy fashion before quickly returning to the security of the crèche. Brown (1958) estimated that young Greater Flamingos could fly at Lake Elmenteita, Kenya, by the age of 75–78 days, but gave the range 62–75 days elsewhere (Brown 1959); and Uys *et al.* (1963) reported young in South Africa able to fly 85 days after the first eggs hatched. In the Camargue, birds are able to fly well at an average age of 80 days (range 71–98, n = 19) (Appendix 4), although they do not immediately leave the breeding site. Chick growth rates vary considerably from one site to another and from one year to another, as the range given here indicates. At Fuente de Piedra, chicks which hatched in 1994, a year of low water levels in Andalusia and poor breeding success, fledged much later than those hatched in a normal year of breeding (M. Rendón-Martos pers. comm.).

The juveniles, prior to and shortly after taking wing, have a soft, almost plaintive, high-pitched call quite different from the voice of the adults, described by Brown *et al.* (1982) as 'kewick-kewick' when begging for food.

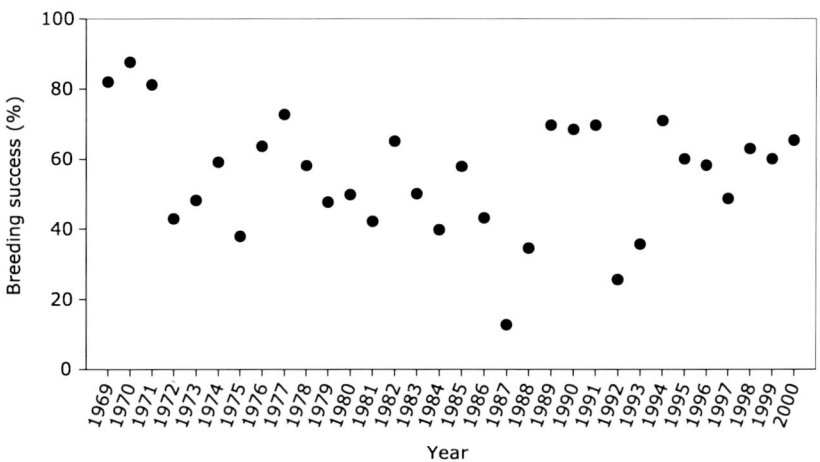

Figure 47. Inter-annual variations in breeding success at the Camargue colony 1969–2000.

1. Three-week old chicks under guard, one of the parents is in the 'chrysanthemum' posture.

2. Flamingo Island and the observation tower in the Camargue.

3. In some places, such as here at Sale Porcus in Sardinia, flamingos drink where freshwater streams enter the lagoon.

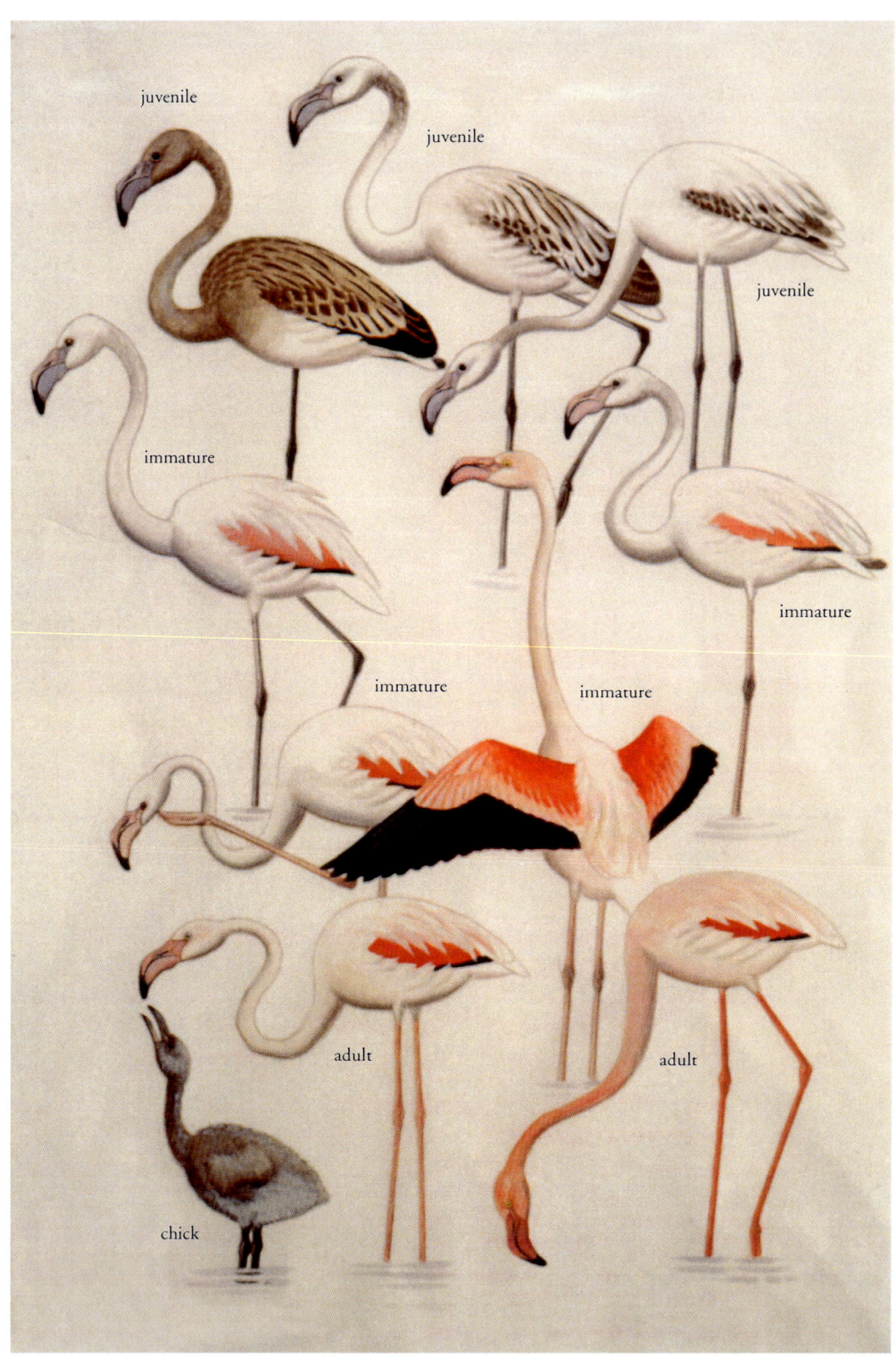

4. Sequence of plumages and coloration of bare parts in the Greater Flamingo. From Johnson *et al.* 1993. Painting by Serge Nicolle.

5. Flamingos wintering in northern parts of the range sometimes encounter severe cold spells; these can be fatal.

6. Flamingos filter their food mostly from shallow brackish or saline waters.

7. Squabbles are frequent, particularly during displays and breeding.

8. Wing salute – note the well-defined incubation patches.

9. An inverted wing-salute.

10. (a) Wing-salute; (b) marching, with some birds in the 'hooking posture'.

11. Copulation takes place in the Mediterranean region from January to May.

12. Breeding in full swing.

13. The single egg is incubated for 28–30 days by both partners.

14. The newly hatched chick, still wet, has 'swollen' pink legs and a pink bill base.

15. At 2–3 days this chick is almost able to stand.

16. Small chicks are fed several times each day on a protein-rich secretion from the parent's crop.

17. A half-grown chick about 40 days old.

18. Crèche of small chicks on Lake Tuz in Turkey. These birds have trekked 17km from the nest site.

19. Juvenile aged 4–5 months.

20. The crèche in the Camargue remains near the breeding site.

21. An immature flamingo 12–15 months old.

23. Restoration of the flamingo island in the Camargue.

22. The first batch of immitation nests. Three months later flamingos occupied this site at the Etang du Fangassier.

24. The colony, observation tower and floating hide at Fangassier.

25. Most migratory movements and foraging flights are crepuscular or nocturnal.

BREEDING SUCCESS

INFLUENCE OF ENVIRONMENTAL FACTORS

We base colony breeding success on the number of chicks fledged per year relative to the maximum number of pairs of flamingos incubating simultaneously during the season. Marked birds were considered to have bred successfully if they were seen feeding a chick of >30 days of age (see Chapter 3), because most chick mortality in the Camargue occurs soon after hatching. Because the true number of pairs attempting breeding is usually greater than the number counted during maximum occupation of the breeding site, this figure will be a slight exaggeration, but it will allow inter-annual comparisons to be made. Inter-annual variation in breeding success at the Fangassier colony is shown in Figure 47. Although breeding success varies widely among years, there is a direct relationship between the number of chicks in the crèche and the number of breeding pairs (Figure 48; $r = 0.739$, $n = 29$, $P < 0.0001$). The fit of that relationship is improved ($r = 0.852$, $n = 28$, $P < 0.0001$) by removing data from 1987, when the colony was considerably disturbed by a child's drifting balloon and 3,000–4,000 pairs abandoned their nests (Boutin & Chérain 1989). Some variation still remains around the regression line, suggesting that environmental factors can influence the productivity of the Fangassier colony.

Tables 13a and 13b show the major annual events influencing breeding success, or causing breeding failure in the Camargue. Heavy rainfall and floods, which vary in intensity from year to year and from one breeding site to another, are serious threats for eggs and young. Predation, both on eggs and on chicks, reduces breeding success (see Chapter 10), but the true impact of predation on colony productivity remains difficult to evaluate.

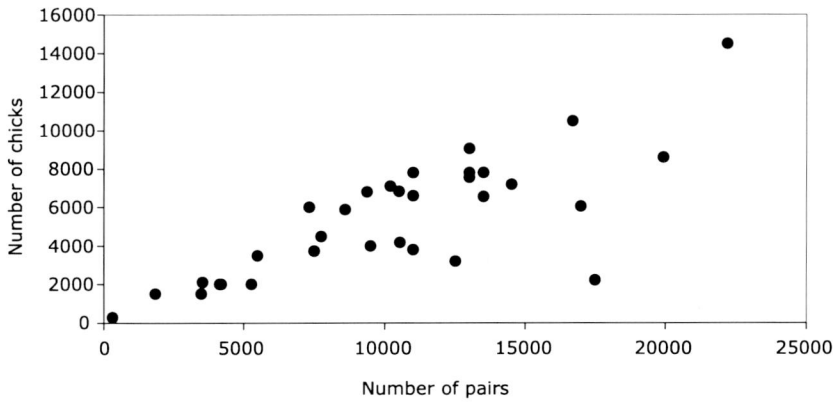

Figure 48. The number of chicks in the crèche in relation to the number of flamingos breeding each year in the Camargue.

Table 13a. **Some of the major annual events influencing breeding success, or causing breeding failure in the Camargue (1950–68).** The colony was established near Aigues-Mortes (Etang de l'Arameau) 1950–52, 54, 55, and near Salin de Giraud (Baisses du Pavias) 1952, 1954–61.

Year (pairs)	(Young) success	Disturbance — remarks — reference
1950 (3000)	(1500) 50%	Predation by 1–2 Red Foxes, disturbance by Badger, Wild Boar and aeroplanes (Guichard 1951; Lomont 1954a).
1951 (3000)	(2000) 66.7%	Erosion of breeding island provokes division of colony into small groups; disturbance by aircraft (Lomont 1954a).
1952 (2400)	(1,800) 75%	Disturbance by aeroplane causes colony to shift site (Lomont 1954a).
1953 (3400)	(2350) 69%	(Hoffmann 1954, Lomont 1954b)
1954 (3000)	(2500) 83.3%	
1955 (3500)	(400) 11.4%	Eggs stolen from two sites in succession (Hoffmann 1955).
1956 (3500)	(750) 21.4%	Severe cold spell in February kills >1000 flamingos in France. Disturbance by photographers and aeroplanes (Hoffmann 1957a).
1957 (4000)	(2350) 58.7%	Some chicks stolen for local collection but main predation by Yellow-legged Gulls, which was estimated at 20 chicks per day during one and a half months (Hoffmann 1959).
1958 (2500)	Complete failure	Unfavourable weather, predation by gulls (Hoffmann 1962).
1959 (3645)	(585) 16%	Predation and disturbance by gulls (Hoffmann 1962).
1960 (8000)	(1600) 20%	Salt foam swamps the colony causing the death of 719 chicks (Hoffmann 1963).
1961 (2500)	(240) 9.6%	Disturbance and predation by Red Fox and gulls (Hoffmann 1963).
1962 (4000)	Complete failure	Island eroded, nests flooded during strong winds, disturbance by gulls and aeroplanes (Hoffmann 1964b).
1963 (3300)	Complete failure	Island eroded, nests flooded during winds, disturbance by gulls and aeroplanes (Hoffmann 1964b).
1964 (0)		Dyke constructed to protect nests but no breeding attempted (Johnson 1966).
1965–68 (0)		No breeding attempted (Johnson 1966, 1970). Spring 1965 was characterised by a long spell of gale-force, northerly winds

Table 13b. Some of the major annual events influencing breeding success, or causing breeding failure in the Camargue (1969–2000). The colony was established in the Etang du Fangassier I 1969, 1972–75, in the Etang du Vaisseau 1970, 71, and in the Etang du Fangassier II 1974–2000.

Year (pairs)	(Young) success	Disturbance—remarks—reference
1969 (7330)	(6000) 81.9	Colonies at Pavias and Fangassier both successful (Johnson 1970).
1970 (320)	(280) 87.5	Breeding at Vaisseau. No dead chicks seen (Johnson 1973).
1971 (1850)	(1500) 81.1	Breeding at Vaisseau (Johnson 1973).
1972 (3500)	(1500) 42.9	Breeding Fangassier I (Johnson 1975a).
1973 (4160)	(2000) 48.1	(Johnson 1975a)
1974 (3560)	(2100) 59.0	New island colonised as annexe site (Johnson 1976).
1975 (5280)	(2000) 37.9	(Johnson 1976)
1976 (5,500)	(3500) 63.6	Island in Fangassier I eroded. Colony moves to new island Fangassier II (Hafner et al. 1979).
1977 (9370)	(6800) 72.6	(Hafner et al. 1979)
1978 (7,750)	(4500) 58.1	High rate of predation by gulls (Hafner et al. 1980).
1979 (4200)	(2000) 47.6	High rate of predation by gulls (Hafner et al. 1980).
1980 (7500)	(3730) 49.7	Gannet takes up residence at colony from 28 April to 23 July and occupies nests. Breeders in the vicinity are scared and abandon nests (Hafner et al. 1982).
1981 (9500)	(4000) 42.1	Island eroded and during heavy rains many egg losses in depressions (Hafner et al. 1982). High rate of predation by gulls—the steep shores prevent the smaller chicks from returning to the island. No wardening and important egg/chick losses as a result of two journalist-photographers visiting the island during hatching.
1982 (10500)	(6825) 65.0	First recorded occurrence of some nest mounds being used by two successive breeding pairs (Hafner et al. 1985).
1983 (14400)	(7200) 50.0	Pairs on dyke (5,600) fail (Hafner et al. 1985). Predation by gulls.
1984 (10535)	(4180) 39.7	Pairs on dyke fail. Predation by gulls (Réserve Nationale de Camargue 1987).
1985 (13500)	(7800) 57.8	Pairs on dyke fail. Predation by gulls (Réserve Nationale de Camargue 1987).

Year (pairs)	(Young) success	Disturbance—remarks—reference
1986 (19926)	(8590) 43.1	Pairs on dyke (7260) fail. Several pairs laid eggs on bales of hay placed on dyke. Egg losses due to gulls and wet and windy weather (Boutin & Chérain 1989).
1987 (17500)	(2200) 12.6	3000–4000 eggs/chicks lost because of disturbance by child's balloon. Pairs on dyke fail. Heavy predation by gulls (Boutin & Chérain 1989).
1988 (11000)	(3800) 34.5	Flamingo island degraded. Many clutches in depressions lost during heavy rains 22–25 April. Pairs on dyke fail (Boutin et al. 1991).
1989 (10200)	(7100) 69.6	Fangassier II dry until 14 April (strike), so late breeding. Birds on (restored) island lay eggs on flat ground but dry spring. c. 300 clutches lost because of Black Swan disturbance (Boutin et al. 1991). Pairs on dyke and mounds of earth nearby (c. 400) fail.
1990 (8600)	(5886) 68.4	Many clutches lost during heavy rains on 24 May (Thibault et al. 1997; Johnson in Seriot 2000).
1991 (13000)	(9050) 69.6	Many eggs lost during heavy rains in mid-April but many replacement clutches laid (Cézilly 1993; Thibault et al. 1997; Johnson in Seriot 2000).
1992 (12500)	(3200) 25.6	Thousands of clutches in depressions lost during heavy rains 21–23 May; this attracts gulls which afterwards predate heavily on colony (Thibault et al. 1997; Johnson in Seriot 2000).
1993 (17000)	(6050) 35.6	Failed breeding on dyke by c. 10000 pairs, some later moving to island to occupy nests of earlier breeders. Predation by gulls, particularly of later breeders, most noticeable in mid-May (Thibault et al. 1997; Johnson in Seriot 2000).
1994 (11000)	(7800) 70.9	Failed breeding on dyke by c. 10000 pairs, some later moving to island to occupy nests of earlier breeders, some of which in depressions between mounds (Thibault et al. 1997; Johnson in Seriot 2000).
1995 (13000)	(7800) 60.0	Many clutches in depressions lost during heavy rains 22–25 April. Some mounds (c. 1300) of the earlier breeders reoccupied by later breeders (Johnson in Seriot 2000; Kayser et al. 2003).
1996 (13000)	(7560) 58.2	Heavy shower in April submerges c. 2400 nests and Flamingo island is abandoned for tern island and dyke nearby (Johnson in Seriot 2000; Kayser et al. 2003).
1997 (13500)	(6563) 48.6	Heavy predation by gulls mostly of chicks in June (Johnson in Seriot 2000; Kayser et al. 2003).
1998 (16700)	(10500) 62.9	Colony in four parts (Flamingo and tern islands and dyke facing each). Pairs successful in all parts (Johnson in Seriot 2000; Kayser et al. 2003).
1999 (11000)	(6600) 60.0	Flamingos prevented from colonising tern island, which was remodelled for terns over winter. Breeding attempts on dyke fail. Heavy predation by gulls in June (Johnson in Seriot 2001; Kayser et al. 2003).
2000 (22200)	(14500) 65.3	Colony in four parts (Flamingo and tern islands and dyke facing each). Pairs successful in all parts (Johnson & Barbraud in Seriot 2002; Kayser et al. 2003).

The island was restored prior to breeding in 1986, 1989, 1996. No successful breeding on dyke 1990–95.

Among the environmental factors known to influence colony productivity, that of water levels around the colony has already received detailed attention. Water levels not only influence the number of breeding birds, but also directly affect the outcome of reproduction. Drastic changes in water levels are known to be one of the major causes of nest desertion, and can occur at any time during the incubation period. Nests may be flooded by rising water levels and important losses sometimes occur in the Camargue following heavy rains or high winds. Peripheral nests and those in depressions are most vulnerable to swamping. Most problems arise, however, when lagoons dry out and scavengers and/or terrestrial predators move in, but these predators are the consequence of drying out and not the direct cause of colony desertion. Parts of the colony may thus be progressively or collectively abandoned, with or without further breeding attempts being made. Occasional massive breeding failure has been reported from practically all sites where flamingos are known to breed. Breeding success can be influenced by moderate fluctuations in water levels, although sometimes in subtle ways. For instance, Cézilly et al. (1995) found that, in the Camargue, colony productivity (the ratio of the number of fledged chicks to the number of breeding pairs) was not directly related to water levels in the Vaccarès system, one of the major foraging areas, during the period 1984–91, and the inclusion of more recent date, for 1992–97 did not change these results (F. Cézilly unpublished results). This may indicate that climatic factors, such as rain and strong winds, rather than water levels per se, are particularly important to colony productivity, either directly, by hampering feeding by adults, flooding nests or forcing adults to abandon their egg during incubation (Cézilly 1993), or indirectly through facilitating predation by gulls (Salathé 1983). However, colony productivity might not be the best index of reproductive success, as it does not guarantee recruitment levels in the following generation. Therefore Cézilly et al. (1995) considered variation in body condition of flamingo chicks at fledging, relative to water levels in the Vaccarès system during the breeding period (March–July) between 1984 and 1991. The average body condition of chicks in the crèche for each year was determined from the slope of the regression line of log (body weight) on log (tarsus length) (see Chapter 3). There was a strong positive relationship between water levels and the average body condition of chicks. We update this analysis here, now taking into account a longer time period, from 1984 to 1997. The relation still holds. The slopes, our estimate of the average chick body condition for a given year, are highly correlated with the mean monthly water levels in the Vaccarès system for the March to July period (Figure 49; $r = 0.82$, $P = 0.0003$). Thus, although water levels do not influence colony productivity, they have a direct effect on breeding success by influencing the average body condition of chicks, which in turn may have important consequences for post-fledging survival and dispersal (Cézilly et al. 1995; Barbraud et al. 2003).

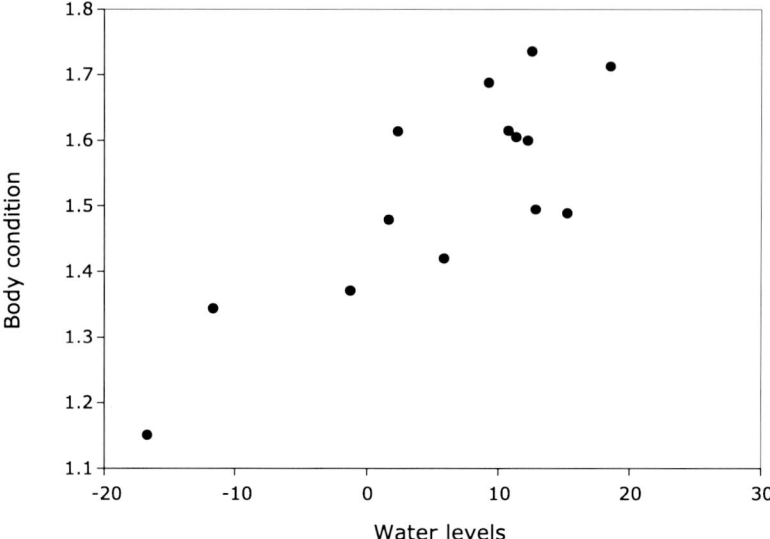

Figure 49. Body condition of flamingo chicks, weighed and measured when caught for banding in July–August, in relation to the level of the waters of the Camargue National Nature Reserve (Vaccarès-Impériaux complex).

AGE-SPECIFIC BREEDING SUCCESS

The relative age of breeding pairs is one important source of variation in breeding success. Reproductive performance is known to vary with age in many bird species (Saether 1990). The typical pattern shows steady improvement in breeding efficiency in the early years, followed by a plateau and, eventually, a decline in older age classes. Various compatible explanations have been advanced for such a pattern, such as differential survival, delayed breeding, individual and pair breeding experience, and foraging ability (Saether 1990; Forslund and Pärt 1995; Cézilly and Nager 1996).

Long-lived colonial waterbirds are particularly good biological models for testing the various hypotheses on age-related improvement in breeding performance (Nelson 1978; Weimerskirch 1992). Observations of marked birds breeding at the Etang du Fangassier have allowed us to study breeding success relative to age over consecutive breeding seasons. Data on hatching success were obtained by recording those nests with a chick, or, if hatching went unnoticed, by recording banded adults seen feeding a chick away from the nest. Fledging success was determined as the percentage of adult birds seen feeding a chick in the crèche among those known to have hatched an egg. However, observations from the tower allowed us to see only a percentage of the birds that bred at Fangassier. In addition, observations at the nursery in the evening did not reveal all successful breeders because it is not possible for observers to identify all birds that raise chicks at Fangassier (many feedings take place late in the evening or at night). Therefore, our estimates (Figure 50) should in no way be seen as reflecting true breeding success. In spite of these limitations, however, we can compare breeding success between different age classes because there is no bias in favour of any particular age group at any stage during the breeding season.

We further analysed data on birds hatched between 1977 and 1996 and re-observed at the Fangassier colony between 1983 and 1999. Ages of breeding birds in the analysis ranged from 3 to 21 years (only one bird hatched in 1977 was observed breeding in 1999 and was not included in the analysis). In order to avoid pseudo-replication, each bird contributed only once to the dataset. That is, for birds recorded breeding at the Fangassier in more than one year, only one observation (corresponding to the breeding performance of that individual at a given age and in a given year) was randomly chosen. Because both hatching success and fledging success (conditional on hatching) can be treated as binomial variables, data were analysed using logistic regressions. Age (continuous variable) and year of observations (nominal variable) were entered as explanatory variables. A total of 5,291 observations were analysed to assess the effect of age on breeding success. Hatching success differed between years (Wald statistic, $\chi^2 = 215.34$, df = 16, $P < 0.0001$). Age had a significant and positive influence on hatching success, independently of variations between years (Wald statistic, $\chi^2 = 33.03$, df = 16, $P < 0.0001$). A total of 1,213 observations were analysed to assess the influence of age on fledging success, for birds that were successful at hatching an egg. Again, there were significant differences in fledging success between years (Wald statistic, $\chi^2 = 81.58$, df = 16, $P < 0.0001$). However, age had no significant influence on the fledging success of birds that were successful at hatching an egg (Wald statistic, $\chi^2 = 0.24$, df = 1, $P = 0.6259$). Our results indicate that the two main components of breeding success, hatching success and chick-raising success, are not equally affected by age. Older flamingos were more successful in seeing an egg

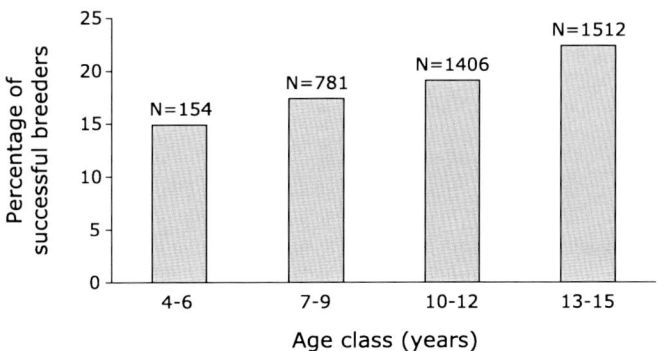

Figure 50. Age-specific breeding success. The proportion of successful breeders for flamingos aged 4 to 15 years recorded in the Camargue 1992–97 (one 3-year-old bird not shown).

through to hatching than younger birds. However, once an egg hatched, birds of all ages were equally successful at raising their chick to fledging.

Part of the lower hatching success of the younger individuals may actually be linked to nest desertion. Nest desertion by some members of a colony is probably a regular phenomenon, even when the colony as a whole has been successful. However, nest desertion seems to involve essentially young breeders. We also observed that females were away from the nest for significantly longer periods than males; the latter, because of their greater size and hence higher competitive ability (Schmitz & Baldassarre 1992), presumably being more efficient at foraging. However, a similar division of duties has been observed in captive birds (see below). In the Camargue, detailed observations of marked individuals throughout the breeding season have revealed that in some cases the off-nest bird does not return to the colony, and in one instance the incubating parent waited nine days on the nest before abandoning it. It is interesting to note that females were much more likely to desert a nest than males (Cézilly 1993). Presumably, females, especially younger ones, experience physiological stress following egg-laying. A combination of physiological stress and adverse weather conditions affecting foraging efficiency may often trigger nest desertion among weak individuals. Such behaviour may be expected in long-lived species such as flamingos, where young individuals should not stake their survival and future reproductive success on their first breeding attempts.

Two main phenomena can explain the observed pattern of variation in breeding success with age. Firstly, the pattern may correspond to either the progressive disappearance of low-quality individuals (see Espie *et al.* 2000; Cam & Monnat 2000) or the progressive appearance of high-quality individuals in the breeding segment of the population (Forslund & Pärt 1995). Secondly, older individuals may benefit from experience or be in better condition than younger individuals at the time of breeding. Experience may result from previous breeding attempts or familiarity with the foraging environment. To test this idea, we looked at evidence for improvement of the breeding performance with age *within* individuals. We found no such evidence in a group of 17 individuals (9 males and 8 females) that had been recorded breeding at the Fangassier colony in more than 10 different years. We also checked whether the probability of improvement increased with increasing time between two reproductive attempts in a larger sample of individuals recorded breeding at least twice, and found again no evidence for improvement within individuals. The absence of improvement of individual breeding performance over the course of the reproductive career of flamingos contrasts with observations in other bird species (Clutton-Brock 1988; Reid *et al.* 2003). One possible reason for this is that in many bird species, pair experience is actually more important than individual age per se in determining breeding success (Black 1996; Cézilly & Nager 1996), for example through better coordination of incubation routines or chick provisioning. As flamingos do not maintain pair bonds over consecutive breeding seasons, there might be little prospect for improvement of the breeding performance with age.

CHAPTER NINE

Survival and recruitment

All population changes are a result of births, deaths, immigration, and emigration. In previous chapters, we have reviewed the factors that influence fecundity and breeding and natal dispersal, and discussed their sources of variation in space and time. In this chapter we focus on two other essential demographic parameters, namely survival and recruitment, i.e. the establishment of new individuals in the breeding component of the population. The estimation of survival and recruitment is particularly important because, for long-lived species such as flamingos (see below), the rate at which a population can grow is far more sensitive to those parameters than to fecundity (Birkhead & Furness 1985; Lebreton & Clobert 1991). That is to say that changes in adult survival and age of access to reproduction will have more profound and more rapid consequences on population size and population dynamics than changes of comparable magnitude in fecundity.

Over the last twenty years, demographic studies have become rather sophisticated and now rely on powerful statistical tools which permit the estimation of survival rates and age-specific breeding probabilities, as well as the assessment of specific biological questions such as comparisons between individuals (e.g. adults vs. young, males vs. females) or among populations (Lebreton et al. 1992, 1993;

Pradel 1996). This chapter starts by reviewing the main known causes of mortality among juveniles and adults. We then provide a detailed summary of our major results concerning the estimation of juvenile and adult survival, and that of age-specific breeding probabilities, and of the various factors affecting these parameters. Because the methodology and statistical jargon may on some occasions become confusing or difficult to follow, we have done our best to convey in this chapter the basic rationale behind the analysis of recovery data and resightings, and refer the reader to the original literature for more details.

CAUSES OF POST-FLEDGING MORTALITY

Recoveries of marked birds reveal some apparent causes of post-fledging mortality, but we cannot yet assess the relative importance of these because the sample is biased, i.e. the rate of recovery of birds varies according to the different causes of death. A bird that dies from a collision with a power-line may be more likely to be recovered than a bird having died of natural causes. One might suppose that the hazards of long-distance flights, over land or sea, take their toll, yet there are few recoveries to support this idea; flamingos are hardly ever reported washed ashore, perhaps because the corpses are devoured by carnivorous fish and/or because the Mediterranean is generally not tidal and they decompose at sea. Some of the commoner causes of death, in our experience, are given below.

Extreme-weather events: Vagaries of the weather claim many victims, whether from drought, severe cold spells or storms. In the Göksu delta, southern Turkey, hundreds of flamingos were among the 10,000 birds reported to have been killed during a hail storm of exceptional intensity in December 1990 (Albrecht 1991). This phenomenon had previously been reported from the Lake of Tunis, in June 1849, when Guyon (1856 in Allen 1956) attributed the death of an enormous number of flamingos to hail. Casualties during severely cold spells, however, clearly account for most documented cases of weather-related mass mortality of full-grown Greater Flamingos.

On 2 January 1985, the temperature plummeted to −11°C at the National Center for Scientific Research (CNRS) meteorological station in the Camargue (Tour du Valat) and to −14°C in Montpellier. It was the start of an exceptionally cold period over much of western Europe. Spells of cold weather are not normally severe or of long duration in the south of France. On this occasion, however, practically all the lagoons along the French Mediterranean coast froze, including the saltpans at Salin de Giraud and Aigues-Mortes. Temperatures remained below zero for 15 successive days until a general thaw set in on 18 January, but there was very little open water before 22–23 January. Waterbirds were hard hit, in particular ducks and Coots, Little and Cattle Egrets and Greater Flamingos (Cézilly 1985; Hafner *et al.* 1992, 1994).

There were 16,000–20,000 flamingos in the south of France before the start of the cold weather (Johnson 1985). Some birds tried to escape along the coast towards Spain and others reached Corsica, but most remained in the Languedoc since it was not the usual season for movements. The first dead flamingos were reported on 5 January, and the number of victims increased daily. A truly impressive effort was made to feed flamingos on rice grain put down where holes in the ice had been maintained, while at least 1,000 birds were saved from starvation in one of the seven centres hastily set up along the coast in response to what was referred to by the press as a catastrophe. It is not known exactly how many flamingos died, but 2,945 corpses were collected during the cold period. The proportion of banded flamingos (n = 141) amongst those recovered was half the number estimated to have disappeared, and it is believed that the true number of casualties must have been *c.* 6,650 (Johnson *et al.* 1991). Even the first of these two figures represents the greatest mortality of flamingos ever recorded in the Mediterranean region.

Mass mortality through starvation is, however, rare. Large numbers of flamingos were reported trapped in the ice and captured by hunters in the Camargue as long ago as 1789 (Glegg in Allen 1956), again in 1839 (Crespon 1840) in 1913–14 (Glegg in Allen 1956) and in December 1931 and 1933 (Trouche 1938). More recently, some hundreds of flamingos died during drastically cold spells in 1947, 1956 (Hoffmann 1957a) and 1962 (Blondel 1964), and Penot (1963) refers to about 60 flamingos dying in the cold spell of January 1960.

Trapped in the mud: In the Camargue only small numbers of flamingos are found dead during the breeding season. They die not only as victims of collisions with overhead cables or barbed wire fences, into which they presumably fly mainly at night, but also when incubating. Birds which lay their eggs in flood-prone depressions between mounds become completely coated with mud during heavy rain or when subjected to wind-blown spray. This happens when they scoop up the surrounding mud to augment their mounds, and then preen, spreading the mud over their plumage to such an extent that they become stuck in the mud and cannot stand. Starvation resulting from such behaviour led to the death of at least 52 birds in one year at the Fangassier colony (Johnson 1991).

Avian botulism: Avian botulism also claims some victims (Van Heerden 1974). Flamingos often feed on the high concentrations of invertebrates in evaporating marshes. Outbreaks of botulism (type C *Clostridium botulinum*) occur when the oxygen content of a water body is low and the temperature high. Such outbreaks may occur practically anywhere throughout the species' range yet, surprisingly, flamingos seldom seem to be the species most affected when there are instances of high mortality in birds (see Van Heerden 1974). It has even been suggested (Allen 1956; Blaker 1967) that flamingos may have acquired a degree of resistance to the botulism toxin type C, as has been reported, for example, in the Turkey Vulture.

Epidemics: Like other bird species, flamingos are susceptible to a number of pathogens which are transmissible to other birds and, eventually, to human beings. Although in several cases the disease agent can survive in the infected flamingo without causing illness or death, some pathogens can be particularly virulent and cause massive mortality in the case of epidemics. One pathogen of increasing concern is the avian influenza virus (AIV), a member of the Orthomyxoviridae family. Because influenza viruses have a high rate of genetic recombination, they constitute a permanent risk for both feral and captive populations of flamingos in various parts of the species' range. In addition, they are widely distributed throughout the world. Flamingos may be particularly vulnerable to infection with AIV. Arenas *et al.* (1990) found that 43% of flamingos were infected with AIV virus in southern Spain. Highly pathogenic H5N1 avian influenza occurred in captive Greater Flamingos in Hong Kong in late 2002, following several consecutive outbreaks of the disease (Ellis *et al.* 2004).

Newcastle disease virus (NDV), an avian paramyxovirus with a very broad host spectrum, is another potential threat to flamingos. It can be transmitted through various vectors such as wind, insects or humans. Data from captive flocks indicate that Newcastle disease can be fatal to flamingos (Kaleta & Marschall 1981), although its virulence in the wild is unknown.

Flamingos may also suffer from infection by avian mycobacteriosis (causing avian tuberculosis), as observed in free-ranging Lesser Flamingos inhabiting Lake Nakuru in Kenya (Cooper & Karstad 1975). Flamingos excrete the virus both in faeces and through respiration. The excreted bacteria can remain in muddy soil for long periods of time and easily spread inside a colony. Flamingos get infected when ingesting or inhaling the bacilli. It seems, however, that they might be less susceptible to avian tuberculosis than other waterbird species, at least from what has been observed in captive flocks at Slimbridge (Brown & Pickering 1992).

Predation: In East Africa Greater Flamingos occasionally fall prey to eagles and Marabou Storks (Brown 1975), the latter when the flamingos are breeding, and presumably also to terrestrial predators such as Spotted Hyenas, although it seems that these usually pursue the more abundant Lesser Flamingos.

Hunting: Hunting claimed many victims before flamingos were protected by law, and in some countries they are still shot, legally or illegally. Hunting was a major cause of mortality in Spain during the 1960s and 1970s and half the recoveries in that country were of birds shot in Andalusia.

Lead poisoning: Flamingos, like ducks and geese, ingest lead shot as grit when feeding and may subsequently die of lead poisoning (Bayle *et al.* 1986). In Spain, Ramo *et al.* (1992) examined 22 flamingos which died of lead poisoning in March 1991 at the Doñana National Park. The gizzards of five birds contained over 100 lead shot and one bird had ingested 328 of the toxic pellets! About 50 flamingos found dead or ill at the Santa Pola saltpans near Alicante, during the winter of

1992/93, were also victims of lead poisoning (Johnson 1995), and at least 34 birds died at the neighbouring El Hondo wetland in January 1998 (*Quercus* 146: 47), some of these birds carrying up to 70 lead pellets in their gizzards. We suspect that ingestion of lead shot causes more deaths than either recoveries or studies show, and is of equal concern with the nominate race in Mexico (Schmitz *et al.* 1990).

Collision with wires: Many flamingos are found dead, or injured, under overhead electric lines, which cross most industrial saltpans. Birds are maimed or killed by collision with the wires rather than by electrocution, but there must be a strong recovery bias here, because of the likelihood of discovery. Flamingos also occasionally fly into barbed-wire fencing. Two Camargue-ringed birds were recovered in this way (Johnson 1983) and in Botswana flamingos have been reported flying at night into cordon fences erected across the Makgadikgadi saltpans to prevent spread of foot-and-mouth disease (McCulloch 2002).

Collision with aircraft: Flamingos are occasionally reported to be in collisions with aircraft. In France this is hardly surprising, because the approaches to two major airports (Marseille-Marignane and Montpellier) cross wetlands which are widely used throughout the year by flamingos.

ESTIMATING SURVIVAL

Results presented in this section concern mainly the analysis of capture-mark-recapture data based upon the resightings of individuals banded as chicks at the Fangassier colony. Because so much of the information we report below relies on models using resighting data developed by Cormack (1964), Jolly (1965), and Seber (1965), it is probably worth a brief explanation of how these models, now commonly referred to as the Cormack-Jolly-Seber (CJS) models, work. For each year of data, each individual that has been marked is assigned a 1 if it has been seen and identified, or a 0 if it has not. This alone does not tell us whether a bird has survived, because not all birds that are alive are seen, as we have illustrated in Figure 51a. When a bird is seen alive during a given year, we know that it has survived up to that time. However, it is not so easy to know whether a bird that has not been seen is still alive. The data look the same in a given year for birds that have died and birds that are alive but not seen (both are 0s). This is where having extensive data becomes important. We have shown a small example in Figure 51b of data for six birds over an eight-year period. None of these birds was seen in year 2, but as data accumulate over several years, we can begin to distinguish those birds that were alive and not seen from those that either died or permanently emigrated from the 'observed' population. For example, from these data, we now know that at least four of the six individuals not seen during year 2 were in fact alive because they were observed in later years. However, two birds (ABB and ACA) were never seen

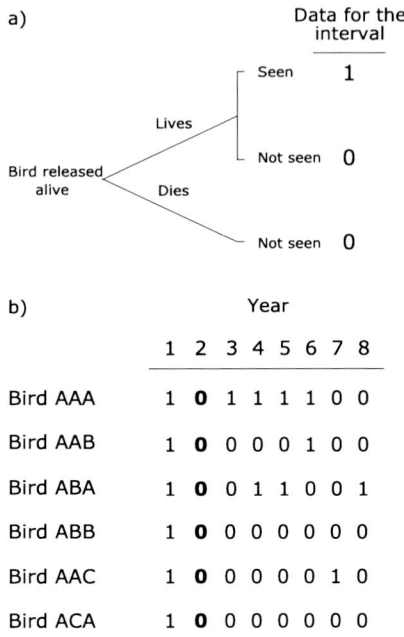

Figure 51. Example of how a resighting model works: a) simple resighting diagram and b) 'local' or 'apparent' survival.

again after the first year. Thus, as far as the models are concerned, they are apparently dead. As the study progresses, we will be able to identify these alive but unseen birds better and better. At the time of writing, we have over 270,000 resightings of about 16,000 individual flamingos over a 22-year period. Finally, we should explain why we refer to 'local' or 'apparent' survival (both terms are used in the literature). Birds that *permanently* emigrate from the observed population, and are never seen again, cannot be distinguished from birds that are dead. However, birds that *temporarily* emigrate, and return later, provide evidence of 'local' or 'apparent' survival, data which actually increase our ability to estimate survival. Flamingo AAC in our example is a case in point.

JUVENILE AND IMMATURE SURVIVAL

It is difficult to obtain precise estimates of juvenile and immature survival because emigration is greater in younger birds than in adults, and the different cohorts of banded birds return to the colony gradually over a period of years. However, a first attempt to model juvenile and immature survival was made by Johnson (1983) using recovery data of birds ringed as chicks. Estimates of survival rates from this analysis were 60% from ringing to first year, 79.7% from first to second year, and

82.9% from second to third year. However, Anderson *et al.* (1985) have since shown that survival estimates from recovery data can be severely biased when birds are ringed only as chicks, particularly for long-lived species. Only 40 years ago flamingos were far more persecuted in the Mediterranean than they are today, and perhaps half of the reported recoveries (n = 415) up to the end of the 1970s were victims of hunting (Johnson 1979). Consequently, survival rates have probably changed since that era, while methods for estimating survival have substantially improved.

More reliable estimates were provided by Johnson *et al.* (1991), who applied the CJS model to resightings of younger age classes and estimated an average survival for cohorts as 83.7% ± 2.1% from ringing to first year, 71.5% ± 4.1% from first to second year, and 88.8% ± 4.0% from second to third year (arithmetic means of estimates +1 sd for the cohorts 1977–85).

However, the estimation of juvenile and immature survival still remains unsatisfactory. Better estimates might be obtained in the future by combining the

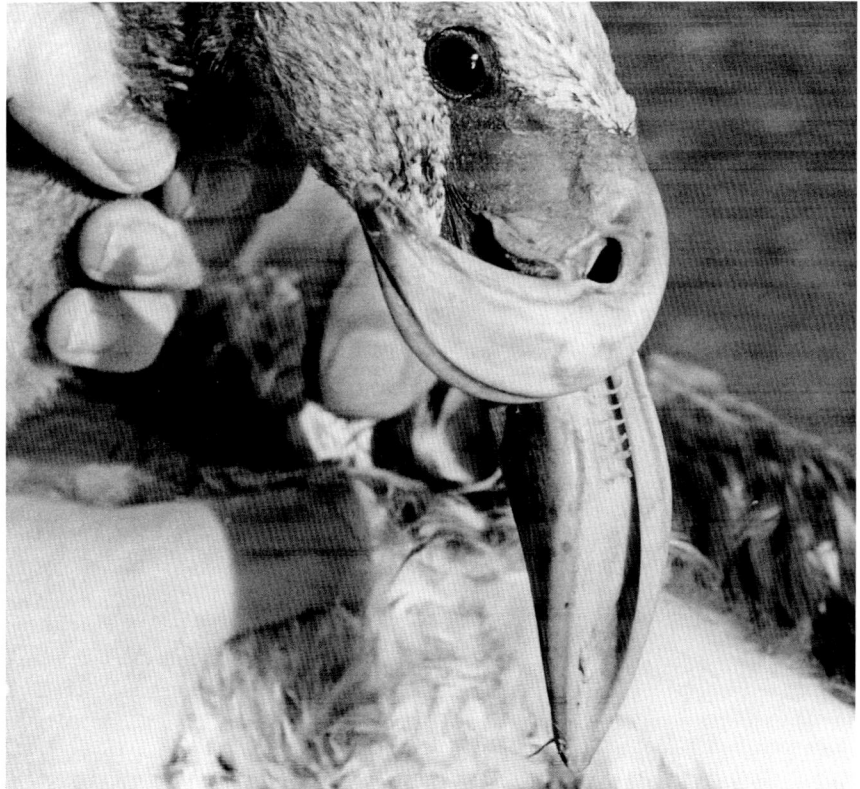

Figure 52. **Flamingo chick with deformed bill.** Of more than 20,000 chicks caught in the Camargue for ringing (1977–2000) only three individuals have been seen with deformed bills. Etang du Fangassier, July 1987. (*F. Mesleard*)

information obtained from resightings with that obtained through the analysis of recoveries. More specifically, differences between cohorts (i.e. birds born in the same year), or between populations, as well as the relation between survival in the first year and chick body condition at ringing, remain to be examined.

ADULT SURVIVAL

Adult survival rates in Greater Flamingos were first estimated by Johnson *et al.* (1991), treating each cohort of banded birds separately. Lebreton *et al.* (1992) provided a more detailed analysis, covering the period 1977–87, which reveals a high heterogeneity in capture rates and indicates a drastic reduction in survival during the exceptionally cold winter of 1984/85. Survival rates over time, other than during this cold winter, were found to be constant and high, with wide confidence limits (mean: 97%; 95% confidence interval: 81.3–99.7%). Progress in methodological approaches continued to allow finer analyses to be carried out, the database having of course increased in size (Cézilly *et al.* 1996; Tavecchia *et al.* 2001). Such analyses have been restricted to resightings of live birds at the Fangassier colony, local survival thus corresponding to uninterrupted breeding in the colony. To avoid confusion between adult survival and the survival or dispersal of juveniles and immatures, the first resighting as a breeder at the Camargue colony was considered the initial capture. The dataset was based upon sightings made each year from April to August, thus satisfying the basic assumptions of capture-mark-recapture models that inter-occasion time intervals are regular and longer than the capture sessions (Lebreton *et al.* 1992).

Our latest analysis (Tavecchia *et al.* 2001) was restricted to the cohorts banded from 1977–91 and resighted in 1983–96. Out of 9,531 birds banded, 2,000 (1,017 males, 983 females) are known to have bred at least once in the Camargue during this period, so our analysis was based on the recapture histories of those 2,000 individuals. We first tested the goodness-of-fit of the CJS model separately for each cohort and each sex, using the programme RELEASE (Burnham *et al.* 1987). Subsequent models were fitted by maximum likelihood estimation using the programme SURGE (version 4.2, Pradel & Lebreton 1991). Model selection was based on the principal of parsimony (Lebreton *et al.* 1992), keeping only significant parameters to describe the data and providing precise estimates of those parameters.

With the exception of females younger than seven years breeding for the first time, the estimates of survival were high (above 0.93 and 0.97 for males and experienced females respectively in normal years), and lay well outside the confidence interval of a previous estimate based on the assumption of constant survival across sex and age (Cézilly *et al.* 1996). Our estimations of adult survival rates (Cézilly *et al.* 1996; Tavecchia *et al.* 2001), however, are based on breeding birds returning to their natal colony, and probably underestimate the adult survival rate of flamingos throughout the whole of their range. Indeed, it is quite likely that some flamingos born in the Camargue leave the western Mediterranean region and never return to their natal colony. Whatever the case, adult mortality in the Greater

Flamingo, at least in the Mediterranean region, is low compared to values obtained in similarly detailed studies of the Ciconiiformes (Freeman & North 1990; Kanyanibwa *et al.* 1990; Hafner *et al.* 1998).

FACTORS INFLUENCING SURVIVAL

As found in previous analyses (Lebreton *et al.* 1992; Cézilly *et al.* 1996), annual survival was strongly affected by the severe winter of 1984/85, and recapture probability varied according to year, sex and age of the bird. Males had a higher resighting probability than females, independently of age. This can be partly explained by the greater probability of males than females making a second breeding attempt (with a new partner) after failure early in the season (Cézilly & Johnson 1995; see Chapter 7).

Interestingly, we found a significant effect of the interaction between age and sex on survival. While young females had a lower survival probability compared with males of similar age, they survived better after seven years old. This, together with previous observations of nest desertion by females (Cézilly 1993) and differences between sexes in feeding young (Cézilly 1994a), led us to suggest that the observed sex-related difference in survival corresponded to asymmetric costs of reproduction, particularly in young age classes. We tested this hypothesis, by including in a model a cost of first-observed reproduction on survival in young females only. This model provided the best fit to the data, explaining the majority of the sex-related differences in survival before females were seven years old.

Yom-Tov & Ollason (1976) suggested that because of their smaller size, females are not as efficient at foraging as males. Indeed, Schmitz & Baldassarre (1992) reported that in the Caribbean Flamingo, size difference between males and females was the principal determinant of contest roles during aggressive interactions on the foraging grounds. One may therefore have expected a marked difference between sexes in survival when food is scarce, as was the case in the south of France in January 1985. However, overall, there was no significant effect of the interaction between sex and winter severity on survival rates (Cézilly *et al.* 1996; Tavecchia *et al.* 2001). This result suggests that competition for food between sexes may have little consequence for survival.

Age generally had a significant, positive influence on survival in breeding birds. However, there was a significant effect of the interaction between sex and age on annual survival. In females, survival probability increased with age regardless of winter severity, whereas in males younger individuals experienced higher mortality during severe winter weather.

Longevity and senescence

Our estimates of adult survival suggest that flamingos may be extremely long-lived. Indeed, data from captivity confirm this. A Greater Flamingo resident in Basel Zoo in Switzerland since at least 1938 (Schenker 1978), and that was still alive (aged at least 68 years) when this book went to press in 2006, must be the oldest-known flamingo in the world (A. Studer-Thiersch pers. comm.). Actually, it is even suspected that this bird, and another one which died in December 2003, may have arrived in Basel Zoo in 1932, already in adult plumage, meaning that the surviving bird could now be over 70 years of age. Three other old birds from this same collection were killed by a Red Fox, one aged 58 years (or perhaps 64) and the two others aged 54 (or perhaps 60). Three of the flamingos at Basel Zoo bred regularly up to 1995, at which time they were at least 57 years old. Another old captive bird was a Caribbean Flamingo which lived in Chicago Zoo from 1934 to 1995, when it died at the age of 60 years and four months (C. King pers. comm.). For comparison, other long-lived bird species and their maximum ages in captivity are Siberian White Crane, 62 years; Eagle Owl, 68 years; Andean Condor, 77 years; and Sulphur-crested Cockatoo, 80 years (Bird 1999).

As the research studies using metal rings reach maturity, data are becoming available on the potential life span of flamingos in the wild, not only from recoveries of ringed birds but also from flamingos observed alive, the ring codes having been read in the field at very close range. Such remarkable data are the fruits of the ringing programme begun in the Camargue by Luc Hoffmann in the late 1940s, and pursued until the early 1960s. Figure 53 shows the frequency distribution of ages at which flamingos marked as chicks in the Camargue between 1947 and 1961 were later recovered dead. Although a majority of individuals died before they were 20 years old, several individuals survived over 30 years. The oldest-known wild flamingo was found freshly dead in Sardinia, 40 years and 23 days after being ringed in 1957. It had probably collided with high-tension cables, as had another bird recovered in France at the age of 38 years and 12 days. The third-oldest wild flamingo known was observed feeding a chick at Fuente de Piedra in 1996, at the age of 36 years, 11 months and 13 days. One other wild bird was recovered after more than 35 years; it died in Tunisia, also after colliding with high-tension cables, aged 35 years and 10 days.

If we consider that survival rates remain constant over the entire adult life, our estimates of survival rates (Cézilly *et al.* 1996; Tavecchia *et al.* 2001) suggest that in the wild, flamingos may well live up to 50 years. However, longevity in the wild might be reduced if flamingos experience a phase of senescence, i.e. a decline in survival rate with advancing age. We found no evidence for a senescence effect on survival rate in our analyses (Cézilly *et al.* 1996; Tavecchia *et al.* 2001). However, this may simply be a consequence of the age samples in the analyses where the oldest individuals were only 19 years old. Senescence therefore is one aspect of the Camargue data to be examined as the banded individuals grow older in order to give a better estimate of average longevity in the wild.

RECRUITMENT

Recruitment has important implications as a factor regulating population size and affecting individual life histories. Most long-lived bird species start to breed only

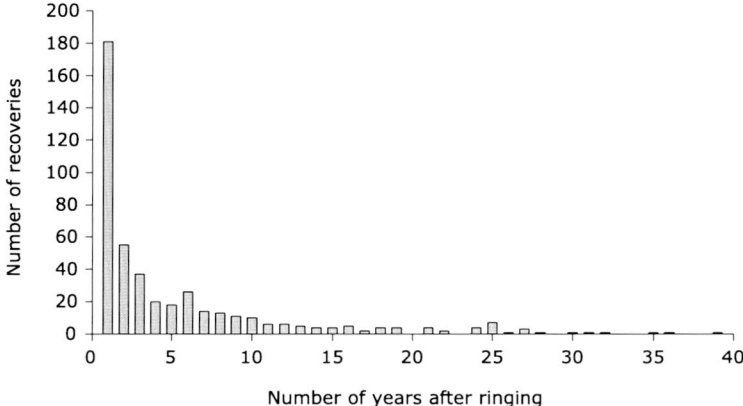

Figure 53. The number of recoveries per year after ringing (resightings not included) of flamingos marked as chicks in the Camargue 1947–61 (Museum Paris alloy rings). N = 453; see Appendix 20.

after they reach maturity and stop growing (Dunnet 1991). Some authors (e.g. Lack 1968b, Orians 1969, Ashmole 1971) have stressed the importance of improved foraging efficiency with age. Others suggest that competition at the nest site plays a major role, since the number of potential breeding sites is limited (Coulson 1968). Finally, several authors (Lack 1954; Mills 1973; Wooler & Coulson 1977) have considered the cost in terms of survival, or the reduction of breeding potential, associated with early breeding.

In spite of its importance, the estimation of recruitment rates remains approximate in the absence of an adequate statistical method of analysis (Clobert & Lebreton 1991; Pradel et al. 1997). Thus, in most studies (e.g. Williams & Joanen 1974, Lloyd & Perrins 1977, Wooler & Coulson 1977, Finney & Cooke 1978, Harris 1981, Thompson et al. 1994), recruitment has been equated with age at first breeding, without allowing for variation in survival and capture rates. In most studies the probability of capture of a breeding bird is less than 1, and some birds recorded breeding for the first time may have already bred without being detected. Models have thus been produced taking such problems into consideration (Lebreton et al. 1990; Clobert et al. 1994), but they still rely partly on hypotheses which have not been adequately tested, such as survival rates of breeders and non-breeders being the same, and capture rates not being age-dependent. In the case of the Greater Flamingo, we know that these hypotheses do not withstand scrutiny (Cézilly et al. 1996; Tavecchia et al. 2001).

AGE-SPECIFIC PROPORTIONS OF BREEDERS

Among the thousands of banded flamingos which are seen in any given year at the Fangassier colony, only a small proportion is able to actually secure nesting places and breed on the island. The proportion of breeding birds among the banded cohorts increases with age (Figure 54). Although we know that Greater Flamingos have bred in captivity when only two years of age (Studer-Thiersch 1975a), and this is also the age of first breeding given by Cramp & Simmons (1977), in the wild they do not attempt breeding when so young, at least not in the Camargue. During 1982–2000, 2,688 banded flamingos of Camargue origin were recorded breeding for the first time at the Fangassier (1,339 males and 1,349 females). Of these only eight birds (one male, seven females) bred at three years of age, before acquiring full adult plumage. Most banded birds are observed breeding for the first time at between four and eight years of age. However, this pattern may not reflect true recruitment accurately, because we know from the information obtained during estimations of survival (Cézilly et al. 1996; Tavecchia et al. 2001) that capture rates increase with age, even among birds having bred at least once. One reason for this might be that younger, inexperienced birds are more likely to fail early during their breeding attempt and thus be undetected by observers.

AGE-SPECIFIC PROBABILITIES OF RECRUITMENT

When studying recruitment in flamingos, we favoured a new method where the estimation of recruitment is independent of survival rate (Pradel et al. 1997). We used the cohorts of flamingos banded in 1977–88 and observed breeding during

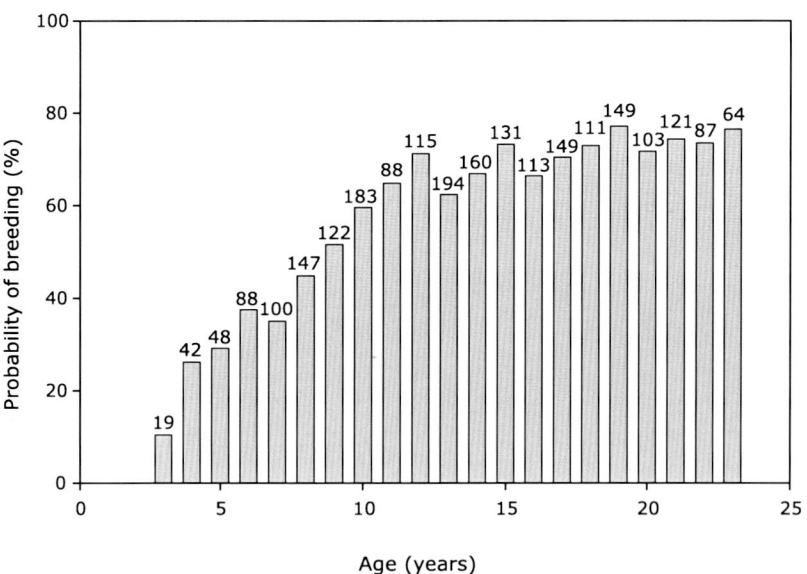

Figure 54. Age specific proportions of breeders among the banded cohorts in 2000 at the Fangassier colony.

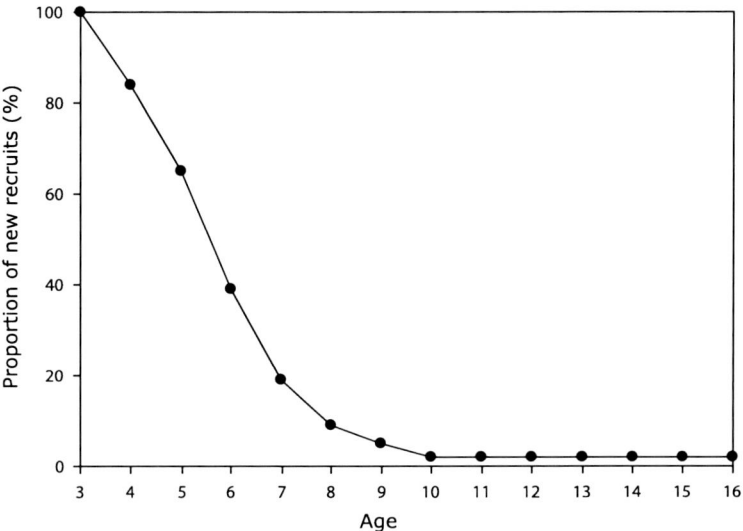

Figure 55. Stabilisation of recruitment with age in Greater Flamingos.

the period 1983–94. The sample concerned 856 males and 824 females. The influence of year of birth, year of first sighting, sex, and age on probability of recruitment and capture rates were all studied using a variety of models. The model that proved most appropriate showed a stabilisation of recruitment when birds were 10 years old (Figure 55). This suggests that birds first detected breeding at the natal colony at nine years old or older might well have bred earlier elsewhere. As mentioned earlier, some individuals are capable of breeding at the age of three years (Johnson *et al*. 1993) and therefore this raises the question of why some birds do not attempt breeding until much later in life.

Delayed recruitment has been related to the need to improve foraging skills through experience (Orians 1969; Recher & Recher 1969) prior to engaging in breeding. Flamingos are filter-feeders (Jenkin 1957) and experimental studies have recently revealed that complex mechanisms underlie filter performance and discrimination capacity (Zweers *et al*. 1995). However, age variables in efficiency of filter-feeding remain unexplored. Apart from limited foraging skills, competition among age classes on foraging grounds (Bildstein *et al*. 1991; Schmitz & Baldassarre 1992) may also impair foraging efficiency of young individuals and thus delay breeding. Alternatively, delayed reproduction may be the consequence of social constraints. As already mentioned, hundreds or thousands of individuals attend the Fangassier colony each year but are unable to secure a nesting site on the breeding island. Reduced mating opportunities owing to directional mating preference for older individuals (see Chapter 7) and costs associated with early reproduction may also force younger individuals to postpone breeding (Tavecchia *et al*. 2001).

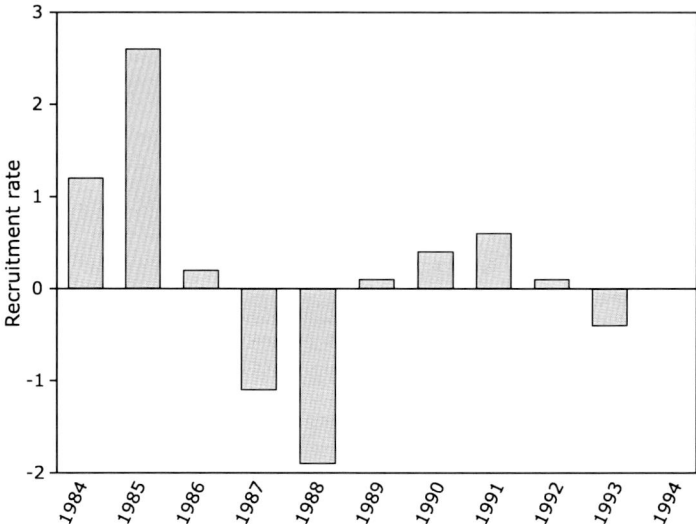

Figure 56. Interannual variations in recruitment rate at the Camargue colony.

Figure 56 shows variation in the recruitment rate among years at the Camargue colony, with a peak in 1985 followed by a marked decline over the following three years. This pattern can be interpreted relative to the wintering strategies of flamingos and to the high mortality that occurred in January 1985. If a majority of birds that breed at the Fangassier colony tend to spend the previous winter in southern France, the deaths of several thousands of flamingos in this region during the cold spell could have dramatically reduced competition for breeding space the following season. Such favourable local conditions may have prompted the recruitment in 1985 of younger age classes and/or inexperienced older individuals. The way in which high adult mortality was, at least partially, compensated for, confirms the existence of a pool of potential recruits which is usually prevented from breeding (Cézilly & Johnson 1995). This pool may contain not only natives from the Camargue but also birds from other colonies (Nager et al. 1996). The control of recruitment to the Fangassier colony therefore appears to be at least partially density dependent. In addition, the cold spell particularly affected the survival of first- and second-year birds (see above), i.e. the 1983 and 1984 cohorts, thus producing a deficit in the recruitment of birds aged 4–6 years in 1987 and 1988 (Figure 56). The situation returned to normal in 1989, with the recruitment of individuals hatched after the cold spell. Pradel et al. (1997) therefore showed that recruitment could indeed be considered constant during the study years, notwithstanding the events of 1985.

CHAPTER TEN

Conservation and management

This chapter covers a wide range of topics. We begin with a brief look at the Greater Flamingo's conservation status in the world before presenting a more detailed picture of the threats the species faces and the wetlands of greatest importance to it. We have included here the problem of a conflict with farming—specifically rice-growing—which is serious enough to have necessitated annual scaring campaigns in the Camargue (France) and in Spain and to have influenced our decision to reduce the size of the flamingos' island at the Etang du Fangassier. The construction and periodic restoration of this island, and other conservation actions carried out in France and elsewhere, are presented here, including the rescue and rehabilitation of both chicks and adults. Finally, we stress, in the light of an uncertain future, the influence of the salt industry on the present numbers, distribution and breeding by flamingos in the Mediterranean region and elsewhere.

CONSERVATION STATUS IN THE WORLD

The Greater Flamingo is in Annex I of the Wild Birds Directive of the European Union, in Appendix II of the Bern Convention (strictly protected species) and of the Bonn (CITES) Convention and is in Column A of the African-Eurasian Migratory Waterbirds Agreement (AEWA) Action Plan. The species is not globally threatened (Collar *et al.* 1994) but, like many other large, colonially nesting waterbirds, is and probably always will be considered of conservation concern because of its dependence on a limited number of wetlands, particularly for breeding. For example, Brown *et al.* (1982) thought that the species may be decreasing in East Africa because of interference during breeding by Great White Pelicans, but increasing in southern Africa owing to a growth in man-made habitats such as saltworks, sewage lagoons and large dams. Brooke (1984) gave the flamingo's conservation status in South Africa as indeterminate while Simmons (1996) referred to numbers having declined across Africa, particularly since the 1970s. The Greater Flamingo is classified as vulnerable in Namibia's Red Data Book (Brown *et al.* in Simmons 1996).

In Europe, although there has been a marked increase in recent years in the number of breeding pairs of Greater Flamingos, and of breeding sites occupied (Spain, Italy), the species remains of conservation concern (Tucker & Heath 1994; Hagemeijer & Blair 1997). This ranking is based on the flamingo's localised distribution, particularly during breeding (Rendón 1997b; Rocamora & Yeatman-Berthelot 1999). Similarly, in Asia, the Greater Flamingo is listed in the Red Data Book of Kazakhstan (Khrokov 1996).

Of the 35 more important breeding sites (Chapter 12) one (Iriki) has been drained and six others, as far as we know, have not been used during the past 20 years. In contrast, however, during this same period, a further six colonies have been established and are regularly occupied. The overall picture of the flamingo's status is one of stability, with possible declines in some parts of the species' range compensated for by increases in others. Saltpans play an important role in determining the flamingo's present-day conservation status, because at one-third of all the breeding sites flamingos either nest or forage in saltpans.

THREATS AND CONFLICTS

Because they breed in only a few places, one of the major threats to flamingos has been disturbance at colonies, by egg-collectors, photographers, aeroplanes and the like. In France, for example, the flamingo's future in the Camargue was a question of major concern in the 1960s, because breeding attempts were thwarted by disturbance from aircraft and predation of eggs and chicks by marauding gulls, aggravated by erosion of the breeding island. Large-scale developments were taking place in the neighbouring regions, to the west for tourism and to the east for industry. Breeding attempts by flamingos in 1962 and 1963 failed, and there followed a five-year period with no nesting. Fortunately, conservation efforts have since been rewarded and the flamingo today has a much stronger hold in the region, although its dependence upon just one breeding site in the country, the Fangassier lagoon, makes it still vulnerable (see Chapter 11). In recent years, improved protection, education and surveillance have played a major role in assuring a more secure position for this species as a breeding bird in France.

Although colonies will always be vulnerable to disturbance and predation, the main threats to flamingos today come from pressures on the wetlands which harbour them, rather than from any particular form of persecution. In this chapter we look at some of the threats facing flamingos and their more important wetlands, and we discuss some of the major conservation actions being carried out for their benefit.

WETLAND LOSS AND DEGRADATION

Although the multiple functions of wetlands are more widely recognised today than they were a few decades ago, thanks to campaigning and conventions (i.e. Ramsar), the threats to these areas persist and even some of the most remote wetlands have recently become threatened by projects which would alter water levels, change water quality or modify wetland use. As human populations grow and develop, so do the threats to the integrity of wetlands, and very few sites, even those which seemingly benefit from protective measures on a regional, national or international level, can be considered totally secure. Sites of major importance to flamingos as either breeding or foraging areas are threatened (or have been lost) by, for example, pollution (Lake Nakuru, Lake Tuz), disturbance through exploitation of their natural resources (Lake Natron), lowered water levels through upstream damming (Iriki depression, Lake Kelbia, Etosha Pan) or reclamation (Lake of Tunis). If flamingo populations have not declined, it is because of improved legislation covering both the species and their remaining wetlands, and also because not all changes have been detrimental to flamingos, the construction or extension of saltpans being one obvious example. Every wetland has not only its own specificity and attraction for flamingos, but its own problems. The conservation status of the Greater Flamingo's main breeding sites is given in Table 14. A global flamingo action plan is presently being prepared by the Flamingo Specialist Group (Wetlands International, Species Survival Commission of IUCN).

ENVIRONMENTAL POLLUTION

Some aspects of the biology of colonial waterbirds have been considered convenient bioindicators, indicating changing environmental conditions (Kushlan 1993).

Table 14 Conservation of the main Greater Flamingo breeding sites.

Site	Flamingo breeding	Ramsar	MAB	Reserve	Park	Other
WEST AFRICA						
Kaolack saltpans	1976	1984	1980	–	Sine-Saloum	–
Chott Boul	1987	–	–	–	–	none
Kiaone Islands	1957	1982	–	–	PNBA	–
Ilot des Flamants	1959	1982	–	–	PNBA	Heritage
MEDITERRANEAN						
Iriki depression	1965	–	–	–	–	none
Doñana	1976	1982	1982	nature	national	IBA
Fuente de Piedra	1963	1983	–	Natural	–	IBA
Alicante wetlands	1973	1989	–	–	natural	IBA
Ebro delta	1992	1993	–	–	regional	IBA
Camargue	1914	1986	1986 ?	–	regional	IBA
Molentargius	1993	1976	–	–	–	IBA-SPA
Margherita di Savoia	1996	1979	–	Natural	–	IBA-SPA
Sidi el Hani	1972	–	–	–	–	none
Sidi Mansour	1963	–	–	–	–	none
Djerid and Fedjaj	1948	–	–	–	–	none
El Malaha	1970	–	–	–	–	none
Camalti saltpans	1982	–	–	hunting	–	IBA
Tuz Gölü	1970	–	–	–	–	none
Seyfe Gölü	1992	1994	–	des.nature	–	SPN–NV
Sultansazligi Gölü	1971	1994	–	nature	–	SPN
Eregli Marshes	1987	–	–	–	–	none
ASIA						
Lake Uromiyeh	1964	1975	1976	–	national	–
Lake Tengiz	1958	1976	–	–	–	protected
Dasht-e-Nawar	1969	–	–	–	–	sanctuary
Ab-e-Istada	1966	–	–	–	–	sanctuary
Lake Sambhar	1995	1990	–	–	–	int.import
Rann of Kutch	1935	–	–	–	–	sanctuary
EAST AFRICA						
Lake Shalla	1988	–	–	–	national	–
Lake Elmenteita	1951	–	–	–	–	private
Lake Magadi	1962	–	–	–	–	none
Lake Natron	1954	–	–	–	–	none

Site	breeding by flamingos	Ramsar	MAB	Reserve	Park	Other
SOUTHERN AFRICA						
Makgadikgadi Pan	1976	–	–	game	–	sanctuary
Lake St Lucia	1967	1986	–	nature	–	–
De Hoop Vlei	1960	1975	–	nature	–	–
Etosha Pan	1957	1995	–	game	national	–

Ramsar list to date 1 May 1998
MAB = UNESCO Man and the Biosphere programme
PNBA = Parc National du Banc d'Arguin
IBA = Important Bird Area
SPA = Specially Protected Area
SPN = Site for the Preservation of Nature
NV = site of natural value

Population decreases of some species have been linked to changes in ecosystem function caused by water management, as with American Wood Storks and other species in the Florida Everglades (Kushlan 1987). Others have been associated with excessive pesticide contamination of food resources, as indicated by decreased numbers of Brown Pelicans in North America (Anderson & Risebrough 1976; Schreiber 1980). The increase in flamingos, which can be observed today throughout the Mediterranean region (Wetlands International 2002), should not be seen as an indication that these birds are living in a healthy environment. Flamingos forage where aquatic organisms abound, and this may well be in polluted areas, as for example at Molentargius in Sardinia and, in Tunisia, on the Lake of Tunis, at Sejoumi, and in the Bay of Sfax. These areas receive, or have received, large quantities of domestic and/or industrial effluents, which flamingos, at the end of the food chain, concentrate in their organs and tissues. The levels of contaminants found in the tissues of these birds, or in their eggs, may thus be evidence of environmental pollution somewhere within their range. Pollutants are considered a threat to flamingos in southern Africa (Williams & Velásquez 1997).

So far, there have been few cases of unusually high mortality of Greater Flamingos where pollutants have been suspected, other than from lead poisoning (see Chapter 9). There have, however, been cases of flamingos with bill malformations which may well be attributable to contamination by heavy metals or polychlorinated biphenyls (PCBs). The most striking of these occurred in the Doñana National Park in Spain in 1988, when eight out of 18 chicks taken into captivity had deformed bills (Mañez 1991). In the Camargue, of 20,000 chicks captured for ringing (1977–2000), only three individuals have been seen with deformed bills. In two cases the upper mandible was shorter than the lower, and in the third (Figure 52), it curved sideways at right-angles to the lower.

Many species of birds have been recorded with bill deformities (see Pomeroy 1962). Possible causes were discussed and 'industrial contamination' was evoked, though something fairly new at the time. In North America, deformities in waterbirds have been attributed to contamination by selenium, boron, mercury and arsenic, which occur in agricultural run-off from irrigated farmland (Anderson 1987). Cormorants, herons and Larids seem to be particularly vulnerable to congenital malformations, and defects at hatching are reported from locations where there are known to be elevated levels of persistent lipophilic contaminants (PCBs and dioxins) or high selenium levels in the aquatic food chain (see Gilbertson *et al.* 1976, Ball 1991, Boudewijn & Dirksen 1995, Volponi 1996).

Very little research in this field has been carried out on flamingos. Cosson & Metayer (1993) examined 22 flamingos which died in southern France during the cold spell of January 1985. Cadmium, lead, copper, zinc and mercury levels were measured in their liver, kidneys, breast muscles, lungs, breastbone, stomach and feathers. The trace elements found could either have their origins in atmospheric pollution from industrial complexes at Fos-sur-mer or in the Rhône valley, or have been obtained by direct contact with, or ingestion of, pollutants. Samples showed great individual variation, with the highest levels well above average, but below those of seriously contaminated birds. These results show that flamingos carry a variety of heavy metals in their tissues and organs, but no harmful effects to the population are apparent as yet (Cosson *et al.* 1988; Amiard-Triquet *et al.* 1991).

More recently, the presence of residues of organochlorine pesticides (OCPs) and PCBs has been investigated in 53 unhatched eggs from Greater Flamingos (Guitart *et al.* 2005). Eggs were collected in 1996 from the National Park of Doñana (Guadalquivir Marshes, south-west Spain), following colony abandonment due to predator attacks. Levels of organochlorine residues indicated only a mild degree of exposure, causing no particular threat for the flamingo population. In particular, eggshell thickness was not correlated with levels of OCPs or PCBs.

EXOTIC FLAMINGOS AND FERAL POPULATIONS

The three species or subspecies of flamingo which are most often kept in captivity are naturally those which most frequently escape, and are seen either living in groups with native species or forming feral groups outside their normal range. In north-western Europe, Greater, Caribbean and Chilean Flamingos established a mixed-species flock in the 1980s, and have since bred quite regularly at Zwillbrocker Venn in Germany (Treep 2000). The colony, which now numbers *c.* 50 birds, is established on a bed of reeds in a freshwater marsh. The birds winter in the North Sea estuaries, particularly in the delta region of The Netherlands, and many have withstood spells of severely cold weather. There is as yet no evidence of any exchange of birds between the North Sea and the Mediterranean.

In the Mediterranean region, Caribbean, Chilean and Lesser Flamingos have all been sighted in various parts of the Greater Flamingo's range. A pair of Caribbean flamingos attempted breeding in the Camargue in 1974 and 1976 (Johnson 1976; Hafner *et al.* 1979) but subsequently one bird and then the other disappeared. The

Chileans, of which up to eight have been seen, have formed monospecific pairs most years, but on at least two occasions have cross-bred with Greaters, presumably successfully, since hybrids have been seen in the wild (Cézilly & Johnson 1992).

Of 12 Lesser Flamingos reported to have escaped from the Balearic Islands in 1989 (Cézilly & Johnson 1992), two or three have since been regularly observed in the Camargue, and in 2001 a pair nested amongst the Greater Flamingos at the Fangassier colony and successfully raised a chick.

Because the different species/subspecies of flamingos are closely related, exotic flamingos can be a threat to the native species in the Mediterranean through the introduction of diseases and parasites, through hybridisation and thus genetic pollution, and through competition. The recent proliferation of the American Ruddy Duck over part of the native White-headed Duck's range in the Western Palearctic is a more extreme example of this problem, which has given rise to much discussion and to eradication efforts (Hughes 1996).

DISTURBANCE AND PREDATION DURING BREEDING

Disturbance by humans

Once they are adult, flamingos have few predators besides humans. When breeding, however, they are very vulnerable because they nest on the ground in large, dense colonies. These attract the attention of both humans and predatory or scavenging animals, which can pose a threat to the well-being of a colony. Records of disturbed or abandoned breeding attempts were common until quite recently (Castan 1963; Hoffmann 1964b; Brown 1973; Mendelssohn 1975; Blasco *et al.* 1979; Grussu 1999). Today, protection measures are in place at many breeding sites, particularly those in the Mediterranean (see 'Wardening' later in this chapter), and egg-stealing or disturbance by humans apparently occurs much less frequently than it did through the first half of the 20th century (Gallet 1949). There is no evidence of any European colonies having been pillaged during the past 50 years. This is not the case in North Africa, however, where we ourselves saw the breeding attempt at El Djem in Tunisia in 1976 thwarted by someone collecting eggs. The same thing had happened at Sidi Mansour a few years earlier (Castan 1963). Globally, however, there is little evidence to indicate that egg-stealing has a major impact on Greater Flamingo populations today.

In the Camargue in 1981, the only year since 1972 that the colony was not under surveillance, two journalists visited the breeding island at the Etang du Fangassier at the peak of hatching, causing the nesting birds to panic. The following day the island and nearby dykes were strewn with corpses of very small chicks (*c.* 2,000) which had been killed by gulls or had starved (Perennou *et al.* 1996). Some years later, in May 1987, we witnessed a freakish but major and quite unforeseeable event which caused the loss of a great many eggs and chicks, also at the height of the breeding season. Alerted by the sudden panic of the whole colony, observers in the tower at the Etang du Fangassier saw a shiny Mylar balloon, shaped like a fish, blow swiftly across the surface of the water trailing a short piece of string and heading directly for the flamingo colony. The observers were unable to intervene and the balloon was blown onto the island, where it became entangled on one of the nests. All the adults fled, leaving behind them mostly well-incubated eggs or small chicks. The Yellow-legged Gulls, less afraid of the silver fish, were quick to benefit from the situation. The balloon remained wavering in the wind from mid-afternoon through the night, before continuing its journey downwind. Only the flamingos nesting furthest from the balloon returned to their nests, the others leaving their clutches exposed to the mercy of the gulls. None of these survived and it is estimated that about 4,000 eggs or chicks were lost because of the balloon (Johnson in Boutin & Chérain 1989).

Disturbance by low-flying aircraft has been and still can be a major form of disturbance at flamingo colonies. This was a very serious problem in the Camargue from the 1940s to the 1970s, and in 1992 the late breeding attempt in the Ebro Delta, Spain, was thwarted by a low-flying aircraft (Luke 1992). In the Camargue, flamingos have largely become accustomed to planes and helicopters near the colony and now panic only at the onset of breeding, if at all. This seems not to be the case in Botswana, however, where planes are a form of disturbance to flamingos on the Makgadikgadi Pans (McCulloch 2002). In this same area, where flamingo chicks trek considerable distances in search of water as their natural wetlands dry, the construction of foot-and-mouth disease veterinary cordon fences across the Sua Pan has proved to be a serious obstacle for the chicks as they follow the receding water (Herremans in Williams & Velásquez 1997).

Mammalian predators and scavengers

A variety of predatory animals and scavengers forage at flamingo colonies if they are accessible. Breeding attempts in the Camargue were formerly jeopardised or failed because of disturbance by Red Foxes, Badgers and Wild Boars (Lomont 1954a; Hoffmann 1963) and those in Doñana in Spain have been destroyed by Wild Boars (Mañez 1991). The presence of terrestrial mammals at colonies may be due to the drying out of the surrounding lagoon, in which case the eggs or chicks lost to predation would in any case probably not have survived. Colonies which are neither established on islands nor in remote places constantly face this threat.

Three species of birds in particular can cause considerable mortality of flamingo eggs and/or chicks, either through disturbance at colonies or by predation. In East Africa these are Great White Pelicans and Marabou Storks, and in the Palearctic region, the Yellow-legged Gull. Marabou Storks forage largely on refuse tips, and since the 1960s their numbers have increased considerably in East Africa, in particular at Lake Elmenteita in Kenya (Brown 1958, 1975). The storks prey both on flamingo chicks and on incubating adults. However, just their presence in a colony is sufficient to cause flamingos to desert their nests, and whole colonies have been wiped out by a few storks (Brown 1973).

Great White Pelicans are another danger to breeding flamingos in Kenya, and are as serious a threat to flamingos as Marabous, particularly at Lake Elmenteita. The pelicans, which are larger and heavier than the flamingos, nest on the same islands, and cause so much disturbance that the flamingos abandon their nests. They are not particularly aggressive but Brown *et al.* (1982) reported flamingo colonies nevertheless being destroyed by pelicans. The presence of very large numbers of pelicans at Lake Elmenteita stems from the introduction into neighbouring Lake Nakuru of the fish *Tilapia grahami* in 1960 (Brown 1973). Since there are no islands on which pelicans can breed at Lake Nakuru, the birds have colonised Lake Elmenteita, and displaced the flamingos.

In the Mediterranean, particularly in the Camargue, Yellow-legged Gulls have for many years taken a heavy toll of flamingo eggs and chicks. Predation and disturbance by these birds have been described by Salathé (1983) (Figure 57). Gulls attack an incubating flamingo either by approaching it from behind and pecking at the protruding leg joints until the flamingo is forced to stand, or from the front, by pulling the flamingo off its nest by its bill. In the former case, when the suffering flamingo eventually stands, the more agile gull snatches the flamingo's egg or small chick from the nest, and the parent can do little to prevent this. Flamingos nesting on the periphery of the island are most vulnerable to such attacks, but in windy weather gulls will fly low over the centre of the colony, and when a flamingo raises its head in a threat posture, will seize it and pull it to its feet. Before the flamingo regains its balance, the gull swoops in and steals the egg or chick. Gulls break open eggs either by pecking at them or by dropping them from a height of one or two metres, making as many attempts as necessary. Predation at the main colony at Fangassier is mostly by a few pairs of resident gulls, but when flamingos attempt to breed on the neighbouring dyke large numbers of gulls (sometimes 50–100), many immature, attack those flamingos which have newly laid eggs, and the birds abandon their nests, sometimes hundreds of pairs at a time.

Yellow-legged Gulls (formerly considered a subspecies of the Herring Gull) have harassed flamingos in the Camargue throughout the breeding season for the past half century, and they have been responsible for the loss of a far greater number of eggs and chicks of flamingos than all other forms of disturbance combined. Gallet (1949) observed in the 1940s that these birds constituted a potential threat to the flamingos, and his fears were soon to be confirmed. In both 1957 and 1959, it was estimated that at least 1,000 flamingo chicks were eaten by gulls (Hoffmann 1959, 1962) which not only preyed upon the eggs and chicks but were a permanent harassment to the flamingos during incubation and chick-raising. The number of pairs of gulls breeding in the Camargue increased from the first three in 1937 (Lomont 1938), to 300 in 1956, 1,000+ in 1973, 3,000+ in 1988 (Sadoul *et al.* 1996), and by the year 2000 exceeded 6,300 pairs (Kayser *et al.* 2003).

In spite of the ever-increasing numbers of gulls in the Camargue, very few of these birds prey on flamingo eggs and chicks (Salathé 1983) and, as noted by this author, the majority of gulls are not a threat to the flamingos. The flamingo population is flourishing and the colony is less vulnerable to abandonment now

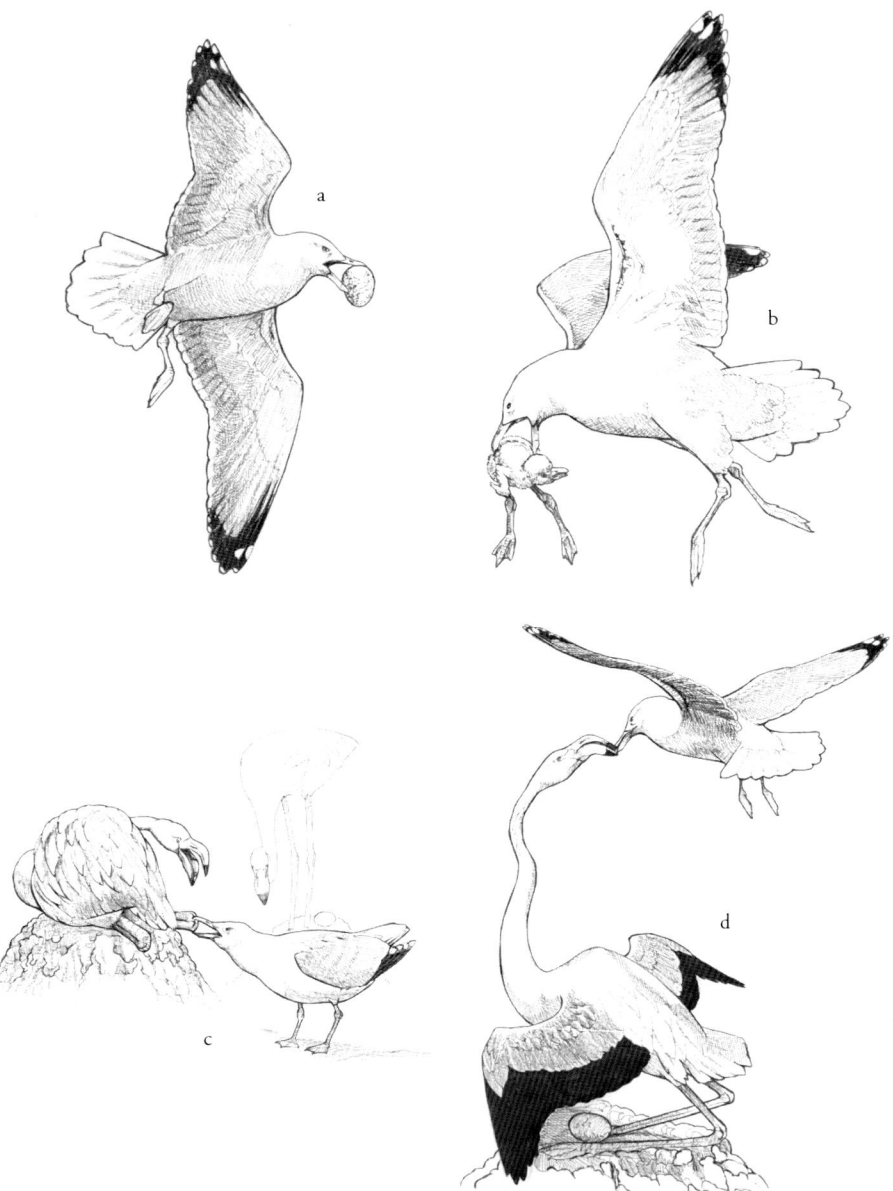

Figure 57. In the Camargue, Yellow-legged Gulls take hundreds or thousands of eggs (a) and small chicks (b) each year by pulling incubating flamingos off their nests, but this predation is no longer a threat to the flamingo population. Most predation occurs on the periphery of the colony, the gulls pecking at the leg joints to make the flamingo stand (c) or pulling the birds off their nests. When it is windy the gulls also fly over the central parts of the colony and jerk birds off their nests (d).

than it was in the 1950s and 1960s. Elsewhere in the Palearctic region, these gulls harass flamingos at Lake Uromiyeh in Iran and in the Ebro delta in Spain, while in Doñana National Park it is the Lesser Black-backed Gulls which have caused breeding failure (Mañez 1991).

Raptors, rather surprisingly, seem to play an insignificant role as predators or scavengers at flamingo colonies. Only Egyptian Vultures have been reported taking chicks which were isolated or at unguarded nests in the Rann of Kutch, India (Shivrajkumar *et al.* 1960), and in East Africa (Brown 1967). In the Camargue, incubating flamingos show no fear of the occasional Black Kite or Marsh Harrier which may pass near the colony.

Exotic or unusual species, on the contrary, can be perceived as a threat by flamingos, and their mere presence may cause panic and nest desertion. In spring 1989, a Black Swan, an Australian native escaped from captivity, visited the Camargue flamingo colony on many occasions (Johnson in Boutin *et al.* 1991). As the swan was unfamiliar to the flamingos and aggressive, its visits to the island caused hundreds of flamingos, even birds half-way through incubation, to desert their nests. In 1980, a Gannet stayed at the colony for several weeks and occupied one of the flamingo mounds, which after all is similar in shape to a Gannet's nest. It even brought in nesting material. Not being familiar with this bird, the flamingos were afraid of it and scattered at the Gannet's approach, abandoning the nests in the vicinity (Hafner *et al.* 1982).

In Europe, the more accessible breeding sites are now wardened, and colony surveillance has generally proved successful against human disturbance (see 'Wardening'). Predation, however, which occurs at most colonies and is quite natural, occasionally reaches such proportions that it may constitute a threat to the colony. Some of the major sources of predation and disturbance, weather included, to which the Camargue colonies have been subjected over the past 50 years are summarised in Chapter 8 (Tables 13a,b).

CONFLICT WITH AGRICULTURE

Half a century ago, Allen (1956) stated that rice farming in the Camargue could be a threat to the flamingo because excessive water from irrigated fields might bring about an increase in the water level of the National Nature Reserve. Indeed, with the advent of rice farming in the 1950s, this is precisely what happened. The Etang du Vaccarès and adjoining lagoons, to which Allen was referring, no longer dry to the extent that they did formerly. This has not, however, resulted in a decrease in the number of flamingos using the Rhône delta, but rather in some birds now foraging in habitats they did not previously use: freshwater marshes, fish farms and rice fields. It is also true that half a century ago the flamingo was less abundant in the delta, and the birds were confined to the brackish lagoons and saltpans. The situation has changed radically since then. Hunters complain that flamingos damage or destroy submerged plant communities which constitute food resources for wintering ducks, and fish farmers have accused the birds of causing turbidity in the waters where they feed, resulting in poorer catches of fish. Both of these claims may be valid, but the main conflict with human interests, and the only one where evaluation of damage has been made, stems from the use of rice fields as foraging areas (Johnson *et al.* 1997).

Flamingos were first observed feeding in rice fields in spring 1978, shortly after the fields had been sown (Hoffmann & Johnson 1991). The number of birds was small and farmers did not complain about crop loss. The following spring, small groups of flamingos again foraged in the paddies of the Camargue and two farmers in the delta expressed concern, but they did not report crop damage. In 1980 the situation seriously deteriorated and thousands of flamingos visited the rice fields. They arrived in the evening in freshly sown fields, where they fed overnight, only leaving when scared away the following morning. Up to 600 birds were seen in a single paddy (*c.* 2 ha.). The problem lasted 5–6 weeks, until the remaining rice plants were about 10 cm in height. Flamingos were not just visiting fields in the vicinity of the colony, but were also flying to paddies up to 20 km inland. Damage was obvious where large numbers of birds were involved; there were large tracts towards the middle of the fields with no plants, while the downwind shoreline resembled a lawn. Farmers reported 450 ha of crops destroyed, representing almost 10% of the total area of rice fields in the delta in 1980. This situation was understandably unacceptable to farmers and the following year a massive scaring campaign was begun. In agreement with the Camargue Regional Park (PNRC) and the Syndicate of Rice Farmers, the Ministry of the Environment in France contracted the Tour du Valat to develop methods to prevent flamingos causing damage to the rice crop (see below).

At first glance it may appear that it is the increase in the number of flamingos occurring in the Camargue that is to blame for this problem, and that the natural marshes and salinas of the Camargue, and of the neighbouring Languedoc lagoons, can no longer satisfy the needs of the colony. This may indeed be true in some years, but other changes have taken place which should not be overlooked, and which are just as striking as the increase in the number of flamingos in the area. Firstly, farming techniques have evolved radically over the past 30 years in a manner which has made the rice fields more attractive to flamingos, and secondly, flamingos have become much tamer than they formerly were and are now quite unafraid of humans, in some places at distances down to just a few metres.

Rice was formerly sown in small fields (of *c.* 1 ha), which it is unlikely that flamingos would ever have visited. Once sprouted, the rice was transplanted by hand by seasonal workers from the rice-growing areas of Spain. Transplanting allowed the weeds, namely rushes, to be eliminated, but this was a labour-intensive method and in the 1970s gave way to mechanisation. The rice is now sown directly into more extensive fields (*c.* 3–7 ha) and the weeds controlled by herbicides. These and other crop treatments are applied by spraying from the air. Trees and hedges have been removed in order to facilitate spraying or fertilising by helicopter, so that the fields have taken on the aspect of extensive shallow marshes. Rice fields are a habitat which flamingos never formerly exploited, even in the Marismas in southwest Spain, where there are regularly large numbers of flamingos, and where large areas of marshland were transformed into paddies in the first half of the 20th century. Although not a major problem every year, this phenomenon has been sufficiently serious for an extensive study to have been undertaken on the use of paddies by birds (Tourenq *et al.* 1999) and for large-scale scaring campaigns to have been organised, which are described later in this chapter.

Being tamer than they formerly were, and foraging to a large degree at night, flamingos now exploit areas which they formerly avoided. The stage has thus been set for a problem which as yet has no solution, and which, on the contrary, has arisen elsewhere, notably in the Ebro delta (Catalonia) since the 1990s.

CONSERVATION ACTIONS

Flamingo breeding sites are few and far between, and consequently they are of high conservation value. The 35 most important are listed in Chapter 12. Many of these sites lie within important bird areas and are legally protected either on a local or national level, or under the umbrella of one or more international wetland or natural heritage conventions (Ramsar, United Nations Educational, Scientific and Cultural Organization (UNESCO) Man and the Biosphere (MAB) Programme). More detailed descriptions of most sites can be found in one or more of the following: *A Directory of Western Palearctic Wetlands* (Carp 1980), *Important Bird Areas in Europe* (Grimmett & Jones 1989), *A Directory of Asian Wetlands* (Scott 1989), *A Directory of African Wetlands* (Hughes & Hughes 1992), *Important Bird Areas in the Middle East* (Evans 1994), *A Directory of Wetlands in the Middle East* (Scott 1995), *Directory of Wetlands of International Importance* (World Conservation Monitoring Centre 1990) and an update of this document (Frazier 1996), *Important Bird Areas in Asia: key sites for conservation* (BirdLife International 2004). Although this legal protection is amply justified, it cannot guarantee the perenniality of a site, and even well-protected wetlands can be degraded by modifications in the surroundings or in the catchment area.

Wetland conservation

The Greater Flamingo's future over the next 50 years is inseparable from that of the wetlands upon which the species depends. Everywhere in the world, wetlands are confronted by drastic changes which can severely affect their characteristics and functions (Dugan & Jones 1993). These changes are the consequences of drainage, various types of pollution, dam-building, agricultural activity and even over-fishing. In the fight to stop and reverse the growing tendency toward the destruction of Mediterranean wetlands, the Greater Flamingo has acquired a symbolic and an exemplary value. Symbolic because the fragile character of the natural resources which it is our duty to manage is perfectly illustrated by the species' ecological requirements. Exemplary because the empirical knowledge necessary for making conservation decisions for long-lived species throughout their range depends on the existence of a long-term research programme of the kind presently being carried out in the Camargue and elsewhere in the Mediterranean. Such studies are very productive and are sorely lacking in ecological research.

The flamingo's future in the Mediterranean basin is closely linked to the future of the marshes, lagoons and salinas of the region, both for breeding and for feeding. These Mediterranean wetlands are threatened in a variety of ways (Garcia-Orcoyen *et al.* 1992; Papayannis & Salathé 1999): through pollution from agricultural developments using massive quantities of chemicals, through pollution from untreated waste water, and by uncontrolled tourism development designed for short-term gain rather than long-term viability. The study of the flamingo's close association with industrial salinas (see Chapter 2 and below) has emphasised the complexity of the species' ecology in this environment, which is managed solely for the benefit of humans. In the Camargue, the wetlands of the delta must satisfy the needs of 50,000 flamingos, including up to 20,000 breeding pairs. According to Rooth (1965) a flamingo must consume about 10% of its body weight per day, which means that 5.5 tonnes fresh weight per day are sieved from the wetlands of the Rhône delta by flamingos (Johnson 1983). These birds are presently able to satisfy their requirements in large part by foraging in the salinas, and the lagoons and marshes of protected areas. Food quality and abundance are dependent upon the water surface area in the Camargue and on management practices. The way in which salt is produced implies a series of lagoons of differing salinities, and this results in a wide range of organisms available for flamingos. Salinas are therefore particularly good places for flamingos to forage and eventually breed. It is difficult to imagine what the consequences might be for breeding flamingos in the Camargue and elsewhere in the Mediterranean region if this industry should one day collapse. Nevertheless, wetland loss in general is very much a reality, while the economics of salt production become increasingly challenging. The constraints and modernisation of this industry have brought about a reduction in the number of employees, and consequently many salinas which adjoin areas heavily frequented by tourists are now subject to disturbance by, in particular, increasing numbers of pedestrians and cyclists. Such a situation could eventually make salinas, many of which are close to towns, less attractive to birds. Another problem linked to the proximity of towns is pollution. At least one salina in the Mediterranean region has recently ceased production because of pollution by waste water, although water is still circulated through the system. The biological richness of these areas is greatly reduced if the salt industry closes down. In some cases, economically viable complementary activities may help to maintain salinas along the Mediterranean coast, eco-tourism for example, and the harvesting of brine shrimps for the aquarium trade, as well as financial compensation for biological conservation through agri-environmental measures developed by the European Union (Sadoul *et al.* 1998). Although the situation is not presently alarming, the threat of a serious reduction of these wetlands hangs in the air. Those salinas which are still active are extensive, or form parts of large estates of extremely high land value. They have tremendous development potential for tourism, which in turn would create employment. Landscape modifications required to produce salt are not too unsightly and they are clearly beneficial to flamingos. They are certainly less damaging to the environment than the changes which would occur if these areas were open to development.

Legal protection

The Greater Flamingo is now protected by law throughout much of its distribution, although it is still, or was until recently, hunted or trapped in parts of its range, for example the Maghreb (Algeria) and Libya, as indicated by ring recoveries, and in Egypt (Mullié & Meininger 1983), Afghanistan (Scott 1995) and India H.S. Sangha pers. comm.), while in Tanzania birds are trapped alive and exported for collections.

The different international conventions are important tools for the conservation of flamingos and, in particular, their habitat. Under the Wild Birds Directive, species listed in Annex I 'shall be the subject of special conservation measures concerning their habitat in order to ensure their survival and reproduction in their area of distribution'. Under the Bern Convention, for species listed in Appendix II, parties agree to take appropriate and necessary measures for the conservation of their habitat, to give special attention to the protection of areas of importance during migration, and to prohibit the deliberate damage or destruction of sites. The Bonn Convention calls for international cooperation in the maintenance of a network of suitable habitat appropriately disposed in relation to migratory routes of species in Appendix II. Such protection measures are vital for the conservation of flamingos, particularly in the Mediterranean, where hunting is practised extensively.

Breeding-site protection

Wardening

Colony surveillance is necessary if flamingos are to breed successfully in areas which are easily visible and accessible to the public. This is the case for most of the European breeding sites which, fortunately, benefit from some degree of protection. Gallet (1949) reported colonies in the Camargue being raided and either partially or entirely destroyed in 1912, 1936, 1939 and 1945. Egg-stealing by humans (for food) has not been reported in Europe for decades, thanks mainly to legal protection and effective wardening. In the second half of the 20th century, threats to the colony changed. In the 1950s foxes became a problem for birds attempting to breed on a dyke when the main breeding island in the saltpans deteriorated through erosion (Lomont 1954a, Hoffmann 1963). Unscrupulous photographers, be they amateur, professional or journalist, are a threat to all species of birds at the nest, and to flamingos in particular. There have, fortunately, been few cases of serious disturbance to breeding flamingos from this source since 1969, and this really is to the credit of those who guard the colonies.

Flamingos are one of the principal attractions for thousands of visitors who flock to the Camargue each year, particularly during the breeding season, and the Rhône delta is no longer the 'wilderness' area that it formerly was. Consequently, since 1969, it has been necessary to warden the colonies established at the Etang du Fangassier. This has been very effective, and disturbance to the breeding birds over the past 30 years has been restricted to a few isolated, mostly minor events. Wardening, and the presence of an information centre, has been assured by conservation organisations and by the salt company, and has proved to be an effective deterrent against people attempting to approach the colony. Colonies are wardened from the start of laying, usually in early April, until the end of June, when chicks are usually united in the crèche and less vulnerable to disturbance.

Over the years, disturbance by low-flying aircraft has been considered by conservationists to be as much of a problem to flamingo colonies as the various forms of terrestrial threats. The Rhône delta, being flat and having a small human population, is surrounded by airports, both civil and military, and planes have been a very serious source of disturbance to breeding flamingos in the Camargue since the 1930s (Bressou 1931). The flamingos have always feared the planes and helicopters, which they first hear and then see as they approach, as opposed to the high-speed jets, to which they hardly react, even when they pass very low over the colony and are deafening to human ears.

Conservationists in the 1960s and 1970s informed the local airport authorities of the impact of low-altitude flights on the behaviour of flamingos in the Camargue. Talks and slide shows were given locally and pilots were requested to avoid flying over the breeding site and to refrain from pursuing flamingos. It has, in fact, been forbidden for many years to fly below 3,000 ft (914 m) over the Camargue National Nature Reserve, and this restriction is enforced. The flamingos, however, breed outside this zone, in an area reserved for military planes and for air traffic crossing the lower reaches of the delta.

The situation evolved gradually over the years and fortunately today flamingos are rarely harassed from the air. At the same time, the birds have learned to tolerate planes to a degree which would have seemed highly unlikely only a decade ago. If they scare at all, it is now only for a short time prior to the start of egg-laying. It is interesting to note that the fear of aircraft appears to have been a local problem. A possible explanation for this fear, given by one of the older workers in the saltpans, is that the birds were pursued and even dive-bombed by aircraft during World War II (1939–45). Indeed, according to Allen (1956), the Etang du Vaccarès was transformed during this period into an aerial target range. During the years when flamingos were scared of aircraft in the Camargue, they were less afraid of them elsewhere. In Tunisia, for example, many birds used to feed in the Lake of Tunis only a few hundred feet below aircraft that were landing or taking off (Allen 1956), and many of these birds were probably of Camargue origin. In addition, when breeding took place in the Sinäi in 1974, Mendelssohn (1975) stated that the flamingos had become accustomed to the movement of aeroplanes and actually bred in an area regularly patrolled by planes.

The flamingo's fear of aircraft in the Camargue persisted well into the 1970s, and it was little short of a miracle that breeding was successful in 1969. Over the past 25 years, there have been no major egg-losses or colony disturbance by aircraft and we can hope that this once major threat has been consigned to the past.

Predator control

As in other parts of the world, gulls of several species have increased in number since the 1940s, and they have extended their range. Population explosions have been linked to the increased availability of food to species which forage at rubbish dumps. Only 10 km east of the Etang du Fangassier is France's largest open-air waste-disposal site, at Entressen, where rubbish from the city of Marseille is still dumped. Tens of thousands of gulls forage at this site.

Authorised Yellow-legged Gull-culling operations were begun in the Camargue in the early 1960s (Blondel 1963), and these were conducted on a wide scale each spring until the early 1980s. Baits containing a strong dose of alpha-chlorolose were placed in nests and at suitable places close by. Most gull colonies were visited twice in the season and some nests were baited twice on the same morning, in an attempt to cull both partners of the pair. Most birds died on the nest, others were collected from the shore of the lagoon. There was an immediate reduction in the number of gulls at the flamingo colony, and this undoubtedly both retarded the current gull population explosion and helped the flamingos to recolonise the Camargue. During the past 20 years, however, there has been a drastic reduction in gull-culling and, not surprisingly, the notable increase in gull numbers mentioned earlier. Similar trends have been recorded elsewhere in southern France and in north-east Spain, with a more than threefold increase over 20 years in the numbers breeding on the Medes Islands in Catalonia (Purroy 1997). Numbers in the Ebro delta also escalated—during the mid-1990s, numbers increased from 2,686 pairs in 1996 (Copete 1998) to 4,166 pairs in 1997 (Copete 2000)—and the flamingos breeding there are now also exposed to harassment by the gulls.

The Yellow-legged Gulls in the Camargue continue to take hundreds or thousands of eggs and chicks each year, but since the flamingos are now also breeding in much greater numbers than they did 40 years ago, little or no culling of gulls has taken place near the flamingo colony since the early 1980s, and would no longer be justified.

BREEDING-SITE MANAGEMENT

In response to local conservation problems, a variety of management techniques have been used to encourage breeding or to improve breeding conditions for flamingos, in both the Old World and the New (Rooth 1982; Castro 1991; Rendón Martos & Johnson 1996). These actions have been carried out in areas already influenced by human presence, notably in saltpans. The most significant of these are discussed below.

Provision and maintenance of breeding islands in the Camargue

Gallet (1949) recognised that flamingos themselves destroy, by trampling and nest-building (Figure 58), the very islands that give them life. The birds themselves, he said, are responsible for the destruction of their own breeding grounds. While new islands were continually forming, this phenomenon was no problem. Islands for breeding were originally created by the dynamic forces of the river and the sea, which during periodic floods left deposits of silt in the lower reaches of the delta. The construction of dykes in the 1850s, however, greatly reduced the influence of both the Mediterranean and the Rhône, and effectively prevented the natural creation of islands.

The four islands used consecutively by flamingos for breeding since the early part of the 20th century are all in permanent lagoons varying in size from 230 ha to 550 ha, having a maximum depth of 110 cm and an average depth of 50 cm. Salinity is in the range 60–100 g/l NaCl for three of them but is only 28–34 g/l NaCl for the fourth, which has been colonised by flamingos only twice since the 1920s. After breeding on the same island in the Baisses du Pavias near Salin de Giraud for nine years, breeding attempts in both 1962 and 1963 failed to produce any chicks (Hoffmann 1963, 1964a). During the previous years (1958–61), breeding success had been poor mainly because the nest mounds, once on an island, now stood in the water and were swamped by salt foam and spray during strong winds. Gallet (1949) had not exaggerated the importance of this phenomenon.

It might seem surprising that flamingos do not nest in the Camargue National Nature Reserve, nor in the adjoining Réserve des Impériaux, a vast area well protected and undisturbed. Indeed, a glance at a map would suggest that there are several suitable sites, but in fact this is not the case. With an extensive cover of vegetation, these reserves are home to terrestrial predators, particularly Red Foxes and Wild Boar, whose presence would prevent successful breeding by flamingos.

The more permanent water levels brought about by the transformation of natural lagoons into saltpans have been beneficial to flamingos for feeding, but in the long term have been detrimental for breeding. During windy weather, islands are subjected to extensive erosion by waves more frequently than they were in the past, and as a result they are sooner or later doomed to disappear.

Since there was an apparent lack of islands suitable for breeding in the Camargue, an effort was made in 1964 to prevent further erosion of the nesting

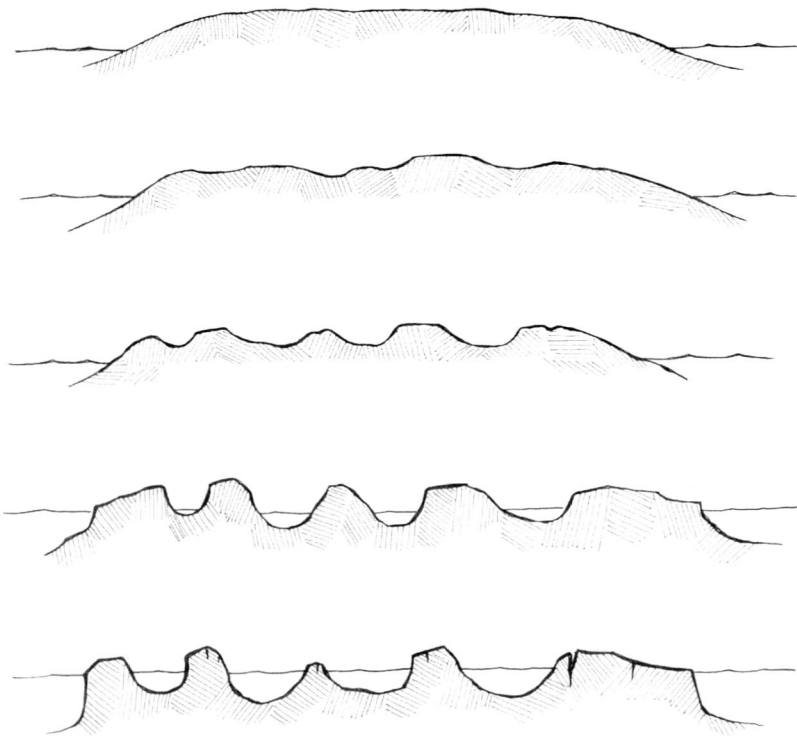

Figure 58. Scheme of the topography of an island following its colonisation by flamingos. Etienne Gallet (1949) observed that flamingos themselves destroy the very islands which gave them birth. It may take several years before a site is totally destroyed, or it may be lost after just one breeding season if consistently exposed to rain and/or spray.

island. Credit must go to Lucas Hoffmann who, in 1964, made what was one of the first efforts to conserve flamingos in the wild, by convincing the salt company to build a dyke around the remnants of the breeding site used in 1952 and again in 1954–61 (Johnson 1966). The intention was to prevent the nests from being submerged during gales. However, flamingos did not attempt breeding over the following four years, either in the Camargue or elsewhere in France, and the effort seemed then to have been of no avail.

In 1969, after a period of seven years without breeding, the species again nested in the Rhône delta (Johnson 1970). Most birds colonised an island in the Etang du Fangassier, a site not previously used for breeding, at least in the recent past. The island, however, was too small for the whole colony and subsidiary groups of mounds were built elsewhere in the Fangassier. Some birds, however, also nested at the old site, on the remnants of the dyke built in 1964 to protect the nests. The return of the flamingo as a breeding species in France was an event which inspired further management actions. Indeed, the new site (Fangassier I) chosen by the flamingos was situated in the middle of the main pan at the Etang du Fangassier (390 ha; Figure 9) and fully exposed to buffeting by the strong winds which are so characteristic of the region. Unless it could be protected or strengthened, it in turn would deteriorate. Since the 1850s, dykes and embankments have prevented flooding of the Camargue by both the river and the sea, and the dynamics of the delta have thus been irreversibly changed. Restoration of this island, however, as became apparent after discussions with the salt company, was not feasible in view of the soft, muddy nature of the lagoon from which the name 'Fangassier' is derived. The director of the salt company was, however, most cooperative and willing to create a new site elsewhere.

The Etang du Fangassier has been one of the flamingo's preferred haunts in the Camargue for a long time. Hughes (1932) reported 10,000 birds at this lagoon in October 1931, and Yeates (1947) also reported large numbers there, in spring 1938. Flamingos have even bred at the Etang du Fangassier in the past (Marc *et al.* 1948). This lagoon became part of the Salins de Giraud in the 1960s, when it was divided into two evaporators by a dyke, the larger western pan being the deeper of the two, and holding water throughout the year. The shallower, eastern pan (145 ha) was, and still is, flooded only during circulation of brine by pumping, from March to August. It is drained in autumn, and in winter is dry except after heavy rain.

Although modern earth-moving machinery can operate in water, an island is best built when the earth is dry. In February 1970, a completely new island (Figures 59, 60; Fangassier II) was constructed in the shallower pan, which was then dry, adjacent to some of the 1969 annex nests (Johnson 1982). It was scraped into existence by a caterpillar tractor, which piled and then compressed an elongated platform of earth (Figure 59), 60 cm above the bed of the lagoon, 220 m in length and varying in width from 20 m to 35 m. It had a surface area of 6,200 m^2 and was situated in the lee of the dyke separating the Fangassier lagoon into two pans. To the human eye, it seemed the ideal nesting site. The flamingos, however, were clearly not as impressed as a few Avocets and gulls were, and they ignored the island for the following three breeding seasons. In February 1974, some artificial mounds were placed on one end of the island in the hope that these would act as decoys and entice flamingos to colonise the site. Artificial mounds have been used to induce breeding in captivity (Duplaix-Hall & Kear 1975). In the wild, it was hoped that the nests would simply attract the flamingos to the site, irrespective of whether they would use them or not. In total, 500 mounds, some genuine flamingo nests transported from the nearby dyke, others moulded from mud-filled buckets (of the type used in France for harvesting grapes), each weighing at least 10 kg, were placed on the end of the island (Figure 61) closest to where flamingos had bred on the bed of the lagoon in 1969, before the island was built.

The first birds to start breeding in 1974 colonised the island in Fangassier I, which was rapidly filled to capacity. Shortly afterwards, prospecting pairs came to the artificial island and settled in amongst the mounds, but only on this part of the

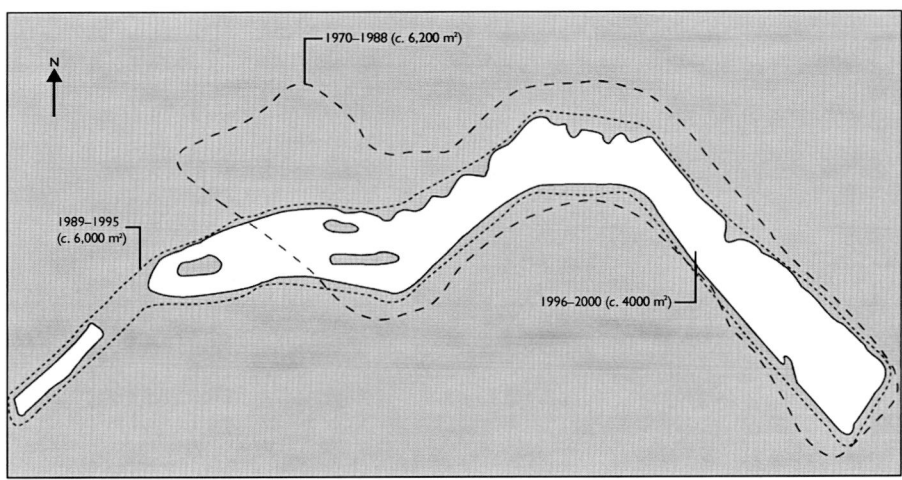

Figure 59. The evolution of the flamingo island at the Etang du Fangassier, Camargue, from its construction in 1970 to the present day. Its narrow shape facilitates observations of marked breeders from the nearby observation tower. It was reduced in size because the nesting population had reached a point where conflicts were developing with local agricultural interests.

Figure 60. Construction of flamingo island at the Etang du Fangassier, Camargue, February 1970. Since its colonisation by flamingos in 1974, over 150,000 chicks have fledged from this site.

island. The mounds were by now sunbaked and cracked, but that was of no importance. About 1,500 pairs of flamingos bred there successfully. A visit to the island after the breeding season revealed that only an estimated 5% to 10% of the false mounds had been occupied, most pairs having preferred to lay their eggs between them, on the flat island or on the scant mound they themselves had made. Was colonisation of the site coincidental to the presence of the fake nests or had the flamingos really been fooled into believing that the mounds were the fruits of their own labours from past years?

The following winter, a further 350 mounds were moulded on the island, 60 m from the first set, and in spring 1975 the sequence of site colonisation was carefully observed. First, the island in Fangassier I was again filled to capacity, but because of erosion, only 2,600 pairs could nest there as opposed to 4,900 pairs in 1969. Next, the artificial island was colonised, at first on the part used the previous year, and then in the area of the second group of mounds, after which the flamingos proceeded to colonise other areas. Our observations left little doubt that the birds had been lured by the mounds.

Since 1976, the last year that the island in Fangassier I was colonised, the Fangassier II island has been the principal if not sole flamingo breeding site in the Camargue (in most years hundreds or thousands of pairs attempt to breed on the dyke nearby but very few are successful). By the end of the 1970s, the Fangassier I island, which prior to being colonised by flamingos had been host in the 1960s to hundreds of breeding gulls, terns and Avocets, was seriously eroded and submerged during windy weather.

The new island was made big enough to accomodate the largest-projected colony of flamingos (8,000 pairs bred in 1960) and some allowance was made for erosion. In 1977, the island was filled to capacity when the number of breeding pairs reached 9,376. Three years later numbers at the Fangassier again increased, and peaked at 20,000 pairs in 1986. This increase was possible because when the number of breeding pairs exceeds $c.$ 10,000, some birds either form a second colony on the neighbouring dyke, or nest in the depressions between the mounds. In years of early breeding, some pairs even reoccupy mounds vacated by the chicks of the earlier breeders. In the first two cases, these are poor-quality nest sites, birds on the dyke being more exposed to high winds than those on the island, to heavy predation by gulls and to disturbance by potential terrestrial predators. Those occupying depressions are vulnerable during heavy rains when they may not only lose their eggs to flooding, but may become so inexorably coated with mud that they are unable to stand or to open their wings and the depression becomes a deathtrap. More than 50 adults have died in this way in one season (1987) in the Camargue (Johnson 1991), and these depressions are more usually deathtraps for many chicks each year.

The island has been restored three times, in 1985/86, 1988/89, and 1995/96 (Plate 23). In 1985/86, it was more seriously eroded on the northern side, where it is most exposed to deterioration by waves when the Mistral blows. The nests on the more sheltered southern side were left undisturbed so that they would act as decoys the following season. The shores of the island were regraded and reinforced with stones. The following spring, the birds were able to settle in even more densely and 14,000 pairs nested that year. Three years later, however, the island was again

Figure 61. Flamingo island at the Etang du Fangassier, Camargue. The first batch of imitation mounds were placed on flamingo island at the Etang du Fangassier, Camargue, in March 1974. Three months after this photograph was taken flamingos bred at this spot, but they preferred to nest between the mounds. (F. Mesleard)

seriously eroded, owing to heavy rains falling before the new earth was compacted, and to a series of wet springs.

During these restorations, the island's shape has been modified slightly (Figure 59) in order to facilitate observations of banded breeding birds from the nearby tower hide, and it has been reduced in size to 4,000 m². This was decided in an attempt to avoid aggravating the conflict between rice farming and nature conservation in the Rhône delta. Restoration in 1988/89 involved flattening the 10,000 natural mounds, which covered the entire surface of the island, thus filling the depressions between the nests. Following this, artificial mounds were once again placed on the island during the winter, in groups of 20–30. The flamingos returned in spring 1989, and the earlier breeders prospecting for a nest site were clearly attracted to those parts with mounds.

If further confirmation of the role played by the mounds was required, observations in 1996 and 1997 were quite convincing. During the third restoration in autumn 1995, the island was again levelled by caterpillar tractor but no fake nests were put down. The flamingos nevertheless returned in spring 1996 and began breeding. However, before they had time to make mounds, there was a very heavy downpour, and 2,500 eggs which had been laid on the flat ground were suddenly floating in wet mud. After one or two days, the island was completely abandoned and the colony moved to a neighbouring island and dyke. Even though the birds bred successfully, this new situation satisfied neither the salt company, who had paid for the restoration of the flamingo island, nor the researchers, who could not observe the colony from the tower hide (and who wanted the flamingos' new nesting site to be occupied by terns, for which it was built, rather than flamingos). Therefore, in February 1997, an effort was made to lure the birds back to 'their' island, and 16 willing helpers from the Tour du Valat moulded a total of 1,625 mounds on the flamingo island, again in groups of 20–30. It was a hard morning's work, for the mud had to be collected from the water, which meant that the mounds were built of slurry, but in spite of the many imperfections the mounds clearly made the surface of the island more attractive, and two months later the flamingos were back.

Over the past 27 years (1974–2000), more than 150,000 chicks have taken wing from this purpose-built flamingo sanctuary. The costs of restoration and maintenance were covered in both 1989 and 1995 by a grant from the owners of the land, the Compagnie des Salins du Midi et des Salines de l'Est (CSME) (now Les Salins). The water levels in the Fangassier, as elsewhere throughout the saltpans, have never been manipulated to favour flamingos in any way, but are controlled solely in the interests of the salt industry.

Providing the Greater Flamingo with a near-perfect home, and restoring it when it suffers from erosion, has allowed the species to nest in the Camargue more frequently and in greater numbers than in the past. However, although this management action has clearly had a major impact not only on the number of flamingos in the Camargue but also in the Mediterranean as a whole, breeding would have been less successful had the colonies not been wardened and the Yellow-legged Gulls not culled at a time when they were a threat, as discussed earlier in this chapter.

Provision and maintenance of breeding islands at Fuente de Piedra

The harvesting of salt and the presence of flamingos at Fuente de Piedra date back to Roman times (Muñoz & Garcia 1983). Prior to the 18th century, however, breeding by flamingos was undoubtedly a very rare event because the only islands were close to the shore and probably surrounded by water only following winters of exceptionally high rainfall. At the end of the 19th century, salt extraction was intensified, and in 1876 the lagoon was divided into a mosaic of small pans separated by canals and dykes (Figure 62). The wetland was managed as saltpans until 1951 (Muñoz & Garcia 1983), when salt extraction ceased and the lagoon was abandoned. The dykes gradually deteriorated and became islands which, following winters of good rainfall, were soon colonised by breeding flamingos (Valverde 1964). The site thus became one of major international importance for flamingos and is host to many other species of waterbirds when flooded.

Management practices at Fuente de Piedra have included both clearance of vegetation to facilitate colonisation by flamingos and reshaping of the breeding islands (the former dykes) in order to reduce erosion and to increase their carrying capacity (Rendón Martos & Johnson 1996). In the Camargue, a colony was simulated by adding mud in the shape of mounds, whereas in Fuente de Piedra, a similar effect was achieved by removing mud from between the mounds. This method has the advantage over moulded nests of allowing rainwater to be retained, so that the flamingos have mud on hand to model their mounds. Broken eggshells (of either flamingos or chickens) scattered around the 'nests' have also proved to be effective in attracting the flamingos.

Fuente de Piedra is a nature reserve and is managed primarily as a flamingo sanctuary but breeding can occur there only following periods of abundant rainfall from October to February (Rendón Martos & Johnson 1996) (see Chapter 8). The lagoon is now protected under several national and international directives.

Irrigation for chicks

Most years when flamingos are able to breed at Fuente de Piedra, the lagoon dries out a few days or even weeks before the chicks are able to fly, generally in June. Since 1978, water has been pumped into the lagoon most years to provide the chicks with a small pool (maximum 5–6 ha) where they can drink and bathe (Vargas Yáñez *et al.* 1983; Réndon Martos & Johnson 1996). The water is taken from a well and piped 3.5 km at 12–15 l/s. Pumping begins only when there is no water left in the vicinity of the breeding island, and continues until the last chicks are able to fly. During dominant south-easterly winds, the water is retained by a low ridge of earth.

212 The Greater Flamingo

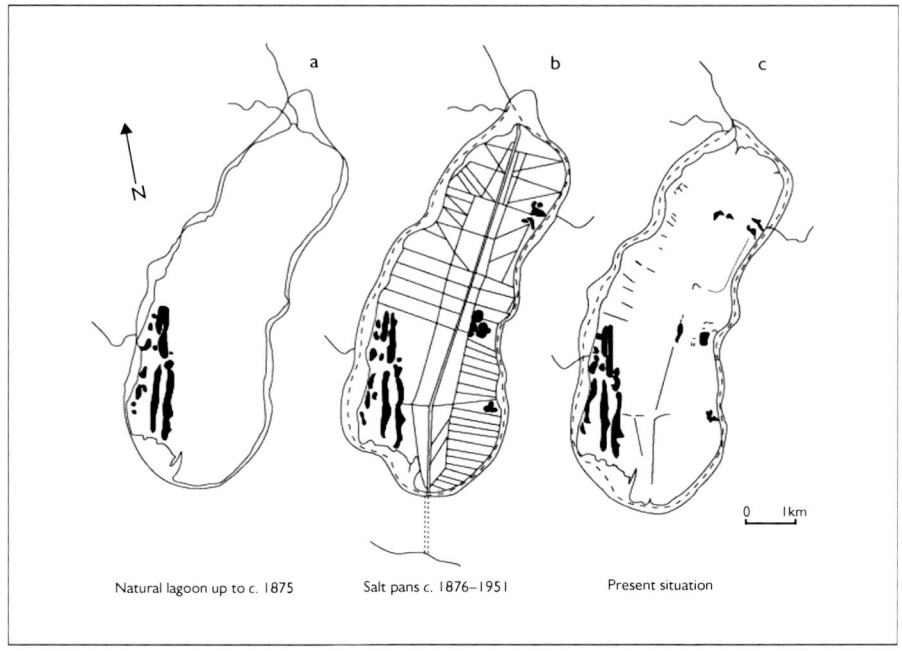

Figure 62. Map of the Fuente de Piedra Lagoon. Flamingos could not formerly breed at Fuente de Piedra because as a natural lagoon (a) there were no suitable islands, and as saltpans (b) it was too disturbed. After the saltpans were abandoned, the dykes eroded (c) and flamingos soon began to breed. A similar situation prevails at Molentargius in Sardinia. Fuente de Piedra is now managed as a flamingo and wildlife sanctuary (from Rendón-Martos & Johnson 1996).

RESCUE AND REHABILITATION

Humans are generally very sensitive when birds are exposed to suffering from illness, pollution, injury or when they are weakened by extremely harsh conditions, and the flamingo perhaps more than other species attracts attention. All flamingos throughout the world can be considered flagship species since their well-being gives rise to concern, and any negative events are widely reported in the press. Extreme weather, poisoning (from botulism or heavy metals) or chicks imprisoned in fatal soda 'anklets' in hypersaline environments are phenomena which can be widely covered by local, national and international press.

Soda 'anklets' and chick rescue operations

When flamingos breed in natural saltpans which are subject to very high evaporation rates, chicks face the danger of starvation if the pan dries before they fledge. In some places they may also be subject to the brine crystallising in deposits of soda on their legs as the pan dries out. This phenomenon has been reported for Lesser and/or Greater Flamingos from East and southern Africa: from Lake Magadi in Kenya (Brown & Root 1971), from the Etosha Pan in Namibia (Berry 1972) and from the Makgadikgadi Pans in Botswana (Liversedge pers. comm.). Chicks have no means of ridding themselves of these salt deposits, which day by day become increasingly cumbersome. Huge numbers of deaths, mostly of Lesser Flamingos, are occasionally reported, as for example at Lake Magadi in 1962 (Brown 1967). On that occasion the Wildlife Society of East Africa paid salt-workers to catch chicks and to break off these apple-sized 'anklets', which they did for a phenomenal 27,000 fledglings. Other handicapped young were driven towards pools of less alkaline water where it was hoped that the 'anklets' would dissolve, and it is estimated that 220,000 chicks were saved from almost certain death. Similar situations have developed at the Etosha Pan, Namibia, in 1963 (Winterbottom 1964 in Simmons 1996), when thousands of chicks were involved, and again in 1969 when *c.* 20,000 stranded young were saved on this pan (Berry 1972). In 1986, 1,200 of 5,220 chicks were rescued but none survived (Archibald & Nott 1987). Greater Flamingos seem to be less affected by soda anklets than Lessers because they tend to breed earlier (Berry 1975; Williams & Velásquez 1997), so that most chicks fledge before the breeding site dries out.

Another chick rescue operation took place in South Africa in 1978. Boshoff (1979) reported 623 Greater Flamingo chicks being captured at Nedersettingsraad dam, Van Wyksvlei (Cape Province) when this site was drained in February. The chicks were transported to Klipsbank dam in the Carnarvon District, where they were released alongside feeding adults. Most are reported to have survived. More recently, in April 1994, 144 Greater Flamingo chicks were captured when hundreds of others died as the Etosha Pan dried out (Fox *et al.* 1997). The birds were hand-reared and 77 of them were released at Walvis Bay. Subsequent observations suggested high mortality soon after release, mainly, it was thought, from predation.

As tragic as it may seem, massive breeding failures are a natural phenomenon in the flamingo's world. The birds have known such events for millions of years, and compensatory adaptations enable them to take such losses in their stride (they are still here to prove it). Flamingos are long-lived and records of exceptionally good breeding seasons are as numerous as reports of large-scale failures. Although it may seem barely acceptable to many people to stand by and not intervene when wild creatures suffer from starvation, well-meaning rescue efforts may even cause long-term damage. From a conservation viewpoint, the usefulness of such rescue operations has been challenged. Simmons (1996) doubts whether many young survive to recruit to future generations, while the trucks used to transport chicks in Namibia have scarred the pans with their deep tracks.

In the Palearctic region, conditions are not quite as extreme as in Africa, and chicks in Tuz Gölü in Turkey trek across the drying saltpan to the nearest stream inflow which is permanent. In Spain's Guadalquivir Marshes, however, the colony surroundings are completely different, with an abundance of higher vegetation. The park authorities have several times organised chick rescue operations, which have involved cutting a path through the vegetation so that guards mounted on horseback could drive the nursery of chicks to the nearest water body a few kilometres distant. Thanks to these efforts 1,500 chicks fledged in 1977, 2,500 in 1978 and 3,500–3,800 in 1984 (Mañez 1991).

Severe weather

Some of the effects of the January 1985 cold spell on the population of flamingos wintering in Mediterranean France have already been discussed (escape movements in Chapter 5, 'Causes of post-fledging mortality' in Chapter 9) and those aspects of this major event are not repeated here. From the onset of that notorious cold spell, people from all walks of life spontaneously tried to help flamingos, by breaking ice to maintain open water, by feeding birds on unsorted rice grain (hunters did so in particular) at the few open water holes at which flamingos were gathered, by catering for weak birds, or collecting corpses. Six nursing centres were hastily set up along the coast, where victims were placed in rooms or shelters. Survivors were placed near buckets of water, or put in troughs, when available, containing rice grain and dried food normally fed to poultry and dogs. Quite remarkably, birds which could stand on their legs fed almost immediately, seemingly unconcerned by their new and bizarre surroundings. Birds which were too weak to stand were administered a solution of sodium chloride (1.5 g per litre of solution), potassium bicarbonate (2 g), sucrose (30 g) and potassium gluconate (7.5 g = 2 g potassium carbonate). This is the solution given in hospitals to children suffering from undernourishment or dehydration. Those birds which were not beyond recovery were very soon back on their feet and feeding themselves alongside stronger birds, and many flamingos were saved from starvation.

While in captivity, 53 adult flamingos were marked with PVC bands (Johnson & Green 1990). They were released at the end of the cold spell and the majority survived. The 15% which were estimated to have died shortly after release were generally lighter and gained less weight while in captivity. Only three months after the cold spell at least 10 of the 45 known survivors bred, while the following year at least 26 did so (Johnson & Green 1990), demonstrating the efficiency of the rescue operations.

In the Mediterranean region, long spells of severely cold weather are rare and quite unpredictable. Those of February 1956, December 1962/January 1963, January 1985 and January 1986 were the four most severe cold spells in southern France during the second half of the 20th century.

Protection of rice fields to reduce crop damage by flamingos

In an attempt to both minimise crop damage and avoid a clash between farming and conservation interests in the Camargue, several partners collaborate each spring in order to discourage flamingos from foraging in rice paddies, which they usually do at night. The first and most intensive scaring campaign took place in 1981 (André & Johnson 1981; Hoffmann & Johnson 1991; Johnson et al. 1997). The complex of rice fields in the Camargue and to the west of the Petit Rhône were all visited by a team of technicians each evening, from an hour before sunset until midnight, and during the whole of the period when the flamingos are most likely to visit the paddies (25 April to 5 June). Flocks of flamingos arriving in the fields, or sometimes already present, were scared by flares or shots fired from a Verey pistol. Problem areas were identified and the following day a bird scarer was placed in the field. Mounted on a tripod, these twin-barrelled propane-powered scarers give a rapid double report, reminiscent of a shotgun, but without ammunition. Firing in opposite directions at about 12-minute intervals, the canon rotates slightly with the momentum from the detonations, thus being heard in all directions. These canons were, and still are to a large degree, effective in scaring the birds.

Many other techniques were tried but for various reasons were found to be unsuitable. Helium-inflated weather balloons, reflecting tape, scarecrows, scarey men and carborundum bird-scarers all proved inefficient for one reason or another (see Hoffmann & Johnson 1991). Orange warning lights have since proved to be quite effective and their flashes can be seen in many rice paddies on spring nights.

Although costly, the 1981 scaring campaign proved to be effective, and little or no damage by flamingos was reported from the rice fields. Over the following years the same 'artillery' was distributed each spring to the farmers and the problem seemed to be contained within limits. Even in 1986, when there were 20,000 pairs of flamingos breeding in the Camargue, twice the number recorded in 1978–80, no crop damage was reported by farmers. The following year, however, some birds again began visiting the rice fields, and although the problem has not reached the proportions of 1980, it has nevertheless become an annual scourge for the farmers.

The phenomenon has received much attention over the years in local, national and international press and is a topic of heated discussion. Scaring devices have proved effective with flamingos because the problem period is relatively short (six weeks) compared to that of the analogous problems of crop damage by geese in northern Europe (up to six months). Scaring is also most effective when there is human presence, but modernisation of farming since the 1970s has of course reduced this presence, and the birds have become less shy.

Flamingos and the salt industry

Saltpans occur throughout much of the flamingo's range and this habitat, as we discussed in Chapter 2 and earlier in the present chapter, is widely used by flamingos for both foraging and breeding. The creation many decades ago of the extensive areas of shallow water necessary for salt production brought about considerable change in the landscape by stabilising, for at least several months of the year, the level of many formerly temporary lagoons. This management, however, is generally well suited to the needs of the species of invertebrates and vertebrates which are adapted to hypersaline environments, and there are very few parts of these modified areas which are too salty or too deep for flamingos to forage in.

Although there may be functional differences among saltpans throughout the world, particularly in the way in which the salt is harvested, they have in common the fact that they lie in areas where evaporation rates are high and exceed precipitation during much of the year. In the tropics, evaporation may exceed precipitation throughout the year, but in the more northerly saltpans in Europe, as in southern France, evaporation generally exceeds precipitation for only six months of the year and salt is produced from March to September. In some of the remaining artisanal saltpans, rainwater falling on the pan absorbs the edaphic salts, and thus within the same lagoon the salinity may vary from almost nil when extensively flooded, to brine and finally to crystallised salt as the pan dries. Most saltworks are comprised of a series of lagoons, each one lying within a much narrower salinity range. The distribution of flamingos has to some extent been influenced by the advent of saltpans, and this habitat, although artificial to varying degrees, has become important to flamingos during much of the yearly cycle.

Saltpans are part of the cultural heritage of the Mediterranean, the oldest-known being Egyptian and dating back to 5000 BC. Sadoul *et al.* (1998) censused 168 saltpans covering over 1,000 km^2 in 18 Mediterranean countries. Although such wetlands are artificial, at least in terms of the flooding regime, they are of great conservation value and the importance of saltpans to wildlife was recognised long ago by Hoffmann (1964a). Water levels stabilised over a period of months provide an abundant and largely predictable supply of invertebrates, and the islands and dykes within these wetlands attract not only flamingos but also Common Shelducks and a variety of breeding gulls, terns and waders, Avocets in particular. Moreover, the lack of tall vegetation on this landscape limits its attraction for foxes, Wild Boar or stray dogs, the main terrestrial predators at flamingo colonies. In this way the salt industry provides a vital habitat for these species. All but the harvesting pans are rich in invertebrates, and therefore ideal habitat for flamingos to feed and also, in some places, to breed. The Fuente de Piedra Lagoon in Andalusia, described above, is a good example of how the salt industry has contributed, albeit incidentally, to the present favourable conservation status of flamingos in the Mediterranean.

In Sardinia, the recently established colony at Molentargius, occupied by 1,000–4,000 breeding pairs of flamingos (1993–2000), is also in inactive saltpans, where dykes have become islands. Here the situation is slightly different, however, because the islands have been created by a rise in water level due mostly to human effluent. Salt production was stopped in 1984–85 because of the polluted water.

Being an ideal habitat for flamingos, saltpans have been a major influence on the status and distribution of the species, particularly in the Mediterranean region, where half the colonies owe their existence to the salt industry. Although saltpans are occasionally reactivated after a period of non-exploitation, and new pans have recently been created in the Mediterranean region, the present tendency is to close them for economic reasons. From 1985 to 1990, 7,000 ha were lost to urban development (Sadoul et al. 1998). As mentioned above, inactive saltpans, as long as they remain flooded and are not unduly disturbed, can be used by flamingos.

CHAPTER ELEVEN

Conclusions: what does the future hold?

The research programme on the biology and conservation of the Greater Flamingo which began in the Camargue more than 50 years ago is unique in many ways. Embarking on such a long-term programme involved tremendous human and material investment and the board of directors of the Tour du Valat has been supportive for many years, patiently awaiting the dividends of this investment. Their patience and encouragement have been rewarded in the long run. Since the PVC-banding scheme started, the number and quality of publications have progressively increased, and within the last ten years several publications have appeared in leading international journals of ecology and ornithology. In other circumstances, the usual criteria under which research projects are evaluated and financed would probably have sounded a death knell for the study before now. Nevertheless, such studies are indispensable to the argument in favour of the effective and coherent management of our natural heritage, particularly in the case of long-lived species. The Greater Flamingo now stands as one of the few long-lived bird species for which extensive analyses of demography, behavioural ecology and life history are available, and the research results achieved in the Camargue are familiar to many population biologists, amateur and professional ornithologists, and conservationists worldwide.

A great deal of our current knowledge of the demographics and behaviour of the flamingo comes from the Camargue study, but there is still much to do. It is therefore essential that the study of the Camargue population continues over at least the next 10 years and that the same quality of data collection is made on banded individuals, and that international cooperation increases within the species' range, in concert with rational conservation-oriented management. This concluding chapter explores possible ways to integrate the pursuit of scientific investigation with a responsible conservationist attitude. We first emphasise the need to improve our understanding of the flamingo's biology and behaviour, and how this can be achieved through maintaining long-term monitoring and fundamental research. Secondly, we discuss how the connection between research and conservation could be strengthened, especially by considering the problem of managing artificial colonies on a larger geographical scale. In both sections, we discuss the importance to flamingo species elsewhere in the world of the research and conservation actions developed in the Mediterranean region.

RESEARCH IN THE FUTURE

There are still many interesting fields of study and research which could be carried out on Greater Flamingos by both amateur and professional biologists. One important contribution by amateur naturalists is the reporting of sightings of banded flamingos. These sightings often provide the raw data for major ongoing studies of many aspects of flamingo biology, for example survival analyses, movements, foraging. The Camargue database is unique, and currently contains over a quarter of a million observations of banded birds. Recently, a particular effort has been made at Tour du Valat to develop new software that allows the joint storage and management of flamingo banding and resighting data in the Mediterranean region. This unique database is permanently available for consultation and analysis as new ideas are formulated, and in the future should facilitate cooperation among research teams working on flamingos. Several areas of research could benefit from such intensified cooperation.

FLAMINGOS AND THEIR ENVIRONMENT

Our understanding of the trophic interactions between flamingos and other constituents of their ecosystems remains limited. The influence on population dynamics of major or subtle changes in wetland function and management is still difficult to predict. At the same time, the real impact of flamingo populations on their environment deserves further consideration. This impact is already cause for some concern for the conservation of the Greater Flamingo. In the Camargue and the Ebro and Guadalquivir deltas, there are conflicts between rice growers and flamingo conservationists because of crop damage caused by feeding birds (Hoffmann & Johnson 1991; Johnson et al. 1997; Tourenq et al. 1999). Such conflicts are not limited to agriculture. It has been suggested, for example, that the activity of flamingos may alter other aspects of natural ecosystems (e.g. invertebrate populations; Glassom & Branch 1997a,b). These concerns lead to the question of how many flamingos an area can support before it suffers adverse effects. Although the definition of 'adverse effects' can be fairly subjective, the better we understand potential environmental impacts, the better we shall be able to address and perhaps minimise any conflicts that arise.

Previous studies have emphasised the importance of commercial salinas as feeding areas for flamingos in the Camargue (Johnson 1973; Britton et al. 1985; Britton & Johnson 1987; see Chapter 6). However, an investigation of the relative importance of other, less stable habitats for both breeding and wintering birds remains to be carried out. Moreover, the mechanisms of foraging-habitat selection by flamingos remain elusive. The importance of prey density (see Tuite 2000) and diversity of prey types on patch exploitation needs further study. It is also important to more fully document the relationship between individual variations in the exploitation of foraging habitat and use patterns observed at the population level. Although the composition of foraging flocks suggests some competition between age classes in exploitation of foraging areas, we do not know whether young birds are forced to exploit marginal habitats or if foraging preferences change with age. Combining resightings of banded birds on foraging grounds with radio-telemetry studies would certainly be of help in the future to better document this important aspect of flamingo ecology. Finally, estimating feeding efficiency in various habitat types and in relation to the size of foraging flocks may lead to a better understanding of foraging decisions in flamingos.

The impact of flamingos on animal and plant communities in their various foraging habitats also needs further study (see Casler & Esté 2000, Esté & Casler 2000). We still do not know the extent of competition between flamingos and other waterfowl species that feed on aquatic invertebrates. We have little idea of how the structure and functioning of freshwater and saltwater marshes might be modified by a reduction in flamingo numbers. This is one line of research for which combining experimental work and field observations is critical (see Bildstein et al. 2000). Captive birds could be used to document the ability of flamingos to distinguish between various prey types or patches of different quality. Where possible, exclosure experiments on foraging grounds (see Hurlbert & Chang 1983) would help us to estimate the effects of foraging by flamingos on prey communities. The same procedure could be used to study the effects of trampling by flamingos on submerged plant communities, particularly in rice fields (Montes & Bernues 1991). Such experimental approaches can be extremely productive and are sorely lacking in ecological research.

METAPOPULATION DYNAMICS

Over the last ten years, the flamingo database has been extensively analysed, using refined statistical techniques developed for the analysis of capture-mark-recapture data (see Chapter 9). Until recently, only single-site models were developed, but progress in modelling now allows the simultaneous consideration of several sites. In particular, it is now possible to directly compare the values of essential demographic parameters, such as survival and recruitment, between breeding sites. In this respect, the establishment of a Mediterranean flamingo database certainly constitutes a valuable advancement for the demographic study of flamingos in a larger spatial context.

We do not know yet if the birds we observe are all part of a single large metapopulation, or if they belong to more discrete sub-units. As a complement to modelling, an important tool for the future will therefore be the use of genetic markers, such as microsatellites, to document the degree of connection between distant breeding populations. Data from banded birds are particularly efficient in documenting movements between sites where a high rate of exchange takes place. However, they perform poorly if the flow of 'migrants' is low. Nevertheless, just a few migrants per generation are enough to homogenise the genetic composition of populations. In this respect, data on population genetics are urgently needed to assess to what extent, for instance, flamingos breeding in the Camargue are isolated from those breeding in East Africa or Iran. The study of Greater Flamingos in a metapopulation framework certainly is a stimulating challenge for the coming years.

AGING AND SENESCENCE

In the future, the flamingo may prove to be an ideal species for studying the process of aging and senescence. The Greater Flamingo stands out among colonial waterbirds because of its exceptional longevity, with which only Procellariiformes can compare (Warham 1990). Data from captivity indicate that female birds aged over 50 years are still able to lay eggs. Estimates of adult survival (see Chapter 9) suggest a life span exceeding 60 years. However, those estimates are based on birds banded since 1977, which means that at the time this book was written, the oldest birds were 'only' 29 years old. We do not know yet if signs which indicate senescence, such as an abrupt decline in survival rate or a sharp decrease in reproductive success, will be observed in flamingos. Recently, a growing number of studies have detected evidence of senescence in natural populations of various bird species (Gustafsson & Pärt 1990; McDonald *et al.* 1996; Møller & de Lope 1999; Nisbet 2001; Robertson & Rendell 2001; Saino *et al.* 2002). However, variation in longevity is extremely wide among bird species, from only two or three years for some passerines, to 20 times longer for flamingos and several seabird species. We do not know at present which factors are responsible for the exceptional longevity of flamingos. One possibility is that their carotenoid-rich diet provides them with an efficient protection against free radicals and oxidative damage that appear to be involved in the process of senescence (Holliday 1995). Future research in that direction would benefit both from long-term studies in the field, in Spain and in the Camargue, and from research of captive flocks in zoological gardens where very old birds can be studied.

DEVELOPING LONG-TERM RESEARCH ON OTHER FLAMINGO SPECIES

The Wetlands International/IUCN-SSC Specialist Group meeting that was held in Miami in 1998 (Baldassarre *et al.* 2000) clearly showed the asymmetry that exists in the extent of scientific knowledge of different flamingo species. While data on movements and areas of importance for foraging and breeding are rather well documented for most species, valuable data on demography, mating system, and breeding biology of wild populations exist only for the Greater Flamingo. For the four other species, knowledge is generally limited to the observation of captive flocks. The use of captive flocks to document flamingo behaviour and reproduction is certainly a valid approach, but one should keep in mind that the conditions under which captive flocks are held are very different from those encountered by wild birds. Birds in zoological gardens are protected from predators and, to a lesser extent, from pathogens. They do not have to cover large distances to find suitable conditions for foraging or breeding. Captive flocks are of limited size (usually under 100 individuals), in marked contrast to the huge aggregations which occur around traditional breeding sites. Some aspects of the behaviour of flamingos observed in captivity might not be relevant to birds in the wild. Studies of the mating system (see Chapter 7) have shown that captive flamingos maintain pair bonds over consecutive breeding attempts, whereas wild birds do not. The discrepancies between the behaviour of captive and wild birds might therefore pose a limit to the usefulness of captive flocks as a surrogate for a deeper understanding of the ecology and behaviour of flamingos. In addition, while there is still much to learn on important subjects such as numbers and distribution, Baillie (1990) suggested that for bird monitoring schemes to be of value to conservation, they must do more than simply chart the fluctuations of bird populations. Our research must be targeted towards helping future management and conservation initiatives to realise their greatest potential, and to understanding the possible effects of these initiatives, both locally and globally. While estimates of demographic parameters may indicate the status of a population, from a conservation standpoint they do little to help us to understand the causes of observed patterns. However, modern techniques using capture-mark-recapture data go beyond the mere estimation of demographic parameters and provide a powerful tool to test biological hypotheses about the various sources of variations that may affect survival and dispersion.

It thus seems particularly important to develop or expand banding programmes and monitoring for other flamingo species. Techniques that have been developed in the Camargue can be exported with success to other areas of the world. An important aspect will be the coordinated training of researchers and technicians, particularly in other countries, to ensure successful capture and marking operations. Tools for data management developed in the Camargue have recently been exported to Mexico, and will be used in research on Caribbean Flamingos. Because flamingos are long-lived, it is particularly important to engage early in long-term research programmes and to make sure that such programmes benefit from sustained funding in the future. Interspecific comparisons will be of great help in understanding the evolution of flamingo life-history traits, and depend ultimately on the collection of long-term data of comparable quality for all five species of flamingos.

CONSERVATION ISSUES

Even apart from the widespread general use of the flamingo as a decorative image, many conservation groups use the bird as their logo, and several countries have issued postage stamps featuring flamingos. Their extreme popularity makes them ideal flagship species for the conservation of wetlands, and it was not surprising that one of the banners displayed at the WWF's 25th anniversary campaign conference at Assisi in 1986 depicted a flamingo. Many wetlands have been protected primarily because of their importance to flamingos, for example the Camargue National Nature Reserve (France), Fuente de Piedra Lagoon (Spain) and Djoudj National Park (Senegal). However, while it might be easier to launch conservation plans for flamingos than for some other vertebrate species, conservation of flamingos in the 21st century is still a challenging task. The utility of conservation actions has sometimes been questioned, and answers must be given to justify these initiatives. The successful conservation actions developed for the Greater Flamingo in the Mediterranean region should not make us complacent about the species' conservation status in other key parts of its range. Neither should we forget that for other flamingo species, conservation issues have only recently been addressed.

IS THE GREATER FLAMINGO AN ENDANGERED SPECIES?

Conservation organisations (e.g. BirdLife International, IUCN) promote action plans, Red Data books and other awareness documents for species whose populations give cause for concern, i.e. because they are declining, or because they have a restricted distribution. As we have seen from population trends (Chapter 4), the Greater Flamingo is not globally threatened (Collar & Stuart 1985; Collar & Andrew 1988), but in parts of its range population declines have been reported, for example in southern Africa (Simmons 1996). Like many colonial birds, flamingos are vulnerable because they nest in so few places and because they are dependent on wetlands which are shallow and coastal. Much of their habitat is therefore at risk, and consequently so are the birds themselves (Johnson 1992).

To determine the species' vulnerability, it is crucial to understand demography at the metapopulation level. While fluctuations in numbers, declines, and even local extinctions might not be desirable, their importance should not be overestimated. What is of prime importance is the stability of the flamingo population throughout the Mediterranean region. Sites that have been abandoned can be colonised again, as was certainly the case when artificial breeding sites did not exist. Again, the establishment of conservation actions and priorities must be based on reliable scientific information. Flamingos might be at risk in some parts of their distribution, and particularly so if local populations are isolated. Documenting the degree of connectivity between populations is therefore a conservation priority for the future. In the short term, it is important to stress that the future of flamingo populations is closely linked to the future of Mediterranean wetlands. Thanks to conservation efforts and public campaigns, public awareness of the value of wetlands is progressively increasing. In many places, however, flamingos exploit habitats that have been modified or are still managed by humans. Economical pressures on certain activities, such as the salt industry, are like the sword of Damocles above flamingos' heads. Large-scale conservation of Mediterranean wetlands, including the maintenance of traditional activities where possible, might well be the best insurance against the decline of flamingo populations.

Are there too many flamingos in the Mediterranean region?

We have seen from data presented in Chapter 4 that there has been a significant increase in the number of flamingos in the western Mediterranean over the past 25 years, and we have seen in Chapter 10 that flamingos now feed in varying numbers each spring in the rice fields of the Camargue and in Spain, where they can cause considerable crop damage. Further, although there is little evidence to support extensive damage to marshes by flamingos, they have been accused of destroying submerged plant communities and consequently to have had a negative impact on marshland ecosystems in Spain (Montes & Bernués 1991). For the same reason, they are unpopular with some hunters in the Camargue, and with some of the fishermen of the Languedoc lagoons, because of the negative effects on fish catches caused by turbid waters around fishing nets. In addition, flamingos now feed in wetlands which are very close to human presence, which they formerly did not. Are these phenomena signs that there are too many flamingos in the Mediterranean region, at least around the western end?

The answer to such a question may well be 'yes, there are too many flamingos', but this surely depends on who is asked. People in general, and visitors in particular, are thrilled to see flamingos when they visit Mediterranean wetlands, and problems linked to the birds' high numbers are few. Other people, such as rice farmers who suffer from damage caused by flamingos, are more likely to express their dissatisfaction and call for regulation of flamingo numbers, at least at a local level. However, the regulation of flamingo populations is a complex problem. Given the generation time of flamingos, their population multiplication rate is much more sensitive to survival after the first year than to fecundity (see Lebreton & Clobert 1991). This means that a reduction in fecundity (breeding success) of, say, 20% would have little consequence on the population demography in the long term, whereas a reduction in adult survival of the same magnitude would have more severe effects. This leaves little opportunity of regulating the population dynamics of flamingos, if such regulation is needed. It would be very difficult to cull eggs without disturbing the whole colony, and the risk of complete desertion would be very high; in any case, a reduction in fecundity would have little effect, if any, on the local densities of flamingos. On the other hand, killing adults is not acceptable on ethical grounds, and regulation devices based on massive culling of adults would almost certainly cause a public uproar. Since flamingos are protected in most parts of their distribution range, only climatic events and severe pandemics are likely to regulate adult survival (see Chapter 9). Cold spells are unpredictable, and no preventive action can be taken to limit their effects on populations of wild birds. It is not clear, however, at what frequency cold spells occur in the geographical range of the species. They are certainly limited to the northern part of the range, and may discourage the flamingo's expansion to the north. Flamingos were severely affected by the cold spell that occurred in the Camargue in the winter of 1984/85 (see Chapter 9). However, the population recovered quickly, at least judging from the number of breeding birds in the following breeding seasons. This was achieved through the recruitment during the 1985 breeding season of young birds (aged three, four and five years old) that typically would have started breeding later in life.

Outbreaks of highly pathogenic viruses are also difficult to predict and control. Experimental studies indicate that the H5N1 virus (see Chapter 9) could infect various avian species and that its virulence varies significantly according to species, even between closely related ones (Perkins & Swayne 2003). For the present, the enhancement of biosecurity procedures on local poultry farms and in retail and wholesale live-bird markets has been sufficient to contain outbreaks of epidemics in south-east Asia. However, the presence of the virus in wild migratory waterbirds causes major concerns for the future. Because of their nomadic movements and their colonial lifestyle, flamingos might be particularly exposed to pandemics.

Because the few problems associated with flamingos are linked to high local densities, the regulation of populations would be best considered on a large spatial scale. It is important to remember that in the 1960s, the situation in the Mediterranean region was not so favourable to flamingos, and it was that which prompted Tour du Valat to take conservation actions such as the creation of an artificial breeding island. Since then, other more or less stable sites have been exploited by flamingos as breeding sites. Obviously, the problem of regulating flamingo numbers is now closely associated with the coordinated management of both natural and artificial breeding sites.

ARTIFICIAL COLONIES AND CONSERVATION: A PARADOX?

The Camargue research and conservation project must surely be considered a success, but at the same time one must draw attention to the other side of the coin. Because the flamingo is an attractive bird, naturalists in several parts of the Mediterranean have tried to copy the Camargue programme in an attempt to encourage flamingos to breed regularly. Such projects generally meet with strong popular support because the spectacle provided by these birds suggests the potential for profit from development of local tourism. The idea of creating breeding islands in different parts of the species' range usually stems from good intentions, but such provision of artificial sites could do more harm than good for the conservation of this species. To illustrate this point, one may reasonably ask whether the use of rice fields as foraging areas is not quite simply a result of overcrowding, with the principal culprit being the island at Fangassier! At the same time, another species which occurs on the same wetlands as flamingos, but for which no measures other than protection have been carried out, is the Common Shelduck. This species also inhabits salinas and, in the Camargue has increased at an even greater rate than the flamingo—from 50 individuals in 1956 to 500 pairs in 1986 (Isenmann 1993), since when the breeding population in the south of France has remained stable (Walmsley 2006). It is hard to imagine an increase in the Common Shelduck population sufficient to have a negative impact beyond its traditional locations, although the species does now appear in small numbers in the rice fields of the Camargue.

The construction of the breeding island at Fangassier has been a conservation action widely appraised in France and elsewhere. The extent to which one should intervene to make the island suitable for breeding every year is a decision that can reasonably be challenged. Flamingos breed in the Camargue every year because on one hand there are many permanent bodies of water providing food, and on the other they have a suitable island. Even in years of low rainfall, water is abundant in the Camargue as a result of irrigation. In no wetlands are water levels managed for flamingos; they are either the result of pumping (saltpans, hunting marshes, rice fields) or partly natural and partly influenced by run-off water (brackish marshes, particularly the Vaccarès-Impériaux complex). The question of maintaining the breeding island at Fangassier each year has often been discussed with colleagues, and opinions have varied. If, from a scientific point of view, the population of flamingos in the Camargue could be maintained by half the number of chicks presently recruiting into the population, then breeding once every two or three years would suffice, as at Fuente de Piedra. To achieve this, however, in the face of abundant water, would require rendering the island periodically unsuitable in one way or another. This would presumably result in birds attempting to nest in other places, perhaps less successfully. Since flamingos are popular birds and have been the subject of a considerable conservation effort by several NGOs, then any effort to impede breeding would need to be accompanied by a strong scientific argument and a clear explanation. At the same time, coordinating the availability of alternative suitable breeding sites in the Mediterranean region could in the long term reduce the local density of flamingos in the Camargue, while guaranteeing the stability of the flamingo population at the regional level. However, such regional coordination might prove difficult to implement, as it relies on cooperation among nations, the interests of which may not be consistently mutual.

The flamingos' behaviour has changed in recent years, at least in the Mediterranean region, to such an extent that is is difficult to imagine just how frightened of aircraft the birds were up to the 1970s. Their fear of humans, thanks to protection, has diminished so much that they can now practically be hand-fed in some places. This relative tameness has also led the birds to establish colonies in sites which would formerly have been considered insecure, and to breed in smaller colonies. We must, therefore, continuously adjust our conservation efforts to the present situation and eventually accept, if necessary, a limit to the potential for breeding in the Camargue. However, the Fangassier colony has provided a unique opportunity to gather important data on flamingos, and the fact that many scientific and conservation issues await further research (see above), in our opinion, contributes to justify the presence of the Fangassier colony, at least in the short term.

Relevance of the Camargue study to other flamingo species

Our studies of many aspects of the flamingo's breeding behaviour and survival are very much 'Camargue-biased'; in other words, regular observations and extensive data collection have been partly facilitated because water levels there are strongly influenced by, or are directly the result of, human management practices. The island on which flamingos have bred for more than 30 years was purposely constructed to enhance breeding, and has been restored for the benefit of the flamingos as the need arises. Therefore, how applicable are the results of the Camargue studies to flamingos elsewhere throughout their range? And has the flamingo 'paradise' at Fangassier influenced the species' historical survival strategies to a point where it is no longer representative of other populations?

To answer the first question, one must consider that what has been observed for the Greater Flamingo in the Camargue is not always valid for other populations and, even more, for other species. As to the second question, it is very unlikely that the nesting conditions prevailing in the Camargue have induced any heritable change in the biology of a species with such a long generation time. The Camargue study has allowed us to collect data which until recently were available only for captive birds. With environmental conditions in the Camargue perhaps closer to those which birds find in captivity than in some of the more natural areas where they breed, for example the salt depressions in Tunisia, Botswana or Namibia, or the Rann of Kutch, it is possible that some of our studies could reveal patterns closer to those observed in captive birds than to those that might exist in wild birds in certain parts of their range. The flamingo modifies its survival strategies according to the opportunities and/or constraints that a wetland area offers. Breeding for some individuals in the Camargue is a near-annual event, but cannot be so in areas which experience drought.

However, what might be applied to other populations and species are the different techniques developed in the Camargue to study flamingos. Some of the natural breeding sites of flamingos offer the possibility of developing research and monitoring programmes similar to those conducted in the Camargue and in Spain. Conservation and management techniques used to favour breeding by flamingos in the Camargue can be applied elsewhere—although we are not suggesting that they always should. They will of course need tailoring to suit the local environment, and it would be a waste of resources to build an island larger than necessary, without considering the trophic resources for flamingos within their foraging range. It would also be wrong to try to attract birds to breed in an area where they would not be secure from all forms of disturbance. Where flamingos breed regularly, marking techniques that were employed in the Camargue, such as dye-marking to assess the foraging range of breeders, could be used and may help to identify the areas of prime importance for conservation around traditional breeding sites. Finally, establishing a regional network of observers and coordinating resightings of marked birds is an important step towards developing conservation plans on a regional scale, based on sound scientific information.

EPILOGUE

As we conclude this book, we realise more than ever how the study of flamingos in the wild is a perfect example of how sound fundamental research and conservation issues are closely linked. We also realise how difficult it is to implement and maintain long-term research without a strong financial and academic background. We hope that developing research and conservation programmes similar to the Camargue flamingo programme will be a priority for future generations, and that other, younger researchers will be as lucky as we have been when observing flamingos in the wild, reading bands, analysing data and contributing to the development of conservation actions. Thirty years after the publication of *Flamingos* (Kear & Duplaix-Hall 1975), we consider the present volume a new progress report in the study of flamingos. We hope that this information will be regularly revised and updated in the coming years, and that scientific knowledge will continue to serve the conservation of wetlands.

CHAPTER TWELVE

An inventory of the more important Greater Flamingo breeding sites

We have attempted to assemble here all breeding reports (up to the year 2000) from the species' entire range. The division into five regions has been done for convenience only and does not infer distinct populations. On the contrary, there is evidence of an exchange of birds among these regions, sometimes on a large scale. Site sheets have been allocated to the 35 most important colonies, where at least one breeding attempt has been successsful since 1950. Neighbouring sites have in some cases been grouped (e.g. Kiaone Islands in Mauritania, Cagliari lagoons in Sardinia).

Minor breeding attempts, as well as nest-building by generally small groups of sub-adult flamingos, are referred to in the introductory pages to each region. In several wetlands that have not been allocated a site sheet, flamingos build nests but do not raise chicks. This behaviour is generally attributable to rather small groups of immature birds, although they may have adult or near-adult plumage. Nests are usually built rather late in the season, with few, if any eggs being laid (see Chapter 8). In some locations, 'false-nesting' has been the precursor of more serious breeding attempts, as observed in the 1990s in the Ebro delta (Spain) and in Apulia (Italy).

230　*The Greater Flamingo*

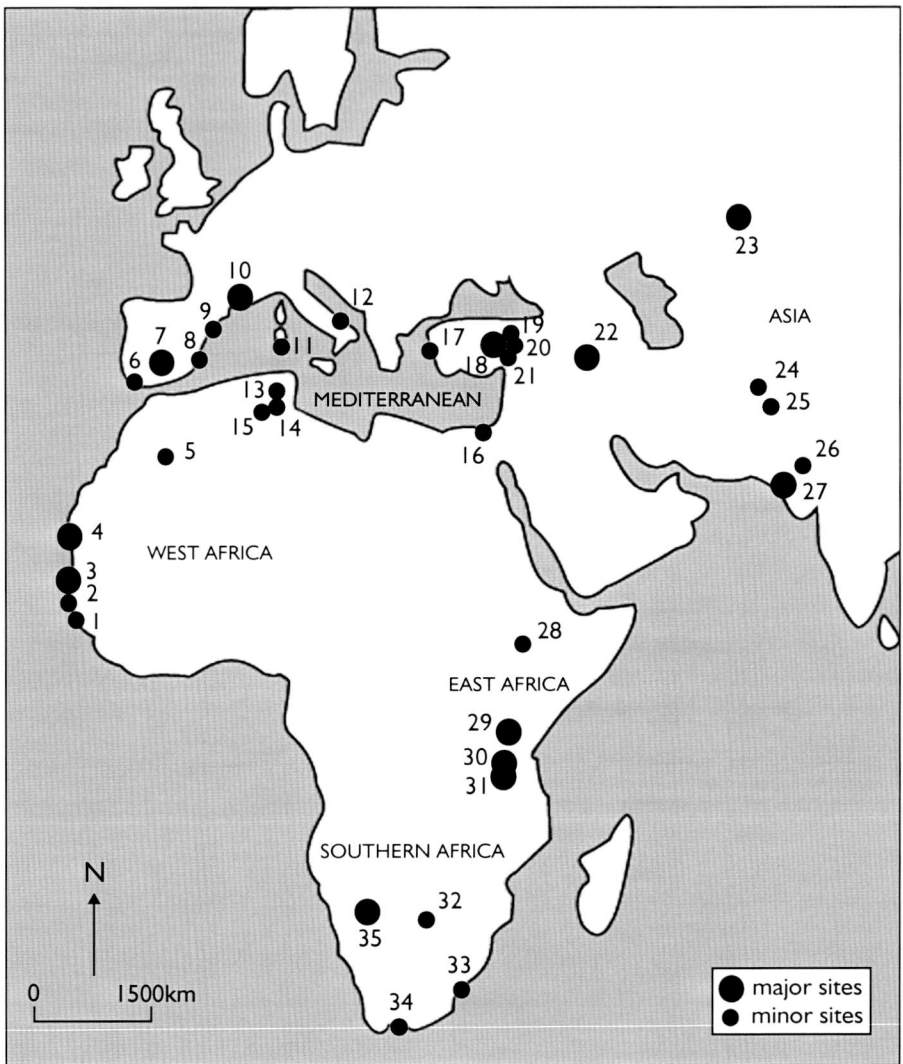

Figure 63. The more important breeding sites of the Greater Flamingo throughout the world.

Almost half (17) of these breeding sites are in playas or salt depressions which flood sufficiently to allow flamingos to breed only at irregular intervals, so that even the most northerly in Europe and Asia are not colonised annually. Five sites are inland lakes which never dry completely, but where breeding by flamingos is influenced by the water level, and only at Lake Uromiyeh in Iran is breeding regular, though perhaps not annual. Three sites are brackish or slightly saline coastal lagoons where breeding has taken place only exceptionally, almost a one-off event. Two sites in the Mediterranean region are in marshes where the colonies are surrounded by higher vegetation. Just as unusual for the species are the two sites (three islands) in West Africa which are in the ocean surrounded by tidal mudflats, conditions which allow flamingos to breed almost annually. Finally, seven sites, mainly in the Mediterranean region, are established within saltpans where water levels are maintained by artificial flooding, providing conditions which are predictable and stable from spring to the end of summer and thus allowing colonies to be established annually.

GREATER FLAMINGO BREEDING SITES

WEST AFRICA

1 Kaolack saltpans (Fatick, Kaolack), Senegal
2 Chott Boul (Trarza), Mauritania
3 Kiaone Islands (Banc d'Arguin NP), Mauritania
4 Ilot des Flamants (Banc d'Arguin NP), Mauritania

MEDITERRANEAN

5 Iriki depression (Ouarzazate), Morocco
6 Doñana (Huelva-Sevilla), Spain
7 Laguna de Fuente de Piedra (Málaga), Spain
8 Santa Pola saltpans and Pantano de El Hondo (Alicante), Spain
9 Salinas de La Trinitat, Ebro delta (Tarragona), Spain
10 Salin de Giraud saltpans, Camargue (Bouches-du-Rhone), France
11 Cagliari (Stagno di Molentargius, di Quartu and Santa Gilla), Sardinia, Italy
12 Margherita di Savoia saltpans (Apulia), Italy
13 Sebkret Sidi el Hani (Sousse), Tunisia
14 Sebkret Sidi Mansour (Gafsa), Tunisia
15 Chotts Djerid and Fedjaj (Tozeur–Gabès), Tunisia
16 El Malaha (Port Said), Egypt
17 Camalti Tuzlasi saltpans (Izmir–Menemen), Turkey
18 Tuz Gölü, (Nigde–Konya–Ankara), central Anatolia, Turkey
19 Seyfe Gölü (Kirsehir), central Anatolia, Turkey
20 Sultansazligi (Kayseri), central Anatolia, Turkey
21 Eregli Marshes (Konya–Karaman), central Anatolia, Turkey

Asia

22 Lake Uromiyeh, Azerbaijan, Iran
23 Lakes Tengiz and Chelkar-Tengiz (Akmola region), Kazakhstan
24 Dasht-e-Nawar (Ghazni province), Afghanistan
25 Ab-e-Istada (Nawar district), Afghanistan
26 Sambhar Lake (Ajmer, Jodhpur, Nagaur), Rajasthan, India
27 Great and Little Rann of Kutch (Gujarat), India

East Africa

28 Lake Shalla (Oromiya), Ethiopia
29 Lake Elmenteita (Rift Valley), Kenya
30 Lake Magadi (Rift Valley), Kenya
31 Lake Natron (Arusha), Tanzania

Southern Africa

32 Makgadikgadi and Sua Pans, Botswana
33 Lake St Lucia (KwaZulu–Natal), South Africa
34 De Hoop Vlei (Western Cape), South Africa
35 Etosha Pan (Kunene), Namibia

WEST AFRICA

Although flamingos probably do not breed annually in West Africa, they do so regularly at several coastal or maritime sites in the region of the Tropic of Cancer. In this part of Sahelian-Senegalese West Africa, rainfall is very low, but the amount of precipitation probably does not determine whether flamingos breed or not, since the main foraging areas are the tidal flats of the Atlantic coast.

The major breeding sites are in Mauritania, on the islands of the Banc d'Arguin and the Bay d'Arguin. Greater Flamingos were first reported nesting on the two **Kiaone Islands** and on the **Ilot des Flamants** in the late 1950s, since when there have been many documented cases of breeding. These islands are in the Atlantic Ocean and surrounded by vast expanses of tidal mudflats. All three islands are sometimes colonised simultaneously, as in 1982, when 17,000 pairs of flamingos bred. Eggs have been laid on another island of the Banc d'Arguin archipelago, the Ilot des Pelicans, two in 1959 (Naurois 1969a) and *c.* 100 in 1974 (Trotignon 1976; Mahé 1985), but this site is obviously sub-optimal for breeding. Lying between the Sahara Desert and the Atlantic Ocean, much of the Mauritanian coastline is quite remote, and flamingos surely breed in this region more frequently than records show, in particular in the Aftout es Saheli, which can be flooded either by the ocean or by the R. Senegal. Here there are just two confirmed reports of breeding, from the **Chott Boul**, where egg-laying has taken place in autumn–winter. Observations of recently fledged juvenile flamingos, in West Africa in May and in the Mediterranean region in early summer, strongly suggest breeding in this area. To the south, in Senegal, there are only two records of breeding at small colonies established in the abandoned **Kaolack saltpans**, in the upper reaches of the Sine-Saloum delta. Nests have also been built (467) in the Geumbeul Reserve near the mouth of the R. Senegal, in April 1995, but they were abandoned when the water level dropped (Ramsar Bureau). Flamingos are reported to have bred on Boa Vista in the Cape Verde Islands, but no longer do so (Naurois 1969b; Naurois & Bonnaffoux 1969).

Flamingos banded as chicks in the Camargue have been observed breeding as far south as the Chott Boul, *c.* 3,750 km from the Rhône delta.

The Greater Flamingo is reported to have decreased on the Banc d'Arguin during the period 1990–95 (Gowthorpe *et al.* 1996). Although it is true that during this period no very large colonies were established in the park, such as those recorded in 1981 and 1985 (see Mahé 1985, Campredon 1987), breeding has been more frequent and we feel that any decrease should not be seen as alarming from a conservation point of view.

1: KAOLACK SALTPANS, SENEGAL

Coordinates: **14° 09′N/16°08′W**   Altitude: **sea-level**

Site description Abandoned saltpans in the upper reaches of the Sine-Saloum delta. Flamingos have bred on three clay islets, the remnants of a dyke.

Wetland status Adjoining Sine Saloum National Park, MAB (1980), Ramsar (1984).

Flamingo breeding Recorded only twice: 1976: on 26 February a colony of *c.* 200 pairs was discovered during an aerial survey by Dupuy (1976). A ground survey on 5 March revealed 146 nests with eggs and seven chicks a few days of age. On 5 April, a crèche of 59 chicks was seen 5 km from the nests. 1979: on 28 February 115 nests were counted, some still with eggs, and *c.* 100 chicks, from one day old almost to fledging (Dupuy 1979). Breeding may also have occurred successfully in early 1977 (Dupuy & Verschuren 1978), but this is not mentioned by Dupuy (1979).

Egg-laying would have started in early February 1976 and in early December 1978.

2: CHOTT BOUL, MAURITANIA

Coordinates: 16°36′N/16°26′W   Altitude: **sea-level**

Site description A coastal depression lying in the Aftout es Saheli (120,000 ha), a vast and remote area between the R. Senegal delta and Nouakchott. Probably only well flooded when storms breach the dunes separating these sites from the Atlantic Ocean, a phenomenon which occurred in March 1985 (Gowthorpe *et al.* 1996). Water also reaches the Aftout from the R. Senegal, formerly when it was in spate and now by sluices upstream from the Diama dam. It is the only site in West Africa where the Lesser Flamingo is known to have attempted breeding, in 1965 (Naurois 1965, 1969a).

Wetland status State ownership. Chott Boul is a Ramsar site but not protected. Water quality is threatened by a proposal to drain irrigation waters from rice fields treated with pesticides and fertiliser to the Chott Boul. Projects exist to supply drinking water to the town of Nouakchott through the Aftout (O. Hamerlynck pers. comm.).

Flamingo breeding 1986: 150 nests in September (Gowthorpe *et al.* 1996). 1987: in early September observers discovered 9,000 nests (Wetten *et al.* 1990–91); they were abandoned but there were 4,500 juveniles and unfledged chicks nearby. 1988: on 4 February J.-L. Lucchesi (pers. comm.) during an aerial survey discovered another colony 50 km to the north of the Chott Boul. On 16 February, a ground survey revealed some birds still on eggs but most with chicks (*c.* 3,000), the oldest aged *c.* 35 days. These observations indicate that in 1987 flamingos bred in spring and again in mid-December. Recently fledged juveniles seen on the Banc d'Arguin, up to 30 aged 3–4 months on 22 May 1986 and another in June 1992 (Cézilly *et al.* 1994b), probably originated from this site, from clutches laid around mid-December. There is, in fact, a reliable report of non-fledged juvenile flamingos being seen in the first half of 1991 in the Diawling Lake to the south of Chott Boul (O. Hamerlynck pers. comm.).

3: KIAONE ISLANDS, MAURITANIA

Coordinates: 20°01′N/16°17′–16°19′W   Altitude: **sea-level**

Site description Two sandstone outcrops about one nautical mile apart and lying 5 km off shore. The larger 'Grande Kiaone' or Kiaone West (1,200 m × 300 m) rises 12–15 m out of the Atlantic Ocean. It has sheer sides but there is a scree slope in the east facing the smaller 'Petite Kiaone' (450 m × 50 m, 6 m high). Flamingos nest on the flat top of the island, down the scree and on the shore above the high-water mark. They also nest on the smaller island, which is separated from the larger only by mudflats at low tide. This has a flat top in the south. Flamingos are unable to make mounds on either island because the ground is stony, so eggs are laid directly on the ground. The spectacle of these 'cliff-nesting' flamingos makes this one of the most impressive colonies in the world, in a remote area of great ornithological importance. Caspian and Royal Terns and the local races of Cormorant, Grey Heron and Spoonbill share these islands with the flamingos for nesting.

Wetland status Within the Banc d'Arguin National Park (1976), Ramsar (1982) and World Heritage site (1989).

Flamingo breeding Tixerant (in Naurois 1969a) discovered hundreds of eggs on Grande Kiaone in 1957 and in early August of that year observed hundreds of chicks able to run. This is the first reference to the species breeding in Mauritania. Since then there have been several records of large, successful colonies as well as failed breeding attempts, and both islands may be colonised simultaneously (as well as the Ilot des Flamants to the north). 1959: 1,500 pairs on Grande Kiaone (GK) with freshly laid eggs on 29 April (Naurois 1969a). 1960: 40 freshly laid eggs on Petite Kiaone (PK) on 11 May (Naurois 1969a). 1974: 20 eggs on PK (Trotignon 1976; Mahé 1985). 1975: no breeding (Trotignon 1976). 1979: breeding (Trotignon & Trotignon 1981). 1980: Trotignon (pers. comm.) counted 383 dead chicks on PK. 1981: 5,000–7,000 nests on both islands (Mahé 1985). 1982: 6,442 pairs on GK and 4,883 pairs on PK (Mahé 1985). 1990: 2,500 pairs on PK (Gowthorpe *et al.* 1996). 1991: 2,500 pairs on PK; on 15–16 June two crèches of 296 and 1,926 chicks aged *c.* two weeks (Cézilly *et al.* 1994a). 1992: 1,240 chicks on GK on 24 June aged 2–4 weeks (Cézilly *et al.* 1994b). 1993: at least 3,000 pairs on GK seen in April–May. On 25 May the chicks, aged 1–2 weeks, were creching. 1994: 5,560 pairs on GK. 1995: 4,730 pairs on GK and PK (Gowthorpe *et al.* 1996). 1996: no data. 1997: *c.* 8,000 pairs bred on GK and raised *c.* 4,650 young. 1998: 2,500 pairs on PK and 4,900 pairs on GK; with 950 and 3,000 chicks, respectively, censused when aged 45 days or less (Parc National du Banc d'Arguin (PNBA), Fondation Internationale pour le Banc d'Arguin (FIBA), Tour du Valat (TdV)). 1999: 13,060 pairs on GK, some birds already incubating on 1 April. On 10 June 9,718 chicks counted in two crèches (PNBA, FIBA, TdV). 2000: 1,500–2,000 chicks aged 10–30 days on GK on 25 June, where many abandoned eggs as well (PNBA, FIBA, TdV).

Egg-laying would have started about 20 March in 1998, 27 March in 2000, end March in 1999, mid-April in 1993, second half of April in 1992, 1997 and end of April in 1957, 1959, 1960, 1991.

4: ILOT DES FLAMANTS, MAURITANIA

Coordinates: **20°35′N/16°40′W**   Altitude: **sea-level**

Site description A low, narrow island of shingle in the Baie d'Arguin, about 200 m in length at high tide when visited in 1986. It was roughly half-moon shaped and just large enough to accommodate the flamingo colony and some hundreds of pairs of Gull-billed, Caspian and Royal Terns which nested alongside the flamingos. Surrounded by mudflats at low tide, it is subject to occasional flooding at high tide (Naurois 1969a).

Wetland status Within the Banc d'Arguin National Park (1976), Ramsar (1982) and Natural Heritage site (1989).

Flamingo breeding First recorded on 6 June 1959, when Naurois (1969a) estimated 9,000 pairs with freshly laid eggs (but later judged this figure to be too high). 1965: 1,000 pairs with eggs (no chicks yet) on 14 June (Naurois 1969a). 1969: 1,000 pairs, young aged 3–4 weeks on 20 July (Westernhagen 1970). 1974: 1,000–3,000 pairs (Kahl 1975b; Mahé 1985) but they were unsuccessful (Trotignon 1976). 1979: c. 1,000 large chicks on 9 August (Trotignon & Trotignon 1981). 1980: 1,500 pairs. 1982: 5,670 pairs. 1983: 3,000 pairs (Mahé 1985). 1985: 12,940 pairs incubating on 20 April, over 5,000 chicks on 29 May (Campredon 1987). 1986: 7,500 pairs, the oldest chicks c. 30 days on 20 May (A. Johnson & P. Campredon pers. obs.). 1988: 4,480 pairs, the first chicks hatching 29 April to 3 May (P. Campredon pers. comm.). 1989: 1,500 pairs with chicks aged up to c. 30 days on 19–20 June (T. v. Spanje pers. comm.). 1993: 150 pairs. 1994: 470 pairs unsuccessful (Gowthorpe et al. 1996). No flamingos bred on this island in 1991, 1992, 1995 and from 1998 to 2000 (PNBA, FIBA, TdV).

Egg-laying would have started around 20 March in 1986, early April 1988, mid April 1985, c. 20 April 1989, on 20 May 1969 and at the end of May in 1959, 1965.

MEDITERRANEAN

The Mediterranean climate has been described as one in which rainfall is much greater in winter than it is in summer. With high evaporation rates and low to zero precipitation, there is a deficient hydrological balance during the summer (see Hobbs et al. 1995, Blondel & Aronson 1999). Autumn rains generally begin in mid-October but there is a profusion of microclimates owing to differences in local terrain (Milliman et al. 1992). Winters are generally mild, with severe frosts occurring only irregularly in the north.

Half of the world's flamingo breeding sites are situated in the Mediterranean region, and at most of them breeding has been well documented over the past 30 years. The main European sites are in France (**Camargue**), Spain (**Fuente de Piedra, Doñana, Ebro delta, Santa Pola saltpans, Pantano de El Hondo**), and Italy (**Cagliari** in Sardinia, **Margherita di Savoia** on the mainland). Flamingos have also bred once at **Orbetello** (Italy) and very recently two other sites have been colonised: the **Laguna Petrola** (Albacete) in Spain, where 98 pairs raised 83 chicks in 1999 (S. Morcillo pers.comm.; Picazo 1999) and c. 300 pairs raised 207 chicks in 2000 (C. Barbraud pers. comm.), and the **Valle di Comacchio** saltpans (Ferrara), Italy, where *c.* 80 pairs raised 68 young in 2000 (N. Baccetti pers. comm.). It is unlikely that there have been any unrecorded cases of flamingos breeding in Europe since the 1960s.

At Lake Messi in Thrace, Greece, three or four breeding attempts have been made and four eggs were laid in June 1992, but these were abandoned (Handrinos & Akriotis 1997). 'False-nesting' has also been recorded in the Camargue, in 1968 (Johnson 1970) prior to the start of regular breeding, and on several occasions since, but rather surprisingly the Etang du Fangassier has remained the only breeding site in France since 1972. There are several reports of false-nesting in Sardinia (Grussu 1999).

In the Maghreb, nesting occurs infrequently at several sites in Tunisia (**Sidi el Hani, Sidi Mansour, Chotts Fedjaj** and **Djerid**). The Tunisian sites are quite well documented, although coverage of the country is not thorough, and breeding attempts are sometimes made on playas other than those having a site sheet here, as in Zougrata in 1963 (Castan 1963) and El Guëttar in 1990. The latter involved a late breeding attempt in mid-June by 1,200 pairs but the colony was destroyed by predators soon after the start of hatching. In May 1976, *c.* 1,000 pairs of flamingos attempted nesting in the Sebkhet El Djem (Morgan 1982) but the colony was raided before our eyes, presumably by local people. In Morocco flamingos formerly nested in the **Iriki depression** in the 1950s–1960s, but they can no longer do so following the construction of an upstream dam in the 1970s. The Knifiss Lagoon, Puerto Cansado, is sometimes referred to as a nesting site (see Naurois 1969a) but there are no substantiated records, either here or at the Merja Zerga, mentioned by Vernon (1973).

In Algeria, the Chott El Hodna (Constantine) has also been mentioned as a breeding site for flamingos (Carp 1980) but there are no confirmed reports, either here or elsewhere in this country (Ledant & van Dijk 1977; Isenmann & Moali 2000). Flamingos are, however, regular visitors and could possibly nest in remote areas bordering the Sahara Desert in years of high rainfall. Elsewhere in North Africa, nesting is occasional in Egypt, at **El Malaha**, though it seems doubtful that many young are produced.

In Asia Minor, breeding is probably annual at one or more sites in Turkey, at **Camalti Tuzlasi, Tuz Gölü**, and **Seyfe Gölü**, and has been reported from **Sultansazligi** and the **Eregli Marshes**. Some of these areas are quite isolated and we may not know of all colonies established in Anatolia, even in recent years.

In Europe, new sites have been colonised in the 1990s in both Spain and Italy and breeding has been attempted in Greece. The separation of colonies between the eastern and western basins of the Mediterranean has thus narrowed. In the Camargue, flamingos have bred annually for over 30 years, making this the most stable and productive colony in the world (over 190,000 chicks raised 1969–2000). In the western Mediterranean there were over 40,000 pairs in seven colonies in 1998, over 30,000 pairs in 1991, 1996, 1997 and 2000, and over 20,000 pairs of flamingos breeding simultaneously in 12 out of the 18 years 1983–2000. The largest colony on record was of 22,200 pairs in the Camargue in 2000 and the greatest number of chicks to take wing from any one site was 15,387 at Fuente de Piedra in 1998. Many flamingos move between the western Mediterranean and West Africa or western Asia.

5: Iriki depression, Morocco

Coordinates: 29°50′N/06°30′W   Altitude: 515 m

Site description An extensive depression 13 km long and 8 km wide (*c.* 20,000 ha) in the valley of the Oued Dräa. This site, 250 km from the Atlantic coast and on the fringe of the Sahara Desert, was formerly flooded irregularly some time between October and January, when the oued was in spate from snow melt in the Atlas Mountains. It dried out again by mid-May. Since the construction upstream of Mansour-ed-Dehbi dam, the Oued Dräa no longer flows continuously (Hughes & Hughes 1992). It seems unlikely that flamingos breed any more at this site, other than perhaps in years of exceptionally high rainfall on the catchment (as possibly in 1979), and Thévenot *et al.* (2003) consider the flamingo to be extinct as a breeding species in Morocco.

Wetland status Unprotected.

Flamingo breeding In March 1957, 500 old nests were discovered (Panouse 1958; Robin 1966). 1965: a colony in two groups, the largest with 524 nests, 30 April to 1 May (Robin 1966). Most had chicks aged <8 days, others were still on eggs. On 6 June there were 400 chicks in a crèche 1 km from the nests. 1966: on 7 May >500 nests, some with chicks, were counted and on 19–20 June >1,500 nests containing eggs or with chicks aged up to 8–10 days (Robin 1968). 1968: at least 500 nests were occupied on 14 April but three days later the colony was raided by local people when the eggs had been incubated for >15 days. 1969–2000: no breeding reported.

Egg-laying would have started at the end of March in 1965, 1966 and 1968.

6: Doñana (Huelva-Sevilla), Andalucía, Spain

Coordinates: 37°00′N/06°25′W   Altitude: sea-level

Site description A vast area (*c.* 50,000 ha) of marshes at the mouth of the R. Guadalquivir, largely dry in summer and inundated in winter, the extent of flooding depending on rainfall. An area rich in wildlife, including Wild Boar, and a major wintering ground for Greylag Geese. The geese make daily flights to the dunes to eat the sand as grit and this inspired hunters in the 1970s to provide the geese with an island of sand in the marshes at Las Nuevas. It is on this island that flamingos have bred several times, but the colony is prone to disturbance by boars. There are extensive saltpans nearby and the whole area is widely used for foraging by flamingos which more regularly breed at Fuente de Piedra, 150 km to the east.

Wetland status Mostly state owned, the flamingo island at 'Baquiruelas' lies within the Doñana Nature Reserve and National Park (1969), which are also UNESCO Man and the Biosphere (MAB) programme reserves and Ramsar (1982) sites.

Flamingo breeding First described by Chapman in 1883 (Chapman 1884). In 1935, 1941 and 1945, breeding attempts seemingly failed (Bernis & Valverde 1954). The 1970s saw the start of a series of mostly successful colonies. 1976: 151 nests built in August (Amat & Garcia 1975) but no eggs. 1977: 1,500 young raised. 1978: 3,300 young raised. 1979: 3,000 pairs failed. 1982: 1,400+ pairs failed. 1984: 3,800 young raised. 1988: 73 incubating in May, 22 young hatched but several with bill malformations (Mañez 1991). No breeding 1980–81, 1983, 1986–87, 1989–94 (M. Rendón Martos pers. comm.). 1997: late breeding attempt by 4,000–5,000 birds (1,250 nests) failed (M. Mañez *in litt.*). 1998: 1,100–1,200 pairs attempted breeding, 155 surviving chicks captured and hand-reared (M. Mañez *in litt.*). 1999–2000: No breeding.

7: Laguna de Fuente de Piedra, Andalucía, Spain

Coordinates: **37°06′N/04°44′W** Altitude: **434 m**

Site description Situated in the southern part of the Iberian Peninsula and lying 50 km to the north of the Mediterranean coast, this is a temporary closed-basin lagoon 6.5 km long × 2.5 km wide (1,364 ha) surrounded by cultivated land. It is flooded by rains falling mostly from September to March and generally dries out in early summer. Rainfall and water depth vary greatly from one year to another and the lagoon may occasionally hold water throughout the year, as in 1990. According to Muñoz & Garcia (1983), the presence of flamingos and salt production both date back to Roman times. Salt production was intensified in 1876, when a series of canals and dykes transformed the depression into a mosaic of saltpans. Salt extraction ended in 1951, after which the dykes gradually eroded (Rendón Martos & Johnson 1996).

Flamingos feed in several smaller fresh or brackish lagoons surrounding Fuente de Piedra as well as in the extensive marshes at the mouth of the R. Guadalquivir, *c.* 140 km to the west.

Wetland status A Ramsar site since 1982, the lagoon was declared a 'Reserva Integral' in 1984 by the Andalusian Regional Government and became a 'Reserva Natural' in 1989. In 1988 it was declared a Special Protection Area under the Birds Directive of the EEC. Since 1985, several islands in the lagoon have been managed for breeding flamingos.

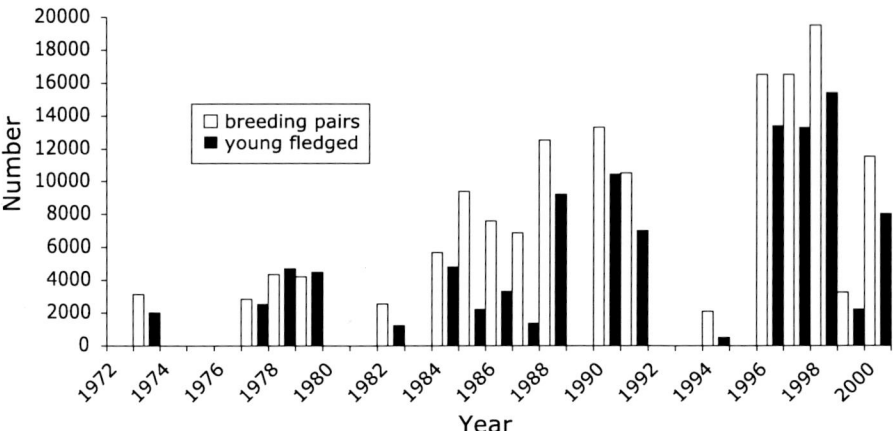

Figure 64. Number of breeding pairs of flamingos and young fledged at Fuente de Piedra. Data from M. Rendón Martos (pers. comm.).

Flamingo breeding Breeding has been documented back to 1963 (Valverde 1964). The size and success of the colonies, which have varied from 800 to 16,500 pairs, are shown in Figure 64. There have been 20 breeding attempts in 35 years (1963–2000) with no breeding in years of low rainfall. It is probable that in historical times breeding was occasionally attempted in years of exceptionally high water, but it is doubtful if flamingos could have bred successfully when saltpans were constructed. Yet it is because of the transformations which took place that the lagoon later became a suitable breeding site for flamingos. Following the collapse of the salt industry, the former dykes gradually eroded, becoming islands, which may periodically have been colonised by flamingos between 1951 and 1963.

A varying proportion of the flamingos which breed at Fuente de Piedra are able to feed in the lagoon. As this dries out, usually before the young are fledged, all the parents are forced to forage in the marshes and saltpans at the mouth of the R. Guadalquivir, 130–160 km to the west. Presumably many of the birds then winter there. Rendón Martos (1996) has shown that the number of breeding pairs of flamingos at Fuente de Piedra is strongly correlated with rainfall in the Guadalquivir Marshes prior to breeding (October–April).

Egg-laying (1977–97, n = 13) has begun between extreme dates of 14 February (1985) and 20 April (1982). In eight out of 13 years breeding has started 1–15 March, average 14 March for the first viable clutches.

8: Santa Pola saltpans and Pantano de El Hondo (Alicante), Spain

Coordinates: 38°13′N/00°35′–00°42′W   Altitude: sea-level

Site description Two managed sites established in previously natural wetlands. Santa Pola is a typical Mediterranean industrial salina (2,700 ha) with shallow lagoons of varying size separated by dykes. The area has been heavily developed for tourism and a busy highway crosses the saltpans. Access is restricted but there is hunting in some parts and large numbers of flamingos have died from lead poisoning. The second site lies 10 km inland from these saltpans, and is the Pantano de El Hondo (2,387 ha), two shallow dams which retain water for irrigation, surrounded by small brackish and freshwater ponds with extensive reedbeds.

Wetland status The saltpans are privately owned but within a natural park (1988). El Hondo is also a natural park (1988) and both are Ramsar sites (1989).

Flamingo breeding 'False-nesting' or unsuccessful breeding attempts have been reported from both sites (Robledano & Calvo 1991).

Santa Pola: Successful breeding in 1973 when 115 pairs produced at least 20–30 chicks (Ibañez Gonzalez et al. 1974; Ferrer et al. 1976).

El Hondo: 1996: 700 birds began breeding in May (160 nests) but later abandoned the colony. 1997: 500–600 pairs raised 491 chicks, the first of these hatching at the beginning of June (Aragoneses & Echevarrias 1998). 1998: c. 1,000 pairs raised 700 chicks (M. Rendón Martos pers. comm.).

No breeding at either site in 1999 and 2000. Egg-laying would have started at the end of April or in early May in 1997.

9: Salinas de La Trinitat, Ebro delta, Spain

Coordinates: 40°35′N/00°40′E   Altitude: sea-level

Site description The colony is established on a sandy island within the saltpans (7,736 ha). This island is situated towards the centre of one of the largest evaporators (603 ha) of medium salinity, which is rich in invertebrates. The island, named *Tora dels Cornills* (c. 1,000 m^2) has an abundant vegetation of glassworts and Egyptian Evergreen shrub. Flamingos nest close to the vegetation, on top of the island 2 m above water as well as along the shoreline. Large numbers of gulls, terns and waders breed in the neighbourhood of the colony and some pairs of Yellow-legged Gulls nest on the same island as the flamingos. Flamingos feed both in the saltpans and in the natural lagoons of the delta and some birds also forage in the rice fields in spring.

Wetland status Privately owned saltpans (Infosa) within a natural park (1983) and Ramsar site (1993).

Flamingo breeding Breeding attempted in 1989 (A. Martínez-Vilalta *in litt.*) and again in 1992, when egg-laying occurred very late (June–July). This colony was disturbed by aircraft and abandoned soon after the chicks hatched in August (Luke 1992). From 1993 to 2000 several hundred pairs bred successfully most years.

	1993	1994	1995	1996	1997	1998	1999	2000
Pairs	?	?	1,500	1,500	1,273	1,461	0	1600
chicks	319	306	1,294	945	788	476	0	1,041

(A. Martínez-Vilalta and Parc Natural del Delta de l'Ebre *in litt.*; Copete 1998, 2000; Martínez Vilalta 2001, 2002, Aymí & Herrando 2003.)

In 1998, curious false-nesting occurred (*c.* 300 nests) in October–November, possibly as a result of unusually high autumn temperatures (pers. obs.).

Egg-laying began early April in 1995, 1997, 1998 and 2000, probably mid-April in 1996, early May in 1993 and mid-May in 1994.

10: SALIN DE GIRAUD SALTPANS, CAMARGUE, FRANCE

Coordinates: 43°25′N/04°38′E   Altitude: sea-level

Site description The Etang du Fangassier is a medium-salinity lagoon (80–110‰) of 536 ha, and forms part of Europe's most extensive saltpans (11,000 ha). It lies within 3 km of the Mediterranean and is divided into a large western pan, which is flooded throughout the year, and a shallower eastern pan, which is drained outside the salt production season. Flamingos breed on a low clay island in the eastern pan, which was purpose-built in 1969 and colonised from 1974 onwards. It has been restored on three occasions (1985, 1988, 1995). The surface area has been reduced from an original 6,200 m² to its present 4,000 m². The competition for nest space has often led to breeding on the dyke separating the two pans and on another nearby island (1996, 1998).

There are very extensive feeding areas for flamingos in and around the Rhône delta and along the Languedoc coast 50 km and more to the west.

Wetland status The Etang du Fangassier is owned by the Compagnie des Salins du Midi et des Salines de l'Est (Salins). It lies within the Camargue Regional Park (1970), which is both a Ramsar site (1986) and MAB reserve.

Flamingo breeding There are several references to flamingos breeding in the Rhône delta prior to the 20th century; Quiqueran de Beaujeu (1551), Darluc (1782–1786) and Crespon (1844). Gallet (1949) and Lomont (1938, 1954a) gave details of 14 breeding attempts from 1914 to 1943 and Hoffmann monitored

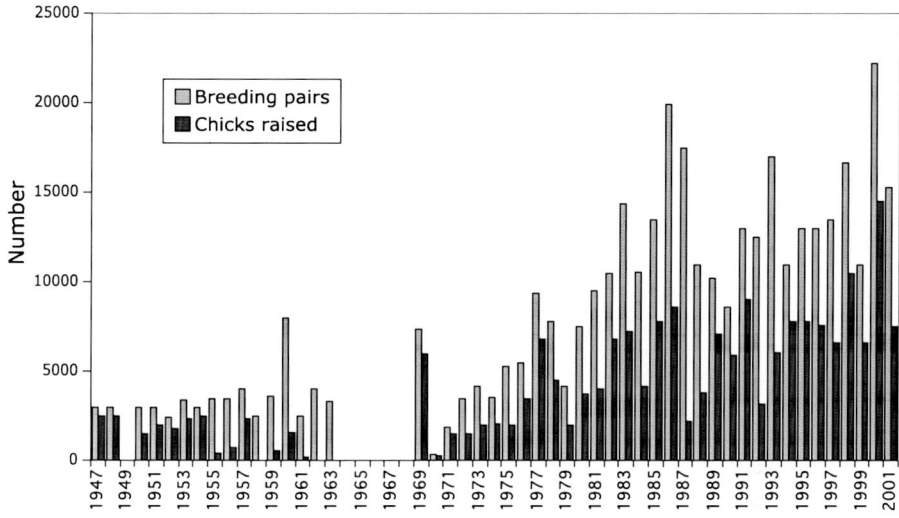

Figure 65. Number of breeding pairs of flamingos and young fledged in the Camargue.

breeding during the 1950s (summary in Johnson 1975b). The size and success of all colonies established 1947–2000 have been closely monitored and are presented in Figure 65. Breeding has occurred annually since 1969, with colonies varying in size from 320 pairs (1970) to 22,200 pairs (2000). The number of chicks raised has fluctuated between 280 and 14,500. Breeding success, measured as a percentage of estimated colony size, has varied between 25% and 87% chicks raised, with an average of 49% (n = 46). This is probably the most frequently used site in the world and also one of the largest colonies, with an annual average of over 10,200 pairs and over 5,400 chicks raised over the 32 years (1969–2000).

Egg-laying has begun between extreme dates of 27 March and 5 May (average 14–15 April, n = 40).

11: STAGNO DI MOLENTARGIUS, DI QUARTU AND SANTA GILLA, SARDINIA, ITALY

Coordinates: *c.* 39°14′N/09°05′E   Altitude: **sea-level**

Site description Molentargius-Quartu lagoon (578 ha) is part of a complex of saltpans at Molentargius-Poetto (1,401 ha), in the suburbs of the coastal cities of Cagliari and Quartu. Salt production came to a (temporary?) halt in 1984–85 because of poor water quality resulting from encroaching urbanisation. It seems that the increasing amount of effluent also brought about a rise in the water level, creating favourable conditions for breeding by flamingos on partly submerged dykes. Flamingos breeding at Molentargius-Quartu feed in the neighbouring saltpans at Capoterra and Santa Gilla and perhaps at several other wetlands in southern Sardinia. They bred at Santa Gilla (Macchiareddu) saltpans in 1999.

Wetland status Ramsar site (1976), Important Bird Area (IBA).

Flamingo breeding Breeding has been successful each year since 1993 and this is a remarkable achievement for the edge of a city. Some birds (banded) breeding in 1993 made a first breeding attempt in the Camargue that same year. In 1994 c. 1,000 pairs began egg-laying on 22 April at the above site (585 nests) and at the beginning of May on another island in the adjoining Quartu lagoon, 2 km distant (350 nests). In both 1995 and 1996 some nests were occupied by two successive breeding pairs. 1998: two islands again occupied after management at the initial site; 850–900 chicks hatching at Quartu and c. 3,300 on the purpose-built island at Molentargius (S. Nissardi pers. comm.).

	1993	*1994*	*1995*	*1996*	*1997*	*1998*	*1999*	*2000*
pairs	1,400	1,100	1,140	2,000	3,330	4,750	1,300	2,500
Young fledged	915	900	700	1,050	2,200	4,000	860	c. 2000
site	M	M+Q	Q	M	Q	M+Q	SG	Q

M = Molentargius, Q = Quartu, SG = Santa Gilla.
(Data from Grussu (1999), A. Atzeni and S. Nissardi, Associazione per il Parco Molentargius Saline Poetto pers. comm.).

Egg-laying began on 2 April in 1995, 1997, the first 10 days of April in 1996, on 22 April in 1994 and early May in 1993, 1999 and 2000 (A. Atzeni pers.comm.).

12: MARGHERITA DI SAVOIA SALTPANS, ITALY

Coordinates: 41°24′N/16°04′E   Altitude: **sea-level**

Site description Typical Mediterranean saltpans and the largest in Italy (3,871 ha). They lie on the coast of the Adriatic Sea in the Gulf of Manfredonia.

Wetland status A Ramsar site (1979), Natural Reserve (1985) and IBA (1988), protected since 1992.

Flamingo breeding A small group of flamingos made a late breeding attempt at the end of summer 1995 and failed (Albanese *et al.* 1997). The island, an eroded dyke, was occupied again in spring 1996, when an unknown number of breeding pairs raised 107 chicks; 88 nests were counted on the island after breeding, suggesting that either some birds laid eggs directly on the ground or some nests were occupied by two successive breeding pairs. More unusual was the second, small wave of breeding birds which colonised the island in September, at least one month after the colony had been abandoned by all the previous breeders. On 31 October, nine adults were seen feeding chicks aged five days. All nine were observed again in November and they took wing at the end of December. This was the first-ever autumn breeding of the Greater Flamingo reported from the Mediterranean region and followed unusually heavy rains (Albanese *et al.* 1997). 1997: 160–200 pairs established on three islets raised *c.* 144 young. 1998: first wave of breeders, *c.* 380 pairs, laid 10 February; second wave in mid-June. A total of 218 young was raised (G. Albanese, N.Baccetti, Istituto Nazionale per la Fauna Selvatica pers. comm.). 1999: *c.* 250 pairs raised *c.* 200 chicks. 2000: spring breeding attempts failed but 122 chicks were raised in autumn–winter (N. Baccetti/INFS pers. comm.).

Egg-laying started around 10 February 1998 for one group but not before mid-June for another. In 1997, it was spread from mid-April through to the beginning of July but with at least nine eggs laid at the end of September. In 2000 the successful breeders laid eggs around 26 August.

13: Sebkret Sidi el Hani, Tunisia

Coordinates: **35°31′N/10°27′E**   Altitude: **about sea-level**

Site description A vast salt lake (*c.* 25,000 ha) with very extensive mudflats at the southern end, where flamingos have been reported breeding on two sandy islands. These are surrounded by water only following abundant rainfall in the area. They have a sparse vegetation and are at times grazed by goats. When there is water surrounding the islands, it can be almost 1 m deep. It is very clear, and rich in brine shrimps. Rainfall is 250 mm per year on average.

Wetland status State owned, unprotected.

Flamingo breeding The first reference to flamingos breeding in Tunisia is by Lavaudan (1924), who mentions finding *c.* 30 old nests in the south of this wetland in February 1924. Breeding was not witnessed at Sidi el Hani, however, until June 1972, when M. Lachaux discovered a very large colony. During a visit on 26 June, it was estimated that there were *c.* 8,000 chicks in the crèche, aged up to two months. Two years later, during a flight over this site on 23 May 1974, Kahl (1975b) and Johnson saw about 1,500 empty nests and a crèche of *c.* 400 chicks (aged 3–4 weeks) nearby.

Egg-laying would have started in mid-March in 1972 and at the end of March in 1974.

14: Sebkret Sidi Mansour, Tunisia

Coordinates: **34°14′N/09°30′E**   Altitude: *c.* **40 m**

Site description Lying in a semi-desert zone (average annual rainfall 1931–60 was 100–150 mm), this depression (*c.* 3,000 ha) is flooded only in years of abundant rainfall, when the water may be >1 m deep, barely brackish and persist throughout the year. In years of low rainfall the site is dry and the land may be ploughed and cultivated with cereals (Carp 1980). Flamingos breed on an island situated in the eastern, less saline, part of the lake.

Wetland status State owned, unprotected.

Flamingo breeding: First recorded in 1963 but unsuccessful because the colony was raided by local people, who collected hundreds of eggs (Castan 1963). 1991: on 5 July a colony of *c.* 4,000 pairs plus 600–700 chicks in crèche (aged <40 days) was observed by P. Pilard and Y. Kayser (pers. comm.). On 8 September successful breeding was confirmed, with fledged young seen and 3,200 chicks still creching.

This was a late breeding made possible by the abundance of water, with egg-laying from end April through to the end of July.

15: CHOTTS DJERID AND FEDJAJ, TUNISIA

Coordinates: 33°18′–34°03′N/07°45′–08°26′E   Altitude: 15–38 m

Site description These two playas together form a closed basin extending 193 km from east to west (495,000 ha). The Djerid is more than 50 km wide and occasionally receives water from the Fedjaj (N. Drake pers. comm.). Precipitation decreases from the coast inland with an average 185 mm per year at Gabès but only 96 mm at Kebili. Rainfall over these playas may be intense, although they are seldom flooded. The Djerid also receives flows of artesian water (Domergue 1949a,b; Hughes & Hughes 1992). Their substrate is of clay or sand, the Djerid being overlaid with gypsum and salt crystals.

Wetland status State owned, unprotected.

Flamingo breeding Since 1948, when eggs were first discovered in the Djerid (Domergue 1949a,b), there have been at least five reports of breeding there. 1949: hundreds of nests. 1950: young on nests on 25 April (Domergue 1951–52). 1951: 600 breeding birds, the egg embryos well developed on 17 April (Domergue 1951–52). 1957: newly fledged young observed at Zougrata on 1 June and in 1959 some chicks already hatched on 5 March, 1,500–1,900 about to fledge on 22 July (Castan 1960). 1991: 3,800 chicks aged 40–70 days on 10 July (P. Pilard & Y. Kayser pers. comm.). No nests were seen, however, and these chicks may possibly have trekked from Fedjaj. Two survey-flights over this latter playa in 1974 (Kahl 1975b) and 1990 revealed crèches of *c.* 8,000 and *c.* 3,000 chicks respectively, the former aged six weeks on 23 May and the latter 6–10 weeks on 19 June.

Egg-laying would have started in early February 1957, mid-February 1959, early to mid-March 1974, 1990 and at the end of March in 1950, 1951, 1991. (Castan (1960) was wrong in assuming that young fledge at age 4–5 months.)

16: EL MALAHA, EGYPT

Coordinates: 31°13′N/32°19′E   Altitude: **slightly above sea-level**

Site description Two hypersaline coastal lagoons separated from the Mediterranean by a sandbar 100–500 m wide. Covering *c.* 3,500 ha they lie in the extreme north-west of Sinai, east of the entrance to the Suez Canal. The extent of flooding varies according to the season and in summer they can be nearly dry. They are, however, connected to the sea by a narrow channel and water levels (max. depth <1 m) can be manipulated to maximise fish production (Baha el Din 1999). Very important for waterbirds, these lagoons have been amongst the least disturbed of the Egyptian wetlands (Walmsley in Meininger & Atta 1994).

Wetland status Unprotected and threatened by an industrial zone (Baha el Din 1999).

Flamingo breeding Until the start of the 20th century, flamingos apparently bred on several Nile delta lakes (Manzala, Bardawil in north Sinai) but by 1929 they had ceased to breed in Egypt (Meininger *et al.* 1986). 1970: 500–600 pairs bred at El-Tina, El Malaha; small chicks visible on aerial photographs taken in April (Mendelssohn 1975; Paz 1987). Breeding is reported to have occurred regularly at this site since then with 350–400 fledglings seen on 7 July 1986 (Goodman & Meininger 1989) and breeding attempts in 1993 and 1994 (Waheed Salama in Baha el Din 1999). When accessible, the colonies are raided by fishermen.

17: CAMALTI TUZLASI SALTPANS, TURKEY

Coordinates: 38°28′N/27°15′E   Altitude: **sea-level**

Site description These are typical Mediterranean saltpans lying on the north side of Izmir Bay on the Aegean Sea. They were created in 1863 and developed in the 1970s–1980s. The saltpans cover 3,300 ha of what was formerly an extensive wetland of 8,800 ha. Flamingos breed on a clay island or on one of the dykes separating the saltpans, but they have also attempted breeding in Homa Lagoon (1,824 ha).

Wetland status State-owned saltpans maintained by Tekel Salt Administration, within a 'Bird Paradise' (1980), an IBA and a Permanent Wildlife Reserve (1980) and protected since 1981 as a site of natural value (Ministry of Culture).

Flamingo breeding Data from M. Siki (pers. comm.) and Eken (1997). Flamingos began breeding here in 1982 with 100–150 pairs egg-laying 22–29 May. 1983: 138 pairs but no eggs laid. 1984–86: 100–150 pairs each year with egg-laying starting the first week of June in 1984 and 1986 and in mid-May 1985. 1988: 50–100 chicks seen in July. 1990: *c.* 600 pairs (egg-laying on 15 April) raised 500 chicks of which five were ringed. 1991: laying from 5 May onwards, 370 pairs raised young.

1992: 93 pairs attempted to breed but failed. 1993: 600 pairs (*OSME Bulletin* 31). 1994: egg-laying from 9 April onwards; 1,752 nest mounds; 750–800 chicks raised. 1995: breeding attempted by 1,450 pairs but colony abandoned (Magnin & Yarar 1997). 1996: late nesting attempts by *c.* 400 pairs at three sites including Homa Lagoon, but no successful breeding. 1997–99: no breeding. 2000: 405 nests but no successful breeding (M. Siki pers. comm.).

Egg-laying began on 9 April 1994, in mid-April 1990, in early May 1991, in mid-May 1985, 22–29 May 1982 and during the first week of June in 1984, 1986.

18: Tuz Gölü, Turkey

Coordinates: **38°43′N/33°22′E**   Altitude: **899 m**

Site description Lake Tuz is a vast, natural and endorheic salt lake (164,000 ha) 80 km × 35 km. It is shallow, having an average depth of only 0.2 m and a maximum depth of only 0.5 m. As it dries, a salt crust forms over its entire surface. There are three saltpans on its periphery. Flamingos have been known to breed here since the late 1960s, on raised ground or on sheets of salt (Warncke 1971) more than 4 km from the shore. Some flamingos which breed here feed in Seyfe Gölü, 100 km to the north-east. As the lake dries, the chicks may trek >15 km from the breeding islands to the mouth of the Konya Channel, which flows into the south-west of the lake.

Wetland status Protected since 1992 as a site of natural value (Ministry of Culture) and since 2000 as an SPA.

Flamingo breeding Discovered by Warncke (1970, 1971) in 1969 when broken eggs were found. 1970: 5,000 nests on 18 May, 20% with chicks aged <10 days, isles of salt crust colonised as they emerged, with laying still in progress. 1971: only 83 nests with eggs found on 31 May, high water level. 1972: 3,500 pairs on 24 May with chicks aged one week (Ornithological Society of Turkey (OST) 1975). 1973: 3,700 pairs, chicks 1–3 weeks on 15 June (OST 1975). 1974: during an aerial survey on 11 June, Kahl (1975b) reported *c.* 1,000 chicks aged 30–40 days. 1991: an aerial survey on 13 June by Tour du Valat/Dogal Hayati Koruma Dernegi (DHKD) revealed 11,000 nests, most of them abandoned, and *c.* 4,500 chicks, the oldest aged 4–5 weeks. 1992: an aerial survey on 22 June by Tour du Valat/DHKD revealed 14,000 chicks (counts on aerial photographs). The oldest were aged *c.* 50 days and had trekked 17 km to the Konya Channel inflow. 1994: evening flights of flamingos out of Seyfe Gölü in the direction of Tuz indicated breeding at Tuz (G. Magnin, DHKD). 1996: large numbers of juvenile flamingos on lakes bordering Tuz in July–August, and in 1997 4,000 juveniles at the Konya Channel inflow suggested successful breeding at this lake both years (G. Magnin, DHKD). 1998: an aerial survey of 30 June (DHKD/Tour du Valat) revealed breeding by at least 12,000 pairs of flamingos. Chicks aged up to six weeks were creching at the breeding site and at the Konya Channel inflow (G. Eken *in litt.*) where on 9 July S. Karauz (*in litt.*) counted 7,000–7,500 juveniles. 1999: on 24 June 1,200 juveniles seen at Konya Channel inflow (S. Karauz pers. comm.). 2000: on 12 July several thousand chicks at this channel inflow (Tour du Valat/DHKD).

Breeding always starts in April, with egg-laying during the first half of the month in 1970, 1974, 1991, 1992, 1998, from 18 April onwards in 1972 and from 26 April onwards in 1973.

19: SEYFE GÖLÜ, TURKEY

Coordinates: **39°12′N/34°25′E**   Altitude: **1,080 m**

Site description An extensive closed-basin lake varying in size from 1,500 ha in summer to 7,000 ha when fully flooded in spring. It is brackish and fed by streams and springs. Drainage of the surrounding land in the 1960s increased the lake surface. It has a maximum depth of 1–2 m. There are many islands, which are colonised by a variety of nesting gulls, terns, waders and pelicans. Evening flights of flamingos out of this lake to the south-west, and the large number of flamingos present (12,000 adults in June 1992) indicate that this is also a foraging site for flamingos breeding at Tuz Gölü, 100 km distant.

Wetland status State owned, a Site for the Preservation of Nature (1970), protected since 1989 as a site of natural value (Ministry of Culture), nature reserve (1990) and Ramsar site (1994).

Flamingo breeding Nesting confirmed in 1992: on 18 June some hundreds of pairs were observed breeding and 300–500 chicks were creching, the oldest aged 15 days. Aerial photographs on 22 June revealed 860 pairs and 114 chicks on one island and 300 pairs on another (Tour du Valat/DHKD). On 25 July the islands were visited and a total of 1,947 nests counted on five islands. The chicks were too distant to be counted (DHKD report). 2000: no breeding.

Egg-laying would have started during the first week of May.

20: SULTANSAZLIGI, TURKEY
(also known as Develi Ovasi and Sultan marshes, sometimes wrongly referred to as Kurbaga Gölü)

Coordinates: **38°20′N/35°17′E**   Altitude: **1074 m**

Site description A vast wetland area (39,000 ha) comprising a series of saline, brackish and freshwater lakes and marshes in the lowest part of Develi Plain. Yay Gölü (3,650 ha), where flamingos have nested, is a closed-basin lake where large numbers of birds also overwinter.

Wetland status State owned, a Permanent Wildlife Reserve (1971), Site for the Preservation of Nature (1988), protected since 1993 as a site of natural value (Ministry of Culture), Ramsar site (1994).

Flamingo breeding 1970: 1,500 pairs (Kasparek 1985). 1974: 8,000 birds with 200 pairs starting to nest 8–10 June (Kahl 1975b).

21: EREGLI MARSHES INCLUDING AKGÖL, TURKEY

Coordinates: **37°30′N/33°50′E**   Altitude: **998 m**

Site description This wetland lies in a steppe zone. Various hydraulic works in central Anatolia transformed this formerly small freshwater wetland into a large marsh of 5,000–7,000 ha (Kilic 1988) which is fed by drainage canals. Extensive reedbeds were reduced in size after an increase in water level in the 1960s. Recently, dams on the inflowing rivers have brought about a reduction in the size of the marsh, which is now smaller than ever (Magnin & Yarar 1997). Flamingos breed on islands of mud with sparse stands of reeds.

Wetland status Protected since 1992 as a site of natural value (Ministry of Culture) and a nature reserve (1995).

Flamingo breeding Warncke (1971) believed breeding was possible here but no nests were seen until 1987, when Kilic (1988) found 35–40 old mounds on 10 May, but no eggs or eggshells. 1991: breeding was confirmed when 217 nests were counted on 1 June, 68 of these with an egg (Magnin & Yarar 1994). However on 13 June the nests had been abandoned. 1993: *c.* 300 pairs attempted nesting on four islands and many chicks had hatched by late June but it is not known whether any young eventually fledged. On one island, which held 90 flamingo nests, 23 pairs of Great White Pelicans were breeding amidst the flamingos (Magnin & Yarar 1994). Egg-laying would have taken place in May 1993.

ASIA

This vast region, which encompasses several very different climates, extends from 50°N latitude in Kazakhstan, south to tropical north-west India, and from sea-level to the highest-known breeding sites of the Greater Flamingo, at 3,200 m altitude. Breeding at all sites may be influenced by rainfall and/or snow melt. The Indian colonies, now apparently far less important than reported around the middle of the

20th century, seem to be largely dependent upon the amount of rain falling during the monsoon.

The most regularly documented breeding sites of Greater Flamingos, Turkey excepted, are **Lakes Tengiz** (Kazakhstan), **Uromiyeh** (Iran) and the **Rann of Kutch** (India), with seemingly less frequent breeding at the high-altitude lakes **Ab-e-Istada** and **Dasht-e-Nawar** (Afghanistan). Data are quite fragmentary, however, and it seems likely that flamingos breed occasionally and perhaps even frequently on other wetlands throughout this region, particularly in north-west India. This must be the case if the species still occurs in the vast numbers reported in the past from Rajasthan. Bharucha (1987), for example, observed brown juveniles begging and being fed by their parents at Bhigwan, Pune-Sholapur in June 1985. Flamingos rarely feed their young after they have left the crèche and Bhigwan is 600 km from the Rann of Kutch! Breeding was confirmed for the first time in 1995 at **Sambhar Lake** (Rajasthan). Another site where Greater Flamingos formerly bred, or have occasionally bred in small numbers, is Thol Lake (Ahmedabad) where in 1980–81, Thakker (1982) counted 70–80 nests. He also saw young flamingos there in June 1981. In Iran, Behrouzi-Rad (1992) reported occasional breeding at lakes Bakhtegan near Shiraz and Qom, central Iran, where 5,000 pairs attempted unsuccessfully to breed in 1988. Breeding also formerly took place on the shores of the Caspian Sea, at Kara-Bogaz Bay until 1937 (Demente'ev *et al.* 1951) and in the north-east where there were 25,000 pairs until 1945–46 (Koning & Rooth 1975).

In the Persian Gulf, Ticehurst *et al.* (1926) found *c.* 500 pairs of flamingos breeding on Bubyan Island off northern Kuwait in 1922 (nests of sand) while the British Museum (Natural History) has eggs from Kuwait collected in 1879, 1884, 1891, 1920, 1921 and 1922 (Platt 1994). Koning & Rooth (1975) refer to Al Fao in Iraq as a breeding site but give no details of nesting there. At Al Ghar, near Abu Dhabi, at least 22 pairs nested in June–July 1993, and some eggs hatched before the colony was abandoned (Platt 1994). This site, now known as **Al Wathba Wetland Reserve**, was colonised again in November 1998. At least 89 nests were built by a group of 1,500 birds and 44 of them were occupied in mid-January 1999 when at least 12 held an egg. Some nests were destroyed soon afterwards by a rise in water level but 10 chicks were seen in a crèche in February and they fledged in April, this being the first successful breeding by flamingos on the Arabian Peninsula in 75 years (Salim Javed pers. comm.). In Saudi Arabia, dozens of flamingos built nests in 1990 at Jeddah South Corniche and port but the colony disappeared after the site was flooded by spring tides (Evans 1994).

In Sri Lanka, flamingos built 635 nests in 1997 at Bundala (Wickramasinghe 1997), where young juveniles are also reported to have been seen. In the Indian Ocean, three complete nests and three incomplete mounds were built on Aldabra Island in April 1995, by a group of 19 flamingos, and a single chick was seen (*Oryx* 29, 1995).

22: Lake Uromiyeh (formerly Lake Rezaiyeh), Iran

Coordinates: 37°15–27′N/45°31′E   Altitude: 1280 m

Site description This closed-basin hypersaline lake has a surface area of 463,000 ha (144 km × 55 km) but is only 5 m deep on average (max. 11 m). It is fed by snow melt, which can raise the level by 2 m. The lake is rich in brine shrimps, an abundant food source which allows many adult flamingos to undergo simultaneous moult (July–August) and become flightless. There are 56 islands, all uninhabited. Flamingos breed intermittently on at least five of the flatter stony or rocky islands: Dowgozlar, Ashk, Espir, Arpatapasy, Kabodan (Behrouzi-Rad 1992). The substrate is of mud or silt, often covered by salt crystals (Scott 1995).

Wetland status Lake Uromiyeh is a state-owned National Park (1967), a Ramsar site (1975) and a Biosphere Reserve (1976).

Flamingo breeding Flamingos are known to have bred most years since 1964 on one or another of the islands but details are fragmentary. 1971 and 1972: 15,000–20,000 breeding pairs (Scott 1975). 1973: 58,500 adults, 20,000 chicks (Scott 1975). 1974: 5,300 breeding pairs, some with small young, but most just hatching on 28–29 June (Kahl 1975b). 1976: 23,000 breeders, 30,000 non-breeders and 9,000 chicks (Ashtiani 1977). Flamingos bred in at least 16 of the years between 1977 and 2000 and hundreds or thousands of chicks have been ringed (Department of the Environment of Iran).

According to Scott (1975) hatching started at the end of June in 1971 and 1972, indicating egg-laying from late May onwards, as would have been the case in 1974, and from 8 June to early July in 1976 (Ashtiani 1977).

23: Lakes Tengiz and Chelkar-Tengiz, Kazakhstan

Coordinates: c. 50°25′N/69°00′E   Altitude: 304 m

Site description Lake Tengiz in western Siberia is the most northerly breeding site of flamingos in the world. It is the most frequently colonised of a series of shallow (3–4 m deep) closed-basin depressions in the northern desert zone of Kazakhstan. It lies in the central Kirghiz steppe about 700 km to the north-east of the Aral Sea and has a surface area of 156,000 ha (74 km × 32 km). Fed by two rivers, the lake floods to a varying degree from one year to another. Annual precipitation is on average 250–300 mm. The mean temperature in July is 20.5°C and in January –17°C. Islands used by flamingos for nesting can be destroyed over winter by ice floes. Adult flamingos undergo simultaneous moult here in summer, when they become flightless.

Wetland status State-owned protected area and Ramsar site (1976). Reported to be gradually drying.

Flamingo breeding: There are many references dating back to the early part of the 20th century, mostly in Russian literature, on the presence and breeding of flamingos on the Tengiz lakes (Volkov 1977). Breeding was recorded in seven years between 1958 and 1972, in the southern group of lakes in wet years (on five islands) and in the northern (Tengiz) group in dry years. The number of breeding birds varied from 7,800 to 30,000–36,000. The maximum number of chicks reported was 10,000–15,000 in 1958.

From 1976 to 2000 breeding was attempted in 17 of the 25 years (A. Andrusenko and A. Koshkin pers.com). In 1986 there were 22,179 pairs in three colonies (E. Stotskaya & V. Krivenko *in litt.*). 1987: 10,500 pairs failed when colonies were flooded (E. Stotskaya & V. Krivenko *in litt.*). 1991: 17,300 nests in six colonies, 2,500 young raised. 1992: two colonies of 1,000 nests (unsuccessful) and 8,000 chicks respectively. 1995: 1,000 chicks. 1998: 15,200 nests, *c*. 12,000 chicks aged *c*. one month on 24–26 July (data for 1992–98 from A. Zhumakan-Uly pers. comm.).

Egg-laying seems to take place in May–June. In 1970 the first chicks hatched 9–10 July (Volkov 1977) indicating laying during the first ten days of June. In 1971 there were 700 nestlings a few days from fledging on 25 August, indicating egg-laying in early May, while in 1998, egg-laying presumably started *c*. 25 May.

24: DASHT-E-NAWAR, AFGHANISTAN

Coordinates: **33°50'N/67°40'E**   Altitude: **3,200 m**

Site description This lake lies in the Hindu Kush, a high desert plateau 50 km × 15 km. It is surrounded by mountains 4,000–4,800 m high and is the world's highest breeding site of the Greater Flamingo. The lake (*c*. 3,500 ha) varies in size according to the season and to the amount of snow melt, and when visited by Klockenhoff & Madel (1970) in summer 1969 was 14 km × 4 km. Flamingos were breeding in the centre of the lake where there were some 40 islets varying in size from 5 m^2 to 500 m^2. The brackish and alkaline water was *c*. 30 cm deep. Nomads graze sheep and dromedaries on the shores of the lake, which is frozen and under snow from October to March.

Wetland status State-owned and declared a wildlife sanctuary for migrating and breeding waterfowl in 1977. The political situation since 1978 has, however, prevented the government from paying attention to the protection of the environment or wildlife.

Flamingo breeding: 1969: Klockenhoff & Madel (1970) visited this site in June–July. On 14 June there were 6,000 flamingos incubating on 27 islets towards the middle of the lake, most of them on eggs but one chick was seen. The following day, they reported 2,000 chicks and by 7 July there were mostly chicks, with few adults still incubating. 1970: almost 12,000 adults with young in August and 1971–72 no breeding because of low water levels (Nogge 1974). *1974:* survey by

Kahl (1975b) on 15 July, who reported 1,200 adult flamingos along with 200 chicks aged 30–40 days. Scott (1995) reports breeding most years except 1971–72.

Egg-laying would have occurred from 7 May onwards in 1974 and *c.* 15 May, or slightly earlier, in 1969.

25: AB-E-ISTADA, AFGHANISTAN

Coordinates: **32°40′N/67°55′E**   Altitude: **2,100 m**

Site description This shallow alkaline lake (16,000 ha) lies in the southern foothills of the Hindu Kush. It was reported to have practically dried out in the 1970s in years of drought. It now fluctuates in size according to rainfall and the functioning of the sluice gates of the Bandeh Sardeh dam, which may be opened during exceptional snow melt, heavy rain and floods. An exceptionally high water level was reported in 1992–93, when the islands customarily used by flamingos for nesting were submerged. There was also a drop in salinity. There are agricultural and pastoral activities around the lake, which freezes over in winter (Scott 1995).

Wetland status State-owned and declared a National Flamingo and Waterfowl Sanctuary (1974) and approved by government as a Waterfowl Sanctuary in 1977. The political situation since 1978 has, however, prevented the government from paying attention to the protection of the environment or wildlife. During the political conflict 1978–91, there was a military base nearby which caused considerable disturbance to wildlife at the lake. Flamingos are still shot for food.

Flamingo breeding Data are very fragmentary; this lake is reported to have held very large numbers of flamingos during the breeding season as long ago as the 15th–16th centuries by Babar the Great (1483–1530), who breakfasted on their eggs (in Paludan 1959). More recently, Akhtar (in Paludan 1959) reported a colony of hundreds of eggs but still no young on 17 July 1946 or 1947. Niethammer (1970) found 500–1,000 eggs and young chicks when he visited the lake in June 1966. 1969: *c.* 1,000 chicks raised (see Scott 1995, p. 19). 1970–72: low water levels and no successful breeding but breeding probably attempted in 1972 (Nogge 1974). Kahl (1975b) visited the lake on 15 July 1974 and observed 8,000 flamingos but found no signs of recent breeding.

Egg-laying would seemingly have taken place in June 1946 or 1947 and in May-June in 1966.

26: SAMBHAR LAKE, INDIA

Coordinates: **27°58′N/75°55′E**   Altitude: **360 m**

Site description This is the shallowest of three salt lakes in Rajasthan and is the largest (23,000 ha) inland alkaline, saline lake in India (Alam 1982). It is 22.5 km long and varies in width from 3.2 km to 11.2 km. The lake is fed by several rivers and streams and the level fluctuates both seasonally and annually, from a few centimetres in depth to a maximum of 2 m, according to drought, flooding and discharge into two reservoirs for salt production. The area has a tropical monsoon climate; temperatures range from *c.* 37°C in summer to *c.* 12°C in winter. The average annual rainfall is *c.* 500 mm. In years of flood the lake can hold water and be host to flamingos throughout the summer. Lesser Flamingos also occur on this lake and are usually more abundant than Greaters.

Wetland status State owned, part of the lake being leased to the Salt Department of the Government of India. There is no habitat protection but shooting is prohibited (Scott 1989). It is a wetland of International Importance and a Ramsar site (1990).

Flamingo breeding First recorded in 1995 by Khumar & Bhargava (1996). In second week of January 1,100 nests were found, many of them with chicks. There were also young birds in various downy and juvenile stages. Egg-laying would seemingly have taken place from about mid-October through to January.

27: GREAT AND LITTLE RANN OF KUTCH, INDIA

Coordinates: *c.* 24°N/69°E   Altitude: **from sea-level to 10 m**

Site description The Great Rann, when dry, is a vast saline plain (700,000 ha) of sun-baked mud and sand with halite and gypsum efflorescences. It is flooded irregularly according to the vagaries of the monsoon rains (average <300 mm July–September) but it also receives wind-driven seawater and tides from the Arabian Sea. Temperatures range from 7°C in winter to >40°C in summer. Flamingos have traditionally bred on Pachham Island, or 'Hanjbet' (Flamingo City) but they may occasionally nest elsewhere in the Great or Little Ranns. These areas are remote and difficult to access, and data are fragmentary.

Wetland status State owned. The Little Rann was declared a Wildlife Sanctuary in 1973.

Flamingo breeding The numbers of birds reported breeding have been quite phenomenal and the Great Rann has been home to the largest Greater Flamingo colony recorded anywhere in the world. The species has been known to nest here since 1896 (Ali 1945) but because of the remoteness of the site, records of breeding are few and far between. 1935: McCann (in Ali 1945) saw large numbers of unfledged young in October. 1945: Ali (1945) saw eggs at various stages of incubation, from fresh to hatching in late March. He reported 105,000 active nests with a total of 500,000 birds at the colony. 1957: breeding occurred with chicks in nurseries in April (Abdulali 1964). 1959–60: on 21 March 1960 Ali (1960) estimated that the colony held twice the number of birds he had seen in 1945 and gave the figure of one million. He saw fresh eggs through to fledged young, as did Shivrajkumar *et al.* (1960) one month later on 19–21 April. They estimated that there were at least half a million chicks at the colony! No such high figures have been reported over the past 50 years and no other colonies anywhere in the world have ever reached one-tenth of these figures. 1970: after three years with no breeding >18,000 nests were occupied (Ali in Koning & Rooth 1975). 1973–74: on 15 November Kahl (in Ali 1974) took aerial photographs of Flamingo City, which held 7,132 occupied nests. It was only the start of breeding since Ali reported eggs at every stage of development as well as chicks up to 40 days on 24 January 1974. The colony held an estimated 10,000 pairs. 1977–87: the Pachham Island colony was not active (Mundkur *et. al.* 1989). 1990–91: breeding was again attempted at Flamingo City by an estimated 25,000–30,000 flamingos, with 10,000–15,000 chicks aged <2 weeks in January (Himmatsinhji 1991; Bapat 1992).

From the above accounts the breeding cycle in the Great Rann seems quite erratic, depending largely on the vagaries of the monsoon. There is no breeding in years of drought. All records of breeding have followed heavy monsoon rains. When the Rann is deeply flooded, however, flamingos either cannot breed or must wait until the water recedes before nesting (Ali 1945, 1974). Ali & Ripley (1978) gave the breeding season as September/October to March/April.

Egg-laying would have taken place from late February to at least March in 1944/45, about August in 1935, from November to April in 1959/60, in November–December in 1990 and from November to at least January in 1973/74. Nests may be used by two successive breeding pairs (Ali 1945).

EAST AFRICA

The Rift Valley lakes of East Africa are among the most renowned in the world for flamingos, in particular for the exceptionally large concentrations of Lessers which occur in Kenya and Tanzania. Greaters are much less abundant but they seem to be present at most sites where the former species occurs, and they may breed in the same place and at the same time as the Lessers. Breeding may take place at almost any time of the year. There are two rainy seasons, the 'long rains' falling March to May and the 'short rains' from November to December. Most records of breeding refer to laying during the latter period. In addition to precipitation, however, which varies greatly from one year to another, Greaters may be triggered into breeding in Kenya by the presence of breeding colonies of Lesser Flamingos (Brown & Britton 1980).

Since the well-documented observations and heroic adventures of Leslie Brown in the 1950s and 1960s, flamingos have received less attention from biologists in East Africa. The areas where they breed remain quite remote, particularly **Lake Natron**, and the data presented here must be considered fragmentary. Breeding by

Greater Flamingos seems to be quite regular on **Lake Elmenteita** (Kenya) and/or **Lake Natron** (Tanzania–Kenya) but has been reported only once from **Lake Magadi** (Kenya) and **Lake Shalla** (Ethiopia). There are records of breeding from L. Nakuru in the more distant past, in 1915 and 1936 (Meinertzhagen 1958) and in 1963 when 6,000–7,000 pairs unsuccessfully attempted to nest on the shoreline (Brown 1973). Breeding has also been attempted at L. Bogoria; in 1978, D. Turner (*in litt.*) reported many nests washed out by heavy rains but saw some chicks in June. Finally, D. Turner (*in litt.*) also reported successful breeding, by 3,000–5,000 pairs, at L. Eyasi (Tanzania) in 1982. Breeding in equatorial Africa can occur in any month of the year.

28: LAKE SHALLA, ETHIOPIA

Coordinates: **07°33′N/38°31′E**   Altitude: **1,558 m**

Site description L. Shalla (43,200 ha) lies in the Rift Valley and is the crater of an extinct volcano. It is separated from L. Abijatta, to the north, by a narrow strip of land, part of the former crater rim. It is deep (max. 266 m) and alkaline with some hot sulphurous springs around the shore. It has nine islands, the one colonised by flamingos in 1988, called Flat Island, being stony and, where the flamingos nested, is devoid of vegetation. The colony was established on terraced slopes of soil and stony ground *c.* 10 m above the lake (Kebede & Hillman 1989).

Wetland status Government land tenure, protected as part of the Abijatta-Shalla National Park.

Flamingo breeding The colony established on Flat Island in 1988 is the only recorded breeding of flamingos in Ethiopia. About 1,000 chicks were discovered on 6 August, the oldest birds estimated to be about two weeks of age. Egg-laying would have begun about 20–25 June.

29: LAKE ELMENTEITA, KENYA

Coordinates: **00°25′S/36°15′E**   Altitude: **1,776 m**

Site description The smallest (1,800 ha) of a chain of alkaline, saline lakes lying in the Rift Valley, and surrounded by farmland. It is a shallow (max. depth 1.2 m), closed-basin lake fed by springs and two small but permanent streams. The water level fluctuates greatly and the lake may be almost dry in years of drought. Much of the flat, muddy shoreline is scattered with stones and rocks of volcanic origin. The western part of the lake has islands of black lava which in dry years are connected to the shore by mudflats, preventing regular use by breeding flamingos.

Wetland status Privately owned and unprotected.

Flamingo breeding Meinertzhagen (in Brown 1958) reported flamingos nesting here in 1903 but it was not until the 1950s that breeding was described in detail (Brown 1955, 1958). Eggs are laid on bare rock or in a slight collection of straw and feathers (Brown 1967). 1951, 1954: no breeding (Brown 1958). 1956: *c.* 3,560 pairs bred, with egg-laying October–December. Disturbance by Marabou Storks but *c.* 1,000 young fledged February–March (Brown 1958). 1957: egg-laying April–June, *c.* 9,000 pairs raised 7,100 young (Brown 1958). 1966: 6,000 pairs raised *c.* 1,100 young (Brown 1973). 1967: successful breeding but no counts (Brown 1973). 1968: 4,500 pairs wiped out by Marabous and flooding in March. A second wave of 8,200 pairs in June disturbed by Great White Pelicans but *c.* 1,200 young raised, and a third attempt by 5,250 pairs in November but most failed because of disturbance by pelicans and humans (Brown 1973). 1970: no breeding (Brown 1973). 1971: colony wiped out by pelicans (Brown 1973). 1977: 500+ pairs successful September–November after failing earlier in the year (D. Turner pers. comm.). 1989: 200–300 pairs successful but colony greatly disturbed by pelicans. 1990: maximum of 200 pairs bred but site taken over by pelicans (D. Turner pers. comm.).

Egg-laying has been reported following the onset of the long rains in April–June and during the period of short rains in October–December.

30: Lake Magadi, Kenya

Coordinates: 01°53′S/36°16′E   Altitude: 579 m

Site description This shallow, alkaline lake (9,700 ha) lies only 25 km north of L. Natron, with which it has much in common. It is a glittering expanse of pink crystalline plates of soda overlying black glutinous mud. It is surrounded and fed by a number of hot springs which maintain permanent lagoons around the main lake. There is a soda factory on the lake shore at Magadi which extracts the trona as brine. The water pH is reported to regularly reach 10.5 (Hughes & Hughes 1992).

Wetland status Unprotected.

Flamingo breeding About 10,000 pairs of Greater Flamingos bred here in 1962 within an enormous Lesser Flamingo colony estimated by Brown and Rooth (1971) to hold *c.* 1,100,000 pairs. It is the only recorded breeding of either species at Magadi. The neighbouring L. Natron (Kenya–Tanzania) is the preferred nesting site, but in 1962 it was flooded all year, preventing breeding there.

Egg-laying, as at the neighbouring L. Natron, can occur at any time of the year but takes place mainly August–November, the end of the cool, dry season and the start of the short rains (Brown & Britton 1980).

31: LAKE NATRON, TANZANIA

Coordinates: 02°20'S/36°00'E   Altitude: 610 m

Site description This is a closed-basin, highly saline lake lying in the Rift Valley. It is 65 km long and has a mean width of 15 km (85,500 ha). The northern extremity of the lake is in Kenya. L. Natron is fed by four rivers and many hot alkaline springs. The water level varies considerably; it can be completely flooded throughout the year (e.g. 1962) but this is rare and it is generally only 20–25% submerged. It lies in a remote area and is the principal breeding site of the Lesser Flamingo in East Africa. The lake bed, composed of a crystalline soda crust overlying alkaline mudflats, is a particularly harsh environment, with midday temperatures of 35°–40°C around the shoreline and 50°–60°C or more on the mud of the lake basin. *Tilapia alcalica* is an endemic fish.

Wetland status Unprotected.

Flamingo breeding L Natron is perhaps the Greater Flamingo's most important breeding area in East Africa, but records are scarce because of the inaccessibility of the colonies, which can only really be observed from the air. Even then it is difficult to distinguish the larger species amongst hundreds of thousands of Lesser Flamingos, but sometimes Greaters nest separately from the Lessers, as in 1968. 1954: two colonies of 100–200 pairs were seen towards the centre of the lake adjoining a much larger Lesser Flamingo colony (Brown 1955, 1958). There were eggs or very small chicks on 20 August but older chicks were also present. 1968: large colony of only Greaters in February (Brown 1975). 1987: 2,000–3,000 pairs bred successfully in June–September with peak in August (D. Turner pers. comm.). 1989: some bred successfully since chicks were seen in December (D. Turner pers. comm.). 1991–92: aerial surveys by R. M. Watson, J. M. Nimmo and G. Theler (*in litt.*) in November and January revealed the presence of Greater Flamingos among a massive breeding by Lessers.

SOUTHERN AFRICA

Two of the most extensive salt depressions in the world are located in southern Africa, the **Makgadikgadi Pans** (Botswana) and **Etosha Pan** (Namibia) and these are clearly the most important sites for flamingo breeding. Both Lessers and Greaters nest according to the flooding of the pans, either separately or in mixed colonies. Successful breeding has also been recorded from **Hoop Vlei** in the Bredasdorp district, at **St Lucia** (Zululand), at the mouth of the Orange River in 1956 and at Calvinia in North Cape in 1997 (Simmons 1997).

Very small colonies or even isolated pairs have been reported attempting to breed in many places in South Africa; for example, in the Orange Free State, a single chick was seen on the Toronto Pan, near Welkom, in October 1951 (Daneel & Robertson 1982). Liversedge (1962) reported 'false-nesting' at St Helena mine in January 1959 and at Allenridge in April of that same year. In this same region, in November 1967, 62 Greater Flamingo chicks were reported on Dankbaar Pan (Daneel & Robertson 1982). In Cape Province, 64 nests were built in January–February 1961 at Rondvlei but they were abandoned (Middlemiss 1961). In 1978 Boshoff (1979) reported 700 unfledged chicks believed abandoned at Van Wyksvlei, Nedersettingstraad dam, when the dam was drained in February. Of these, 623 were caught and released on a nearby wetland used by feeding adults, and they are believed to have eventually fledged. Flamingos are thought to have bred twice previously since 1973 on islands in the Nedersettingstraad dam. In 1996 at Kamfers dam, near Kimberley, Northern Cape Province, 50–100 nests were constructed and six eggs laid in March, the colony later being abandoned (Anderson 1994).

When the Makgadikgadi pan is dry, vast numbers of flamingos, Lessers in particular, must be either on Etosha Pan 980 km to the west, as suggested by Robertson & Johnson (1979), or in the Rift Valley, >2,000 km to the north, as suggested by Tuite (1981a) when referring to the smaller species. In support of movements between Namibia and Botswana is a recovery of a Greater Flamingo on the border of these two countries, 700 km east of Walvis Bay, Namibia, where it was released following an aborted breeding in Etosha in 1994 (Fox *et al.* 1997). There are also reports of nocturnal movements, presumably mostly Lessers, from west to east over Maun (Borello *et. al.* 1998).

Flamingo chicks, Lessers in particular, are known to trek vast distances over the dried salt crust as the water of the Makgadikgadi and Etosha Pans evaporates. Such a phenomenon was first recorded in 1971 at Etosha (Berry 1972) and has since been filmed at Makgadikgadi (Liversedge, *Year of the Flamebird*).

Simmons (1996) reports a decrease in the numbers of flamingos throughout southern Africa, particularly in Namibia.

32: Makgadikgadi and Sua Pans, Botswana

Coordinates: **20°53′S/26°12′E**   Altitude: **910–915 m**

Site description Makgadikgadi is one of the largest (*c.* 1,000,000 ha) salt depressions in the world, covering *c.* 16,000 km². It comprises the Ntwetwe Flats and Sua (Sowa) Pan (*c.* 3,700 km²). The substrate is sandy, alkaline, lacustrine soils overlaid, when dry, by a salt crust. It is shallow, with islands which are remote and completely undisturbed. Flamingos breed in the eastern part, in the Sua Pan (112 km × 72 km), which is flooded by five rivers and by direct precipitation. Mean annual rainfall is 450 mm but can be as low as 75 mm. Most rain occurs

November to April, with great annual variations and irregular flooding of the closed drainage pans. Very large numbers of both Lesser and Greater Flamingos breed on the pan some years. When the pan is dry it is thought that these birds move west to Etosha Pan in Namibia and/or north-east to the Rift Valley lakes. There is a soda ash plant (2,200 ha) at Sua Pan.

Wetland status Makgadikgadi Game Reserve (1970) and the northern part of Sua Pan is protected as the Nata Sanctuary.

Flamingo breeding 1978: in July, Robertson & Johnson (1979) reported 548 occupied nests and 17,183 chicks in crèches. 1988: 50,000–52,000 adults present at the colony and 21,500 chicks raised (Hancock 1990; McCulloch & Irvine 2004). 1996: 5,000–10,000 eggs abandoned (Simmons 1997). 1998–99: no breeding (McCulloch & Irvine 2004). 1999–2000: a record 23,869 pairs nested and raised c. 18,498 chicks (McCulloch & Irvine 2004). 2000–01: 651 pairs nested unsuccessfully (McCulloch & Irvine 2004).

33: Lake St Lucia, South Africa

Coordinates: **27°37′–28°30′S/32°22′–32°34′E**   Altitude: **0.5–1 m**

Site description A shallow (<1 m) but extensive (155,000 ha) coastal dune lake connected to the sea by a long narrow channel (Berruti 1983). The substrate is fine silt but with heavy black vetic clay and fine grey sand where flamingos bred. Porter & Forrest (1974) recorded 15.2‰ salinity in June 1972 and 35.3‰ in December.

Wetland status State owned, a nature reserve and park (1939) and Ramsar site (1986).

Flamingo breeding: 1967: 30 chicks seen on 25 June, the first evidence of breeding at this lake (Porter & Forrest 1974). The nesting area was heavily trampled by hippopotamus prior to breeding. 1972: following exceptionally heavy rains January–March, 6,000 pairs nested and raised 4,000 chicks.

34: De Hoop Vlei, South Africa

Coordinates: **34°27′S/20°22′E**   Altitude: **sea-level**

Site description This lake is the southernmost large permanent body of brackish water in Africa (Uys & Macleod 1967). It is 6.5 km × 0.5 km (750 ha) and lies close to the Indian Ocean coast of Southern Cape. It is separated from the ocean by dunes. It is fed for at least part of the year by the R. Sout and also receives fresh water from springs. Rainfall is 380 mm, mostly March–October. During flooding it has a maximum depth of 8 m and during drought can dry completely. The salinity also varies considerably (3‰–60‰). The substrate is of mud, fine silt and underlying stones. Flamingos bred at Renierskraal Farm, 3.2 km from the sea.

Wetland status De Hoop Vlei is a nature reserve (1956) and Ramsar site (1975) but flamingos bred outside the reserve.

Flamingo breeding 1960: *c.* 800 pairs of flamingos bred in October–November and raised at least 350 chicks, the first mass breeding of flamingos in South Africa (Uys & Martin 1961; Uys *et al.* 1963). 1961: unsuccessful, 120 eggs destroyed by predation. A single pair of flamingos is reported to have bred successfully prior to 1960 (at De Hoop Vlei?), which Middlemiss (1961) refers to only as being in the Bredasdorp district.

35: ETOSHA PAN, NAMIBIA

Coordinates: **19°00′S/16°00′E**   Altitude: **1,073–1,086 m**

Site description This flat, saline depression lying in savanna country is 120 km × 55 km (460,000 ha) and is dry for much of the year. The substrate is of clay, silt or sand overlaid, when dry, with a salt crust in the parts which regularly flood. It is partly flooded when rainfall exceeds 440 mm per year but has not been fully flooded for 50 years (Hughes & Hughes 1992). Water depth does not generally exceed 1 m. Both species of flamingos may breed or attempt breeding, either together or separately.

Wetland status State owned and fully protected, Etosha Game Reserve (1907), National Park (1975) and Ramsar site (1995).

Flamingo breeding Since 1957 known to be an important site but only three major breeding events have been recorded in 40 years (Simmons 1996; Fox *et al.* 1997). 1963: seemingly failed breeding attempt (Winterbottom in Simmons 1996). 1971: one million Lessers and Greaters were present in February. Greaters were breeding February–May, with 27,000 birds at nests in May (Berry 1972). 1974: an aerial census by Kahl (1975b) on 29 April revealed 5,000 pairs, some birds still incubating, others with chicks up to 45 days. 1986: short flooding of the pan, no successful breeding (Archibald & Nott 1987). 1989: 3,000 nests, 700 chicks but the only survivors were captured and hand-reared. 1991–93: no breeding. 1994–95: failed breeding attempts. 1997–98: 15,000–20,000 breeding pairs raised more chicks than during the past two decades (N. Brain in Dodman *et al.* 1997). 2000: several thousand chicks raised at two sites on the pan and fledged young observed from an unknown colony (R. Simmons pers.comm.).

Appendices

APPENDIX 1

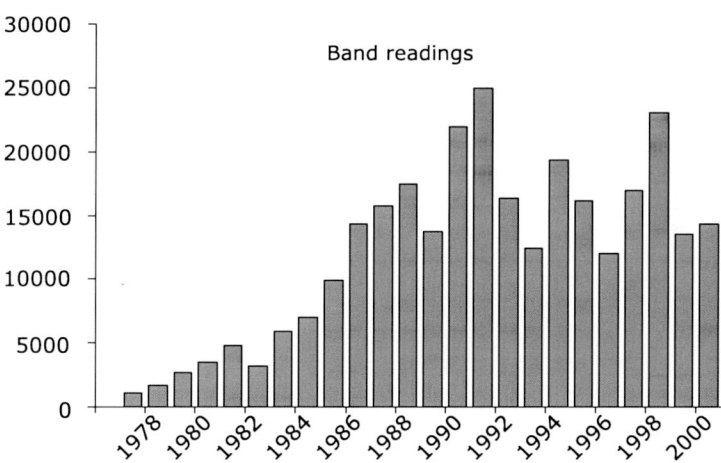

Appendix 1. The number of sightings per year of the flamingos PVC-banded as chicks in the Camargue 1977–2000. N = 289,954.

Year	Readings	Year	Readings
1977	1044	1989	13671
1978	1638	1990	21867
1979	2596	1991	24912
1980	3363	1992	16216
1981	4753	1993	12319
1982	3065	1994	19278
1983	5799	1995	16008
1984	6879	1996	11953
1985	9808	1997	16905
1986	14242	1998	22991
1987	15641	1999	13439
1988	17349	2000	14218
			289954

APPENDIX 2

Appendix 2. Greater Flamingo marking schemes, past and present, in Europe and Asia (casual ringing efforts excluded).

Site	Years	Ring type	Juveniles	Full-grown	Total	Recoveries/ Resightings	Scheme
Caspian Sea USSR	1935–1970	metal only			604	?	Moscow
Rann of Kutch (India)	1945	metal only	192		192	0?	BNHS
Camargue (France)	1947–1961 1977–2000	metal only PVC+ metal	6417 17118	0 59	6417 17177	560 314,008	Tour du Valat/ CRBPO
Fuente de Piedra (Spain)	1964–1986 1986–2000	metal only PVC+ metal	232 9978	127	232 10105	54,185	EBD/ ICONA
Lake Tengiz (Kazakhstan)	1967–1981 1978	metal only collars	8109+	3446+	12040	90	Moscow
L.Uromiyeh (Iran)	1970 1983 1971–1990	collars collars metal	2250 100		2250 100 35000	2 191	Tehran
P.N.Doñana (Spain)	1986	metal only PVC+ metal					EBD/ ICONA
Orbetello (Italy)	1994	PVC+ metal	26	0	26	1144	Bologna
Molentargius - Santa Gilla (Sardinia-Italy)	1997	PVC+ metal	977	0	977	1736	Bologna
Comacchio (Italy)	2000	PVC+ metal	66	0	66		Bologna

APPENDIX 3

Appendix 3. The characteristics of PVC leg-bands which can be read in the field used on Greater Flamingos in the Mediterranean region 1977–2000.

Band type	Band colour	Colour of digits	Site	Band identification
ABCD	yellow or white	black	Camargue (France)	Alphanumerical codes of 3 or 4 letters/numbers. Band on right tibia, metal ring on left tibia.
C I X4	orange or white	black	Fuente de Piedra (Spain)	Black line between first and second digits of 3 or 4 alphanumerical codes. Band on left tibia, metal ring on right tibia.
IAA	blue	white	Orbetello (Italy)	'I' is first of 3 letters. Band on left tibia, metal ring on right tibia.
IBV	blue	white	Comacchio saltpans (Italy)	Series following on from the Orbetello bands.
MABC	red	white	Molentargius and Santa Gilla (Sardinia, Italy)	'M' is first of 4 letters. Band on left tibia, metal ring on right tibia.

APPENDIX 4

Appendix 4. The timing of the annual ringing operation in the Camargue in relation to the start of egg-laying, the age of the chicks and fledging date. Note that the ringing operation takes place on average 107 days (range 91–126, n = 24) after the start of egg-laying (first viable clutches) when the chicks are aged 35–77 days. They can fly on average at 80 days of age (range 71–98, n = 19).

Year	Start of laying	Date of ringing	Time lapse	Age range of chicks	First chicks on wing	oldest chicks
1977	19–20.04	20.07	91	20–63	01.08	74
1978	22.04	02.08	103	44–73	02.08	73
1979	25.04	03.08	99	56–71	03.08	71
1980	18.04	01.08	105	43–76	05.08	80
1981	14.04	27.07	104	52–75	28.07	76
1982	03.04	27.07	115	49–86	–	–
1983	11.04	22.07	102	29–73	–	–
1984	11.04	24.07	104	36–75	–	–
1985	07.04	23.07	106	24–78	25.07	80
1986	03.04	17.07	105	27–76	03.08	93
1987	03.04	22.07	111	21–81	–	–
1988	04.04	20.07	107	27–78	22.07	80
1989	05.05	09.08	96	37–67	16.08	74
1990	06.04	18.07	103	38–74	24.07	80
1991	27.03	31.07	126	24–97	–	–
1992	13.04	29.07	107	38–78	06.08	85
1993	08.04	29.07	112	28–83	25.07	79
1994	28.03	20.07	114	32–85	25.07	90
1995	09.04	02.08	115	34–86	14.08	98
1996	30.04	07.08	99	38–70	12.08	75
1997	18.04	30.07	103	57–74	05.08	80
1998	08.04	29.07	112	55–83	28.07	82
1999	13.04	28.07	106	38–77	29.07	78
2000	03.04	26.07	114	43–85	20.07	79

APPENDIX 5

Appendix 5. The number of flamingos PVC-banded as chicks in the Camargue and recaptured (resighted or recovered) per cohort 1977–2000. Note that 92.6% of flamingos are observed at least once, in the crèche or elsewhere, or are recovered after banding. 77.4% are seen or recovered after fledging.

Year	Number ringed	Seen Fangassier or elsewhere	%	Seen after fledging	%
1977	559	511	91.4	413	73.9
1978	650	607	93.4	533	82.0
1979	651	599	92.0	548	84.2
1980	761	722	94.9	557	73.2
1981	697	685	98.3	575	82.5
1982	650	609	93.7	552	84.9
1983	720	695	96.5	593	82.4
1984	781	739	94.6	622	79.6
1985	552	536	97.1	478	86.6
1986	599	591	98.7	496	82.8
1987	600	599	99.8	527	87.8
1988	600	590	98.3	473	78.8
1989	594	574	96.6	460	77.4
1990	598	595	99.5	567	94.8
1991	518	510	98.5	445	85.9
1992	839	779	92.8	654	77.9
1993	875	727	83.1	632	72.2
1994	850	810	95.3	675	79.4
1995	870	742	85.3	655	75.3
1996	800	751	93.9	595	74.4
1997	954	932	97.7	733	76.8
1998	800	689	86.1	478	59.8
1999	800	691	86.4	531	66.4
2000	800	571	71.4	452	56.5

APPENDIX 6

Appendix 6. The number of pairs of Greater Flamingos recorded breeding throughout the Old World in 1972–2000 (see Figure 14). Data sources are referred to in the site sheets given in Chapter 12.

	W. Africa	Mediterr.	Asia	E. Africa	S. Africa	Total
1972	0	17000	20000	–	6000	43000
1973	0	11090	30000	–	–	41090
1974	2000	16260	5500	–	–	23760
1975	0	5280	–	–	–	5280
1976	160	6650	10000	–	–	16810
1977	0	13690	–	500+	–	14190
1978	0	14605	–	–	18000	32605
1979	1115	11400	–	–	–	12515
1980	1500	7500	–	–	–	9000
1981	6000	9500	–	–	–	15500
1982	16995	14565	–	–	–	31560
1983	3000	14400	–	–	–	17400
1984	0	20100	–	–	–	20250
1985	12940	26050	–	–	–	38990
1986	7500	28080	22180	–	–	57760
1987	9000	24200	10500	2500	–	46200
1988	7480	23700	–	1000	25000	57180
1989	1500	10200	–	250	–	11950
1990	2500	26700	10000	200	–	39400
1991	2500	44820	17300	–	–	64620
1992	1240	28650	9000	–	–	38890
1993	3150	21060	–	–	–	24210
1994	6030	17493	–	–	–	23523
1995	0	17090	1100	–	–	18190
1996	4730	34766	–	–	–	39495
1997	8000	40083	–	–	–	48083
1998	7400	56941	15200	–	17500	97041
1999	13060	17071	750+	–	–	30881
2000	2000	40885	–	–	20000	62885

APPENDIX 7

Appendix 7. The number of pairs and distribution of Greater Flamingos recorded breeding in West Africa in 1972–2000 (see Figure 15). Data sources are indicated on site sheets (see Chapter 12).

	Senegal	Chott Boul	Pet. Kiaone	Grd. Kiaone	Flamant	Total
1972	0	–	–	–	0 ?	0 ?
1973	0	–	–	–	–	0 ?
1974	0	–	20	–	1000	1020+
1975	0	–	–	–	–	0 ?
1976	200	–	–	–	–	200+
1977	0	–	–	–	–	0 ?
1978	0	–	–	–	–	0 ?
1979	215+	–	–	–	1000+	1215+
1980	0	–	–	–	1500	1500+
1981	0	–	c. 3000	c.3000	–	6000
1982	0	–	4883	6442	5670	16995
1983	0	–	–	–	3000	3000
1984	0	–	–	–	–	0 ?
1985	0	–	–	–	12940	12940
1986	0	150 ?	–	–	7500	7650
1987	0	9000	–	–	–	9000
1988	0	3000	–	–	4480	7480
1989	0	–	–	–	1500	1500
1990	0	–	2500	–	–	2500
1991	0	–	2500	–	0	2500
1992	0	?	–	1240+	0	1240+
1993	0	–	–	3000+	150	3150+
1994	0	–	–	5560	470	6030
1995	0	–	2365	2365	0	4730
1996	0	–	?	?	0	0 ?
1997	0	–	0	8000	0	8000
1998	0	–	c. 2500	4900	0	7400
1999	0	–	0	13060	0	13060
2000	0	–	0	c. 2000	0	c. 2000

APPENDIX 8

Appendix 8. The number of pairs of Greater Flamingos recorded breeding in the western Mediterranean in 1972–2000 (see Figure 17). Note that some sites the figure(s) may be the average of a reported range, or the number of chicks fledging. Data sources are indicated on site sheets. In Tunisia, unofficial report of c. 1000 chicks near Kebili May 1992 and 2 juvs.aged c. 4 months at Sfax 10 July 1993.

	Camargue	Tunisia	F. Piedra	Doñana	Alicante	Ebro	Sardinia	Orbetello	Apulia	Petrola	Comacchio	Total
1972	3500	10000	0	0	0	0	0	0	0	0	0	13500
1973	4160	0	3115+	0	115	0	0	0	0	0	0	7390
1974	3560	11500+	0	0	0	0	0	0	0	0	0	15060
1975	5280	0	0	0	0	0	0	0	0	0	0	5280
1976	5500	1000+	0	151	0	0	0	0	0	0	0	6651
1977	9370	0	2820	1500	0	0	0	0	0	0	0	13690
1978	7750	0	4355	2500	0	0	0	0	0	0	0	14605
1979	4200	0	4200	3000	0	0	0	0	0	0	0	11400
1980	7500	0	0	0	0	0	0	0	0	0	0	7500
1981	9500	0	0	0	0	0	0	0	0	0	0	9500
1982	10500	0	2513	1400	0	0	0	0	0	0	0	14413
1983	14400	0	0	0	0	0	0	0	0	0	0	14400
1984	10600	0	5700	3800	0	0	0	0	0	0	0	20100
1985	13500	0	9400	3000	0	0	0	0	0	0	0	25900
1986	19930	0	7600	0	0	0	0	0	0	0	0	27530
1987	17500	0	6700	0	0	0	0	0	0	0	0	24200
1988	11000	0	12500	100	0	0	0	0	0	0	0	23600
1989	10200	0	0	0	0	0	0	0	0	0	0	10200
1990	8600	4200+	13300	0	0	0	0	0	0	0	0	26100
1991	13000	9750	10500	0	0	0	0	0	0	0	0	33250
1992	12500	0 ?	0	0	99	0	0	0	0	0	0	12599
1993	17000	0 ?	0	0	0	1760	1400	0	0	0	0	20160

	Camargue	Tunisia	F. Piedra	Doñana	Alicante	Ebro	Sardinia	Orbetello	Apulia	Petrola	Comacchio	Total
1994	11000	0	2083	0	0	1500	1100	60	0	0	0	15743
1995	13000	0	0	0	0	1500	1140	0	0	0	0	15640
1996	13000	0	16500	1250	0	1500	2000	0	116+	0	0	34366
1997	13500	0	16900	700	200	1273	3330	0	180	0	0	36083
1998	16700	0	19500	1150	1000	1461	4750	0	380	0	0	44941
1999	11000	0	3240	0	0	0	1300	0	250	81	0	15871
2000	22200	0	11500	0	0	1600	2500	0	2300	300	80	40480

APPENDIX 9

Appendix 9. IWC counts of flamingos over the western Mediterranean. Counts of Greater Flamingos made during the Wetlands International mid-January waterfowl censuses in 1965–2000. See Figures 16 and 18. Dash denotes data absent or incomplete.

Year	France	Spain	Portugal	Italy-Sard.	Tunisia	Algeria	Morocco	Total
1965	856	–	–	–	–	–	–	–
1966	662	–	–	–	–	–	–	–
1967	2493	–	–	–	–	–	–	–
1968	1149	–	–	–	–	–	–	–
1969	500	–	–	–	–	–	–	–
1970	2712	–	–	–	–	–	–	–
1971	2222	–	–	–	–	–	–	–
1972	4752	–	–	–	–	–	–	–
1973	8390	8777	–	420	36978	–	–	54565
1974	11131	7684	30	–	9000	–	2029	–
1975	8174	4236	193	1900	38547	781	–	53831
1976	12758	–	30	–	–	–	–	–
1977	13127	–	0	–	25000	1250	–	–
1978	10206	11965	–	8450	38000	1565	–	70186
1979	14370	10668	–	–		1445	1046	–
1980	13616	10036	–	–		400	1223	–
1981	16825	7156	263	5707	25500	541	3098	59090
1982	22414	10575	60	5577	25695	764	1502	66587
1983	14168	8124	5	–	–	–	3225	–
1984	19697	–	0	–	20269	–	573	–
1985	8475	10638	–	–	20365	–	2024	41502
1986	8893	13577	33	–	19831	9243	2650	54227
1987	7769	13278	600	11500	20112	4075	3593	60927
1988	18398	15348	150	–	–	3195	2338	–
1989	22873	17170	620	–	–	1820	2700	–
1990	17929	10574	524	2726	20000	1425	1763	54941
1991	24318	19058	1006	1727	10611	2006	2576	61302
1992	21151	21849	698	6576	14485	5918	1477	72154
1993	27613	28577	3998	8797	9277	2727	4318	85307
1994	25983	26376	3980	13347	12808	6126	3612	92232
1995	27500	21747	2915	10808	10824	300	4506	78600
1996	26580	24344	373	7919	–	5950	–	65166
1997	23656	16895	668	13781	5438	17011	–	77449
1998	23070	24003	1695	10564	–	21060	–	80392
1999	27733	27529	6273	17537	6993	24542	–	110607
2000	26556	–	–	–	27481	–	–	–

APPENDIX 10

Appendix 10. The number of Greater Flamingo chicks fledged in the western Mediterranean in 1972–2000. Data sources are indicated on site sheets. In Tunisia, unofficial report of c. 1000 chicks near Kebili in May 1992 and two juveniles aged c. 4 months at Sfax on 10 July 1993.

	Cam.	F de P	Maris.	Tunisia	Ebro	Sardinia	Orbet.	Apulia	Alicante	Petrola	Comacc	Total
1972	1500	0	0	8000 ?	0	0	0	0	0	0	0	9500
1973	2000	2000	0	0	0	0	0	0	30	0	0	4030
1974	2100	0	0	?	0	0	0	0	0	0	0	2100
1975	2000	0	0	0	0	0	0	0	0	0	0	2000
1976	3500	0	0	0	0	0	0	0	0	0	0	3500
1977	6800	2500	1500	0	0	0	0	0	0	0	0	10800
1978	4500	4700	3300	0	0	0	0	0	0	0	0	12500
1979	2000	4500	0	0	0	0	0	0	0	0	0	6500
1980	3730	0	0	0	0	0	0	0	0	0	0	3730
1981	4000	0	0	0	0	0	0	0	0	0	0	4000
1982	6825	1215	0	0	0	0	0	0	0	0	0	8040
1983	7200	0	0	0	0	0	0	0	0	0	0	7200
1984	4180	4800	3800	0	0	0	0	0	0	0	0	12780
1985	7800	2200	0	0	0	0	0	0	0	0	0	10000
1986	8590	3300	0	0	0	0	0	0	0	0	0	11890
1987	2200	1360	0	0	0	0	0	0	0	0	0	3560
1988	3800	9200	22	0	0	0	0	0	0	0	0	13022
1989	7100	0	0	0	0	0	0	0	0	0	0	7100
1990	5886	10417	0	3000	0	0	0	0	0	0	0	19303
1991	9050	7005	0	7000	0	0	0	0	0	0	0	23055
1992	3200	0	0	0 ?	0	0	0	0	0	0	0	3200
1993	6050	0	0	0 ?	319	900	0	0	0	0	0	7269
1994	7800	478	0	0	306	900	26	0	0	0	0	9510

Appendix 10. continued.

	Cam.	F de P	Maris.	Tunisia	Ebro	Sardinia	Orbet.	Apulia	Alicante	Petrola	Comacc	Total
1995	7800	0	0	0	1294	635–685	0	0	0	0	0	9729
1996	7560	13352	0	0	945	2050	0	116+	0	0	0	24023
1997	6563	13272	19	0	788	2000	0	144	491	0	0	23277
1998	10500	15387	155	0	476	4200	0	218	700	0	0	31636
1999	6600	2205	0	0	0	800	0	200	0	83	0	9888
2000	14500	8019	0	0	1044	2000	0	120	?	212	68	25963

APPENDIX 11

Appendix 11. The numbers of Camargue-banded flamingos recorded breeding in the different Mediterranean and West African colonies in 1982–2000. N = 10,948. These numbers are strongly influenced by annual variations in colony size and accessibility, and by observer effort.

colony and distance from Camargue	Chott Boul 3588 km SW	Doñana 1178 km SW	F. de Piedra 1063 km SW	Alicante 727 km SSW	Ebro delta 455 km SW	Camargue	Cagliari 596 km SSE	Orbetello 546 km ESE	Comacchio 621 km ENE	Apulia 965 km ESE
1982	?	?	?	0	0	4	0	0	0	0
1983	?	?	0	0	0	38	0	0	0	0
1984	?	?	?	0	0	53	0	0	0	0
1985	?	?	?	0	0	195	0	0	0	0
1986	?	?	44	0	0	430	0	0	0	0
1987	?	?	28	0	0	316	0	0	0	0
1988	2	?	200	0	0	276	0	0	0	0
1989	?	?	0	0	0	278	0	0	0	0
1990	?	?	252	0	0	532	0	0	0	0
1991	?	?	20	0	0	972	0	0	0	0
1992	?	?	0	0	0	641	0	0	0	0
1993	?	?	0	0	?	518	12	0	0	0
1994	?	?	5	0	?	747	17	3	0	0
1995	?	?	0	0	?	754	36	0	0	0
1996	?	?	112	0	?	594	2	0	0	0
1997	?	1	73	0	?	756	2	0	0	0
1998	?	?	143	32	27	1026	95	0	0	1
1999	?	?	38	0	0	522	0	0	0	?
2000	?	0	84	0	28	1019	16	0	4	?
total	2	1	999	32	55	9671	180	3	4	1

APPENDIX 12

Appendix 12. Colony of first-recorded breeding of flamingos banded as chicks in the Camargue. N = 3,328. Efforts to resight banded birds began in the Camargue in 1983 but not until 1986 at Fuente de Piedra. These numbers are strongly influenced by annual variations in colony size and accessibility, and by observer effort.

colony and distance from Camargue	Chott Boul 3588 km SW	Doñana 1178 km SW	F. de Piedra 1063 km SW	Alicante 727 km SSW	Ebro delta 455 km SW	Camargue	Cagliari 596 km SSE	Orbetello 546 km ESE	Comacchio 621 km ENE	Apulia 965 km ESE
1982	?	?	?	0	0	4	0	0	0	0
1983	?	?	0	0	0	35	0	0	0	0
1984	?	?	?	0	0	39	0	0	0	0
1985	?	?	?	0	0	167	0	0	0	0
1986	?	?	42	0	0	303	0	0	0	0
1987	?	?	17	0	0	123	0	0	0	0
1988	2	?	121	0	0	70	0	0	0	0
1989	?	?	0	0	0	74	0	0	0	0
1990	?	?	118	0	0	168	0	0	0	0
1991	?	?	5	0	0	350	0	0	0	0
1992	?	?	0	0	?	119	0	0	0	0
1993	?	?	0	0	?	79	8	0	0	0
1994	?	?	1	0	?	130	13	3	0	0
1995	?	?	0	0	?	171	22	0	0	0
1996	?	?	50	0	?	117	2	0	0	0
1997	?	1	45	0	?	128	1	0	0	0
1998	?	?	77	28	18	226	64	0	0	1
1999	?	?	17	0	0	74	0	0	0	?
2000	?	0	35	0	11	237	8	0	4	?
total	2	1	528	28	29	2614	118	3	4	1

APPENDIX 13

Appendix 13. Observations of colour-banded Greater Flamingos seen breeding at Fuente de Piedra (1987–90) and located on their foraging grounds between visits to the colony (birds banded as fledglings in the Camargue 1977–85). Data from Tour du Valat files and from Rendón Martos et al. 2000.

Band code	Sex	Seen at colony	Seen foraging	Location of foraging area	Distance
AAU	♀	24.04.1987 08.05.1987	25.04.1987	Gosque Lagoon	10 km
ACN	♀	11.05.1987	22.05.1987	Guadalquivir Marshes	143 km
		23.05.1987 09.07.1987	19.06.1987	Guadalquivir Marshes	148 km
		12.04.1990 01.05.1990	25.04.1990	Guadalquivir Marshes	143 km
ATB	♀	25.04.1988 29.04.1988	26.04.1988	Guadalquivir Marshes	140 km
		11.05.1988 17.05.1988	15.05.1988	Guadalquivir Marshes	143 km
AXS	♀	05.06.1988 09.06.1988	08.06.1988	Guadalquivir Marshes	145 km
CGU	♀	05.06.1988 09.06.1988	07.06.1988	Guadalquivir marshes	145 km
CKG	♂	25.05.1988 30.05.1988	27.05.1988	Cadiz saltpans	143 km
KAC	♂	21.05.1987 25.05.1987	22.05.1987	Apromasa—Cadiz saltpans	143 km
		10.04.1988 28.05.1988	27.05.1988	Apromasa—Cadiz saltpans	143 km
		08.06.1988 16.06.1988	15.06.1988	Apromasa—Cadiz saltpans	143 km
KFX	♂	07.05.1987 11.05.1987	08.05.1987	FAO—Guadalquivir marshes (Huelva)	143 km
KKT	♂	14.06.1988 18.06.1988	15.06.1988	Cadiz saltpans	143 km
TDS	♀	25.04.1988 16.05.1988	15.05.1988	Campillos Lagoon	10 km
TXV	♂	13.05.1988 20.05.1988	15.05.1988	Cadiz saltpans	143 km
		10.06.1988	28.06.1988	Guadalquivir marshes	145 km
			15.07.1988	Cadiz saltpans	143 km
XFS	♀	01.04.1990 27.04.1990	25.04.1990	Odiel marshes	198 km
XSV	♀	18.04.1988 07.06.1988	26.04.1988	Guadalquivir marshes	140 km

Band code	Sex	Seen at colony	Seen foraging	Location of foraging area	Distance
AD4	♂	20.05.1990 28.05.1990	26.05.1990	Guadalquivir marshes	132 km
BS6	♂	25.05.1988 29.05.1988	27.05.1988	Apromasa—Cadiz saltpans	143 km
CK5	♂	30.04.1988 03.06.1988	15.05.1988 27.05.1988	Cadiz saltpans Cadiz saltpans	143 km 143 km
CS3	♀	01.06.1990 05.06.1990	03.06.1990	Guadalquivir marshes	132 km
KJ7	♂	23.04.1987 27.04.1987	26.04.1987	Bonanza—Cadiz saltpans	143 km
ACBN	♂	11.07.1990 23.07.1990	20.07.1990	Cadiz saltpans	143 km

APPENDIX 14

Appendix 14. The mean egg-laying dates at flamingo colonies in north-west Africa, the Mediterranean and south-west Asia corrected with geographic index for altitude, latitude and longitude. See Figures 37 and 38. Note that the geographical index is the sum of altitude, latitude and longitude corrected. The mean laying date is the Julian Calendar with egg-laying beginning here from 15 December in Chott Boul, Mauritania to 24 May in Kazakhstan.

Alt (m)	Lat (N)	Long	Alt cor	Lat cor	Lon dec	Lon cor	Geo index	Mean laydate	N
L. Tengiz, Kazakhstan	304	50°25'	69°00'E	2.4	50.4	69.0	13.8	66.6	144	3
L. Uromiyeh, Iran	1280	37°20'	45°30'E	10.2	37.3	45.5	9.1	56.6	149	4
Dasht-e-Nawar, Afghanistan	3200	33°50'	67°40'E	25.6	33.8	67.7	13.5	72.9	131	2
Camargue, France	0	43°25'	04°38'E	0	43.4	4.6	0.9	44.3	104	40
Ebro delta, Spain	0	40°35'	00°40'E	0	40.6	0.7	0.1	40.7	106	7
Cagliari, Sardinia	0	39°14'	09°05'E	0	39.2	9.2	1.8	41.0	109	7
Camalti Tuzlasi, Turkey	0	38°28'	27°15'E	0	38.5	27.3	5.5	44.0	132	7
Tuz Gölü, Turkey	899	38°43'	33°22'E	7.2	38.7	33.4	6.7	52.6	103	7
Fuente de Piedra, Spain	434	37°06'	04°44'W	3.5	37.1	-4.7	-0.9	39.7	74	13
Seb. Sidi el Hani, Tunisia	0	35°31'	10°27'E	0	35.5	10.5	2.1	37.6	79	2
Chott Djerid, Tunisia	25	33°50'	08°05'E	0.2	33.8	8.0	1.6	35.6	67	7
Iriki depression, Morocco	515	29°50'	06°30'W	4.1	29.8	-6.5	-1.3	32.6	84	3
Ilot des Flamants, Mauritania	0	20°35'	16°40'W	0	20.6	-16.7	-3.3	17.3	117	7
Kiaone Isl. Mauritania	0	20°01'	16°18'W	0	20.0	-16.3	-3.3	16.7	104	10
Chott Boul, Mauritania	0	16°36'	16°26'W	0	16.6	-16.4	-3.3	13.3	-15	2
Kaolack saltpans, Senegal	0	14°09'	16°08'W	0	14.2	-16.1	-3.2	11.0	5	2

APPENDIX 15

Appendix 15. **The start of egg-laying in relation to the date of flooding of the Etang du Fangassier, Camargue.** Flamingos start laying on average 21 days after the flooding by pumping of the breeding site (range 2–33 days, n = 22).

Year	Lagoon flooded	Start of laying	Time lapse, remarks
1975	April 5	April 18	13 days
1976		April 15	
1977	March 25	April 20	26 days or more
1978		April 22	
1979	April 4–9	April 25	16–21 days
1980		April 18	
1981		April 14	
1982	March 12–16	April 3	18–22 days
1983	March 9–10	April 11	32–33 days
1984	March 19	April 11	23 days
1985	March 18	April 7	20 days or more
1986	March 15	April 3	19 days
1987	March 11	April 3	23 days
1988	March 18	April 4	17 days
1989	April 15	May 5	20 days, late flooding because of strike
1990	March 12	April 6	25 days
1991	March 24–26	March 27	2–3 days, but rainwater around island
1992	March 19–20	April 13	24–25 days
1993	March 18	April 8	21 days or more
1994	March 10–19	March 28	9–18 days
1995	March 27–29	April 9	11–13 days
1996	April 2	April 18	16 days
1997	March 21	April 18	28 days
1998	March 6	April 8	33 days
1999	March 11	April 13	33 days
2000	March 12	April 3	22 days

APPENDIX 16

Appendix 16. Greater Flamingo breeding in the Camargue: egg-laying period, colony size and success.

year	start	delay	end	spread	pairs	chicks	success
1947	04.05	38	?	?	3000	2500	83.3
1948	19.04	23	06.05	18	3000	2500	83.3
1950	09.04	13	12.05	34	3000	1500	50.0
1951	12.04	16	09.05	28	3000	2000	66.7
1952	?	?	?	?	2400	1800	75.0
1953	?	?	?	?	3400	2350	69.0
1954	?	?	?	?	3000	2500	83.3
1955	15.04	19	25.06	65	3500	400	11.4
1956	?	?	?	?	3500	750	21.4
1957	09.04	13	?	?	4000	2350	58.7
1958	21.04	25	?	?	2500	0	0
1959	15.04	19	?	?	3645	585	16.0
1960	15.04	19	15.05	31	8000	1600	20.0
1961	20.04	24	01.06	43	2500	240	9.6
1969	28.04	32	01.06	35	7330	6000	81.9
1970	01.05	35	20.05	20	320	280	87.5
1971	16.04	20	?	?	1850	1500	81.1
1972	10.04	14	10.05	31	3500	1500	42.9
1973	03.05	37	10.05	8	4160	2000	48.1
1974	15.04	19	15.05	31	3560	2100	59.0
1975	18.04	22	15.05	28	5280	2000	37.9
1976	15.04	19	15.05	31	5500	3500	63.6
1977	20.04	24	01.06	43	9370	6800	72.6
1978	22.04	26	21.05	30	7750	4500	58.1
1979	25.04	29	10.05	16	4200	2000	47.6
1980	18.04	22	21.05	34	7500	3730	49.7
1981	14.04	18	07.05	24	9500	4000	42.1
1982	03.04	7	10.05	38	10500	6825	65.0
1983	11.04	15	25.05	45	14500	7200	50.0
1984	11.04	15	20.05	40	10535	4180	39.7
1985	07.04	11	31.05	55	13500	7800	57.8
1986	03.04	7	22.05	50	19926	8590	43.1
1987	03.04	7	02.06	61	17500	2200	12.6
1988	04.04	8	25.05	52	11000	3800	34.5
1989	05.05	39	04.06	31	10200	7100	70.0
1990	06.04	10	12.05	37	8600	5886	68.4
1991	27.03	0	08.06	74	13000	9050	69.6
1992	13.04	17	23.05	41	12500	3200	25.6
1993	08.04	12	02.06	56	17000	6050	35.6
1994	28.03	1	20.05	54	11000	7800	70.9
1995	09.04	13	31.05	53	13000	7800	60.0

Appendix 16. continued.

year	start	delay	end	spread	pairs	chicks	success
1996	30.04	34	31.05	32	13000	7560	58.2
1997	18.04	22	05.05	18	13500	6563	48.6
1998	08.04	12	05.06	59	16700	10500	62.9
1999	13.04	17	22.05	40	11000	6600	60.0
2000	29.03	2	15.05	47	22200	14500	65.3

APPENDIX 17

Appendix 17. Dimensions and weights of 17 freshly abandoned Greater Flamingo eggs from the Camargue.

Length (mm)	Width (mm)	Weight (g)	Date
89.4	56.8	162	30.04.1993
105.0	57.4	180	30.04.1993
95.2	58.5	177	30.04.1993
92.7	56.5	161	30.04.1993
86.1	56.0	153	10.05.1983
91.6	56.0	180	10.05.1983
91.6	58.9	195	10.05.1983
89.3	55.0	160	10.05.1983
84.4	57.3	174	10.05.1983
91.4	55.9	178	10.05.1983
90.6	59.3	194	10.05.1983
88.9	54.7	162	10.05.1983
89.4	56.4	174	10.05.1983
90.1	58.9	192	10.05.1983
89.5	55.8	169	10.05.1983
94.0	54.8	155	14.06.1983
88.5	53.0	?	14.06.1983

APPENDIX 18

Appendix 18. Some stages in the growth of flamingo chicks.

Age	Description
1–3 days	White down, pinkish-orange bill and legs, egg tooth prominent. Brooded for long periods, stands with difficulty day 3 (Plates 14, 15).
4–6 days	Greyish-white down, bill and legs pink becoming greyish, egg tooth still visible. Stands easily day 5 (Plate 16).
7–9 days	Greyish down, bill and legs grey with black joints developing at day 7 and obvious day 9. Actively walks around on nest, pecking at objects, trampling with feet and beating tiny wings. By day 9 wanders out of the nest. In a stretched stance, chick's crown is height of average female's leg joint by day 7 or half-way up average male's tarsus.
10–11 days	Greyish down, bill and legs grey. No longer brooded.
12 days	Legs black, bill slightly curved.
15–16 days	Dark grey down with a noticeably darker patch on back of head. Chick's back is on level with female's leg joint, its crown on a level with male's leg joint.
27 days	Uniform dark grey down. Feather tracts barely visible but form of closed wing discernable above lower back.
50 days	Back and wings are feathered; the leading edge of the closed wing is hidden under feathers on flanks. Belly is white, neck and breast buffish.
70 days	Chick's back is on a level with top of male's tibia.

APPENDIX 19

Appendix 19a. Duration of chick feeding bouts in relation to chick age. The date of hatching of these chicks was known because a parent was identifiable. Data from Camargue 1985–98.

Age of chick	Av. feeding time (min)	Sample size
<20 days	6.0	4
21–40 days	10.1	15
41–60 days	13.7	23
61–80 days	16.5	29
81–100 days	17.4	30
>100 days	24.0	3

Appendix 19b. Duration of chick feeding bouts by male and female parents in relation to chick age. The date of hatching of these chicks was known because a parent was identifiable. Data from Camargue 1985–98. See Figure 46.

Age of chicks	♂♂ (min)	Sample size	♀♀ (min)	Sample size
21–40 days	8.3	10	10.8	5
41–60 days	12.4	7	14.2	16
61–80 days	17.2	19	15.5	10
>81 days	20.2	19	15.1	14

APPENDIX 20

Appendix 20. The number of recoveries per year after ringing (resightings not included) of flamingos marked as chicks in the Camargue in 1947–61 (Museum Paris alloy rings). Data presented in Figure 53.

Year	Recoveries	Year	Recoveries	Year	Recoveries	Year	Recoveries
1	181	11	6	21	4	31	1
2	55	12	6	22	2	32	1
3	37	13	5	23	0	33	0
4	20	14	4	24	4	34	0
5	18	15	4	25	7	35	1
6	26	16	5	26	1	36	1
7	14	17	2	27	3	37	0
8	13	18	4	28	1	38	0
9	11	19	4	29	0	39	1
10	10	20	0	30	1	40	0

APPENDIX 21

Appendix 21.   Scientific names of animals and plants mentioned in the text

Birds

Adelie Penguin *Pygoscelis adeliae*
Andean Condor *Vultur gryphus*
Andean Flamingo *Phoenicoparrus andinus*
Atlantic Puffin *Fratercula arctica*
Banded Stilt *Cladorhynchus leucocephalus*
Barnacle Goose *Branta leucopsis*
Bean Goose *Anser fabalis*
Bewick's Swan *Cygnus columbianus bewickii*
Black-footed Albatross *Diomedea nigripes*
Black Kite *Milvus migrans*
Black Swan *Cygnus atratus*
Blue-eyed Shag *Phalacrocorax atriceps*
Brandt's Cormorant *Phalacrocorax penicillatus*
Brown Pelican *Pelecanus occidentalis*
Caribbean (or American) Flamingo *Phoenicopterus ruber ruber*
Carrion Crow *Corvus corone*
Caspian Tern *Hydroprogne caspia*
Cattle Egret *Bubulcus ibis*
Chilean Flamingo *Phoenicopterus chilensis*
Common Crane *Grus grus*
Common Eider *Somateria mollissima*
Common Gull *Larus canus*
Common Shelduck *Tadorna tadorna*
Coot *Fulica atra*
Cranes *Grus* spp.
Eagle Owl *Bubo bubo*
Egyptian Vulture *Neophron percnopterus*
Emperor Penguin *Aptenodytes forsteri*
Eurasian Curlew *Numenius arquata*
Eurasian Spoonbill *Platalea leucorodia*

Feral Pigeon *Columba livia*
Garganey *Anas querquedula*
Great Blue Heron *Ardea herodias*
Great Cormorant *Phalacrocorax carbo*
Great White Pelican *Pelecanus onocrotalus*
Greater Canada Goose *Branta canadensis*
Greater Flamingo *Phoenicopterus ruber roseus*
Grey-faced (or Great-winged) Petrel *Pterodroma macroptera gouldi*
Grey Heron *Ardea cinerea*
Greylag Goose *Anser anser*
Gull-billed Tern *Gelochelidon nilotica*
Herring Gull *Larus argentatus*
James's (or Puna) Flamingo *Phoenicoparrus jamesi*
Kestrel *Falco tinnunculus*
Kittiwake *Rissa tridactyla*
Laysan Albatross *Phoebastria immutabilis*
Lesser Black-backed Gull *Larus fuscus*
Lesser Flamingo *Phoenicopterus minor*
Little Egret *Egretta garzetta*
Little Penguin *Eudyptula minor*
Marabou Stork *Leptoptilos crumeniferus*
Marsh Harrier *Circus aeruginosus*
Northern Gannet *Morus bassanus*
Northern Mockingbird *Mimus polyglottos*
Northern Shoveler *Anas clypeata*
Osprey *Pandion haliaetus*
Pied Avocet *Recurvirostra avosetta*
Pintail *Anas acuta*
Ring-billed Gull *Larus delawarensis*
Roseate Spoonbill *Ajaia ajaja*
Royal Albatross *Diomedea epomophora*
Royal Tern *Sterna maxima*
Ruddy Duck *Oxyura jamaicensis*
Scarlet Ibis *Eudocimus ruber*
Siberian White Crane *Grus leucogeranus*
Slender-billed Gull *Larus genei*
Sooty Albatross *Phoebetria fusca*
Sooty Tern *Sterna fuscata*

Squacco Heron *Ardeola ralloides*
Sulphur-crested Cockatoo *Cacatua galerita*
Turkey Vulture *Cathartes aura*
White Ibis *Eudocimus albus*
White Stork *Ciconia ciconia*
White-headed Duck *Oxyura leucocephala*
Whooper Swan *Cygnus cygnus*
Wood Stork *Mycteria americana*
Yellow-legged Gull *Larus cachinnans*

Mammals

Badger *Meles meles*
Hippopotamus *Hippopotamus amphibius*
Red Fox *Vulpes vulpes*
Spotted Hyena *Crocuta crocuta*
Wild Boar *Sus scrofa*

Invertebrates and fish

acorn barnacle *Balanus* spp.
brine fly *Ephydra cinerea*
brine shrimp *Artemia* spp.
fairy shrimp *Anostraca* spp.
midges Chironomidae
mullet *Mugil* spp.
ragworm *Nereis diversicolor*
salt fly *Ephydra bivittata*
sand smelt *Atherina boyeri*
snails *Hydrobia* spp.
Soldier Crabs *Dotilla fenestrata*

Plants

ditch grass *Ruppia* spp.
eelgrass *Zostera* spp.
Egyptian evergreen shrub *Thymelaea hirsuta*
Fennel Pondweed *Potamogeton pectinatus*

glasswort *Salicornia* spp., *Arthrocnemum* spp. and *Artemisia* spp.
reeds *Phragmites* spp.
rice *Oryza sativa*
rushes *Juncus* spp.
sedges *Scirpus* spp.

Algae

blue-green algae (Cyanophyta) *Spirulina platensis*
chaetomorph algae *Chaetomorpha linum*
filamentous algae *Cladophora* spp.

Glossary

Ornithological terms are defined mainly as in Campbell & Lack (1985).

AEWA: African-Eurasian Migratory Waterbird Agreement.

assortative mating: a particular pattern of pairing in which pair-members are more similar for one or several characters than expected under the null hypothesis of random pairing in the population (synonymous with homogamy).

attentive period: duration of individual incubation spells between changes of partners.

banding: the process of marking birds on their tibia with a PVC band.

CEEP: Conservatoire—Etudes des Ecosystèmes de Provence—Alpes du sud (Aix-en-Provence).

chott: (Arabic) the area (saline) surrounding a 'sebkha', a seasonal or semi-permanent salt lake.

clade: a group of organisms evolved from a common ancestor.

cladistics: method of classification of animals and plants on the basis of those shared characteristics which are assumed to indicate common ancestry.

CNRS: Centre National de la Recherche Scientifique (National Centre for Scientific Research).

colony: a number of birds breeding gregariously, the term vaguely including the location and the nests.

comfort behaviour: a complex of basic, highly stereotyped maintenance activities concerned with the care of the plumage, such as feather maintenance, bathing, care of skin and bare parts, resting postures.

CRBPO: Centre de Recherche sur la Biologie des Populations d'Oiseaux.

cysts: a dormant stage protected by a thick membrane.

DHKD: Dogal Hayati Koruma Dernegi, the BirdLife Partner in Turkey.

dispersal—natal: movement from the natal colony to the site of the first reproduction;

  breeding: movements between successive breeding attempts.

endorheic: closed basin with no outflow.

evaporator: a shallow lagoon used to concentrate seawater.

false-feeding: during displaying, a bird dips its bill into the water as if going to feed, but does not. This movement is very brief and followed by head-shaking and scratching of the neck.

false-nesting or **false breeding attempt:** nest-building by groups of flamingos, presumed mostly young birds, usually late in the season, and with few, if any, eggs being laid.

false-preening: displacement activity during bouts of display when a bird suddenly and momentarily preens but without any stimuli appropriate to preening being present.

GRIVE: Groupe de recherche et d'informations sur les Vertébrés (Montpellier).

holocrine: producing or being a secretion resulting from lysis of secretory cells. A method of production of digestive juices or other fluids in which the cells of the glands disintegrate to form part of the fluid. Such a mode of secretion occurs in the sebaceous glands of mammals and in the intestines of certain insects.

IBA: Important Bird Area. The IBA Programme of BirdLife International is a worldwide initiative aimed at identifying and protecting a network of sites critical for the conservation of the world's birds.

IUCN: The World Conservation Union (previously International Union for Conservation of Nature and Natural Resources).

IWC: International Waterbird Census

laparoscopy: viewing of the organs of the abdomen by insertion of a fibre-optic instrument through the abdominal wall.

LPO: Ligue pour la Protection des Oiseaux (Aude).

MAB: UNESCO Man and the Biosphere programme.

mattoral shrubland: a plant community or ecosystem characteristic of the regions with Mediterranean-type climates. Also known with the local names of chaparral (California); maquis (Mediterranean), mattoral (Chile), mallee scrub (Australia), fynbos (South Africa).

metapopulation: set of populations living in unconnected habitat patches but linked by movement of individuals between them.

migration: the act of moving from one spatial unit to another.

monogamous: having only one reproductive partner at a time.

obligate: by necessity.

philopatric: faithful to home area.

playa: a desert basin from which water evaporates quickly.

remiges: the main flight feathers (primaries and secondaries).

reverse migration: a phenomenon in which birds fly in a direction opposite to that which they would be expected to take during a particular migration.

ringing: marking birds on their tarsus, or tibia, with a metal ring engraved with a unique code and the address of the ringing scheme.

salt gland: lobes lying on the skull which allow a bird to excrete a concentrated saline solution.

saltpan: a shallow basin used to concentrate seawater and precipitate salt.

sebkha: shallow, salty depression, a semi-permanent salt lake.

semi-precocial: term used to describe young which at hatching have their eyes open, are down-covered, stay at the nest until able to walk, and are fed by their parents.

SPA: Special Protection Area. A strictly protected site classified in accordance with Article 4 of the EC Birds Directive, which came into force in April 1979. SPAs are classified for rare and vulnerable birds, listed in Annex I to the Birds Directive, and for regularly occurring migratory species.

UNESCO: United Nations Educational, Scientific and Cultural Organization.

uniparous: producing but one egg or young at a time.

wind seiche: fluctuations in the water level of a lagoon caused by wind.

References

ABDULALI, H. 1964. On the food and other habits of the Greater Flamingo (*Phoenicopterus roseus* PALLAS) in India. *Journal of the Bombay Natural History Society* 61: 60–68.

AGUILERA, E. 1989. Sperm competition and copulation intervals of the white spoonbill (*Platalea leucorodia*, Aves: Threskiornithidae). *Ethology* 82: 230–237.

AINLEY, D.G., LERESCH, R.E. & SLADEN, W.J.L. 1983. Breeding biology of the Adélie Penguin. University of California Press, Berkeley.

ALAM, M. 1982. The flamingos of Sambhar Lake. *Journal of the Bombay Natural History Society* 79: 194–195.

ALBANESE, G., BACCETTI, N., MAGNANI, A., SERRA, L. & ZENATELLO, M. 1997. Breeding of the Greater Flamingo *Phoenicopterus ruber roseus* in Apulia, S.E. Italy. *Alauda* 65: 202–204.

ALBRECHT, S. 1991. 10,000 birds killed by hailstones. In News and Information. *Bulletin Ornithological Society of the Middle East* 26: 37.

ALERSTAM, T. 1990. *Bird Migration*. Cambridge University Press, Cambridge.

ALERSTAM, T. & BAUER, C.A. 1973. A radar study of the spring migration of the Crane (*Grus grus*) over the southern Baltic area. *Vogelwarte* 27: 1–16.

ALEXANDER, B. 1898. An ornithological expedition to the Cape Verde Islands. *Ibis* 4: 74–118.

ALI, S. 1945. More about the flamingos in Kutch. *Journal of the Bombay Natural History Society* 45: 586–593.

ALI, S. 1960. 'Flamingo City' revisited: nesting of the Rosy Pelican (*Pelecanus onocrotalus* Linnaeus) in the Rann of Kutch. *Journal of the Bombay Natural History Society* 57: 412–415.

ALI, S. 1974. Breeding of the Lesser Flamingo, *Phoeniconaias minor* (GEOFFROY) in Kutch. *Journal of the Bombay Natural History Society* 71: 141–144.

ALI, S. & HUSSAIN, S.A. 1982. Studies on the movement and population structure of Indian avifauna. *Annual Report II*. Bombay Natural History Society, Bombay.

ALI, S. & RIPLEY, S.D. 1978. *Handbook of the Birds of India and Pakistan, with those of Bangladesh, Nepal, Bhutan and Sri Lanka*. Vol. 1. Second edition. Oxford University Press, London.

ALLEN, R.P. 1956. *The Flamingos: their life history and survival*. Research Report No. 5, National Audubon Society, New York.

AMAT, J.A. & GARCIA, L. 1975. Nidificación de *Phoenicopterus ruber* en las Marismas del Bajo Guadalquivir. *Doñana, Acta Vertebrata* 2: 275.

AMETOV, M. 1981. *Birds of Karakalpakia and their Conservation*. Nukus, Karakalpakstan. [In Russian]

AMIARD-TRIQUET, C., PAIN, D. & DELVES, H.T. 1991. Exposure to trace elements of flamingos living in a Biosphere Reserve, the Camargue (France). *Environmental Pollution* 69: 193–201.

ANDERSON, I. 1987. Epidemic of bird deformities sweeps US. *New Scientist* 3 September. 1987: 21.

ANDERSON, D.T. 1998. *Invertebrate Zoology*. Oxford University Press, Melbourne, Oxford, Auckland, New York.

ANDERSON, M.D. 1994. Greater Flamingo breeding attempt at Kamfers Dam, Kimberley. *Mirafra* 11: 45–46.
ANDERSON, D.R., BURNHAM, K.P. & WHITE, G.C. 1985. Problems in estimating age-specific survival rates from recovery data of birds ringed as young. *Journal of Animal Ecology* 54: 89–98.
ANDERSSON, M. 1994. *Sexual Selection*. Princeton University Press, Princeton.
ANDRÉ, P. & JOHNSON, A.R. 1981. Le problème des flamants roses dans les rizières de Camargue et les résultats de la campagne de dissuasion du printemps 1981. *Bulletin Parc Naturel Régional de Camargue* 22–23: 20–35.
ANDRUSENKO, N.N. 1981. On successive moulting of primaries in Greater Flamingo. P. 10 in *Ecology and Protection of Birds*. Shtiintsa, Kishinev. [In Russian]
ANKEY, C.D., AFTON, A.D. & ALISANSKAS, R.T. 1991. The role of nutrient reserves in limiting waterfowl reproduction. *Condor* 93: 1029–1032.
ANON. 1995. Flamingo first from Aldabra. *Oryx* 29: 232–233.
ARAGONESES, J. & ECHEVARRIAS, J.L. 1998. El flamenco vuelve a criar en los humedales del sur de Alicante. *Quercus* 144: 16–18.
ARCHIBALD, T.J. & NOTT, T.B. 1987. The breeding success of Flamingoes in Etosha National Park, 1986. *Madoqua* 15: 269–270.
ARENAS, A., CARRANZA, J., PEREA, A., MIRANDA, A., MALDONADO, A. & HERMOSO, M. 1990. Type A influenza viruses in birds in southern Spain: serological survey by enzyme-linked immunosorbent assay and haemagglutination. *Avian Pathology* 19: 539–546.
ARENGO, F. & BALDASSARRE, G.A. 1995. Effects of food density on the behaviour and distribution of nonbreeding American Flamingos in Yucatan, Mexico. *Condor* 97: 325–334.
ARENGO, F. & BALDASSARRE, G.A. 1999. Resource variability and conservation of American Flamingos in coastal wetlands of Yucatan, Mexico. *Journal of Wildlife Management* 63: 1201.
ARENGO, F. & BALDASSARRE, G.A. 2002. Patch choice and foraging behaviour of nonbreeding American Flamingos in Yucatan, Mexico. *Condor* 104: 452–457.
ARGYLE, F.B. 1975. Report on Bird ringing in Iran, 1970–1974. Unpublished progress report, Ornithology Unit, Division of Parks and Wildlife, Department of the Environment, Iran.
ARGYLE, F.B. 1976. Report on Bird ringing in Iran, 1975. Unpublished progress report, Ornithology Unit, Division of Parks and Wildlife, Department of the Environment, Iran.
ASH, J.S. & MISKELL, J.E. 1998. *Birds of Somalia*. Pica Press, Mountfield.
ASHFORD, H.R., CÉZILLY, F., HAFNER, H. & GORY, G. 1994. Scarcity of haematozoa in some colonial birds in southern France. *Ekologija* 4: 33–35.
ASHMOLE, N.P. 1963. The biology of the Widewake or Sooty Tern *Sterna fuscata* on Ascension Island. *Ibis* 103b: 297–364.
ASHMOLE, N.P. 1971. Seabird ecology and the marine environment. Pp. 224–286 in Farner, D.S. & King, J.R. (eds), *Avian Biology*, Vol. 1. Academic Press, London.
ASHTIANI, M.A.A.Z. 1977. Breeding biology of the Greater Flamingo (*Phoenicopterus ruber roseus*) in Lake Rezaiyeh National Park, Iran. Unpublished MSc thesis, University of Michigan.
ATKINSON-WILLES, G.L. 1969. The mid-winter distribution of wildfowl in Europe, northern Africa and south-west Asia, 1967 and 1968. *Wildfowl* 20: 98–111.
AUSTIN, J.J., CARTER, R.E. & PARKIN, T. 1993. Genetic evidence for extra-pair fertilization in socially monogamous Short-tailed Shearwaters, *Puffinus tenuirostris* (Procellariiformes: Procellaridae), using DNA fingerprinting. *Australian Journal of Zoology* 41: 1–11.
BACCETTI, N., CIANCHI, F., DALL'ANTONIA, P., DE FAVERI, A. & SERRA, L. 1994. Nidificazione de Fenicottero, *Phoenicopterus ruber*, nella Laguna di Orbetello. *Rivista Italiana di Ornitologia* 64: 86–87.

BAHA EL DIN, S.M. 1999. *Directory of Important Bird Areas in Egypt*. The Palm Press, Cairo.
BAILLIE, S.R. 1990. Integrated population monitoring of breeding birds in Britain and Ireland. *Ibis* 132: 151–166.
BAKER, J.R. 1938. The relation between latitude and breeding season in birds. *Proceedings of the Zoological Society of London*, Series A 108: 557–582.
BAKER, R.R. 1978. *The Evolutionary Ecology of Animal Migration*. Hodder & Stoughton, London.
BALDASSARRE, G.A., ARENGO, F. & BILDSTEIN, K.L. (eds). 2000. Conservation biology of flamingos. *Waterbirds* 23 (Special Publication 1).
BALL, D. 1991. A high incidence of bent beaks in nestling Pied Cormorants. *Emu* 91: 257.
BANNERMAN, D.A. 1953. *The Birds of West and Equatorial Africa*. Oliver & Boyd, London.
BANNERMAN, D.A. 1963. *Birds of the Atlantic Islands*. Oliver & Boyd, Edinburgh & London.
BAPAT, N.N. 1992. A visit to the 'Flamingo City' in the Great Rann of Kutch, Gujarat. *Journal of the Bombay Natural History Society* 89: 366–367.
BARBRAUD, C., JOHNSON, A.R. & BERTAULT, G. 2003. Phenotypic correlates of post-fledging dispersal in a population of greater flamingos: the importance of body condition. *Journal of Animal Ecology* 72: 246–257.
BATESON, P. 1983. *Mate Choice*. Cambridge University Press, Cambridge, New York & Melbourne.
BAUER, K.M. & GLUTZ VON BLOTZHEIM, U.N. 1966. *Handbuch der Vögel Mitteleuropas*. Vol. 1. Akademie Verlag, Frankfurt.
BAUER, W., HELVERSEN, O.V., HODGE, M. & MARTENS, J. 1969. Aves. In Kanellis, A. (ed.), *Catalogus Faunae Graeciae*. Privately published, Thessaloniki.
BAYLE, P., DHERMAIN, F. & KECK, G. 1986. Trois cas de saturnisme chez le Flamant rose (*Phoenicopterus ruber*) dans la région de Marseille. *Bulletin Société Linnéenne de Provence.* 38: 95–97.
BEAUCHAMP, G. & MCNEIL, R. 2003. Vigilance in Greater Flamingos foraging at night. *Ethology* 109: 511–520.
BEAUCHAMP, G. 2005. Non-random patterns of vigilance in flocks of the greater flamingo, *Phoenicopterus ruber ruber*. *Animal Behaviour* 71: 593–598.
BEHROUZI-RAD, B. 1992. On the movements of the Greater Flamingo, *Phoenicopterus ruber*, in Iran. *Zoology in the Middle East* 6: 21–27.
BELLROSE, F.C. & CROMPTON, R.C. 1981. Migration speeds of three waterfowl species. *Wilson Bulletin* 93: 121–124.
BENSON, C.W. & BENSON, F.M. 1977. *The Birds of Malawi*. Montfort Press, Limbe.
BENSON, C.W., BROOKE, R.K., DOWSETT, R.J. & IRWIN, M.P.S. 1971. *The Birds of Zambia*. Collins, London.
BERNIS, F. & FERNANDEZ-CRUZ, M. 1965. Actividad del Centro de Migración de la Sociedad Española de Ornitológia, Bienio 1963–1964. *Ardeola* 11: 5–51.
BERNIS, F. & VALVERDE, J.A. 1954. Sur le Flamant rose dans la Péninsule Ibérique. *Alauda* 22: 32–39.
BERRUTI, A. 1983. The biomass, energy consumption and breeding of waterbirds relative to hydrological conditions at Lake St Lucia. *Ostrich* 54: 65–82.
BERRY, H.H. 1972. Flamingo breeding on the Etosha Pan, South West Africa, during 1971. *Madoqua* 5: 5–31.
BERRY, H.H. 1975. South West Africa. Pp. 53–60 in Kear, J. & Duplaix-Hall, N. (eds), *Flamingos*. Poyser, Berkhamsted.
BERRY, H.H. & BERRY, C.U. 1976. Hand-rearing abandoned Greater Flamingoes *Phoenicopterus ruber* L. in Etosha National Park, South West Africa. *Madoqua* 9: 27–32.
BERTHOLD, P. 1993. *Bird Migration: a general survey*. Oxford University Press, Oxford, New York & Tokyo.

BERTHOLD, P. 1996. *Control of Bird Migration*. Chapman & Hall, London, Glasgow, Weinheim, New York, Tokyo, Melbourne & Madras.
BÊTY, J., GAUTHIER, G. & GIROUX, J.-F. 2003. Body condition, migration, and timing of reproduction in snow geese: a test of the condition-dependent model of optimal clutch size. *American Naturalist* 162: 811–816.
BHARUCHA, E. 1987. Some aspects of behaviour observed in the Greater Flamingo at Bhigwan. *Journal of the Bombay Natural History Society* 84: 677–678.
BIBBY, C.J., BURGESS, N.D. & HILL, D.A. 1992. *Bird Census Techniques*. Academic Press, London.
BILDSTEIN, K.L., BALDASSARRE, G.A. & ARENGO, F. 2000. Flamingo science: Current status and future needs. *Waterbirds* 23 (Special Publication 1): 206–211.
BILDSTEIN, K.L., FREDERICK, P.C. & SPALDING, M.G. 1991. Feeding patterns and aggressive behaviour in juvenile and adult American Flamingos (*Phoenicopterus ruber ruber*). *Condor* 93: 916–925.
BILDSTEIN, K.L., GOLDEN, C.B. & MCCRAITH, B.J. 1993. Feeding behavior, aggression, and conservation biology of flamingos: integrating studies of captive and free-ranging birds. *American Zoologist* 33: 117–125.
BILDSTEIN, K.L., POST, W., JOHNSTONE, J. & FREDERICK, P. 1990. Wetlands, rainfall, and the breeding ecology of white ibises in coastal South Carolina. *Wilson Bulletin* 102: 84–98.
BIRD, D.M. 1999. *The Bird Almanac: the ultimate guide to essential facts and figures of the world's birds*. Firefly Books, Buffalo, New York.
BIRDLIFE INTERNATIONAL. 2004. *Important Bird Areas in Asia: key sites for conservation*. BirdLife Conservation Series No. 13, BirdLife International, Cambridge.
BIRKHEAD, T.R. & FURNESS, R.W. 1985. Regulation of seabird populations. Pp. 145–167 in Sibly, R.M. & Smith, R.H. (eds), *Behavioral Ecology, Ecological Consequences of Adaptive Behaviour*. Blackwell, Oxford.
BIRKHEAD, T.R. & MØLLER, A.P. 1992. *Sperm Competition in Birds: evolutionary causes and consequences*. Academic Press, London.
BLACK, J.M. (ed.) 1996. *Partnerships in Birds: the study of monogamy*. Oxford University Press, Oxford.
BLACK, J.M. & OWEN, M. 1995. Reproductive performance and assortative pairing in relation to age in Barnacle Geese. *Journal of Animal Ecology* 64: 234–244.
BLAKER, D. 1967. An outbreak of *botulinus* poisoning among waterbirds. *Ostrich* 38: 144–147.
BLASCO, M., LUCENA, J. & RODRIGUEZ, J. 1979. *Los flamencos de Fuente de Piedra*. Naturalia Hispanica No. 23. ICONA, Madrid.
BLONDEL, J. 1963. Le problème du contrôle des effectifs du Goéland argenté (*Larus argentatus michahellis* Naumann) en Camargue. *La Terre et la Vie* 17: 301–305.
BLONDEL, J. 1964. Compte rendu ornithologique pour les années 1962 et 1963. *La Terre et la Vie* 18: 294–368.
BLONDEL, J. 1966. Compte rendu ornithologique pour les années 1964 et 1965. *La Terre et la Vie* 20: 237–254.
BLONDEL, J., & ARONSON, J. 1999. *Biology and wildlife of the Mediterranean region*. Oxford University Press, Oxford & New York.
BOARD, R.G. 1982. Properties of avian egg shells and their adaptive value. *Biological Reviews* 57: 1–28.
BORELLO, W.D., MUNDY, P.J. & LIVERSEDGE, T.N. 1998. Movements of Greater and Lesser Flamingos in southern Africa. Pp. 201–218 in Lesham, Y., Lachman, E. & Berthold, P. (eds), *Migrating Birds Know No Boundaries*. The Torgos No. 28, Israel Ornithology Centre, Tel-Aviv.
BOSHOFF, A.F. 1979. A breeding record for the Greater Flamingo in the Cape Province. *Ostrich* 50: 124.

BOUDEWIJN, T.J. & DIRKSEN, S. 1995. Impact of contaminants on the breeding success of Cormorant *Phalacrocorax carbo sinensis* in The Netherlands. *Ardea* 83: 325–338.
BOUTIN, J. & CHÉRAIN, Y. 1989. Compte rendu ornithologique camarguais pour les années 1986–1987. *Revue d'Ecologie (La Terre et la Vie)* 44: 165–189.
BOUTIN, J., CHÉRAIN, Y. & VANDEWALLE, P. 1991. Compte rendu ornithologique camarguais pour les années 1988–1989. *Revue d'Ecologie (La Terre et la Vie)* 46: 263–289.
BOWLES, A.E., ELLIS-JOSEPH, S.A. & TODD, F.S. 1988. Re-uniting in three captive penguin species: perspective on the factors promoting long-term pair bonds in the wild. *Cormorant* 16: 121–122.
BOYKO, H. (ed.). 1966. *Salinity and aridity: new approaches to old problems*. Junk, The Hague.
BRADLEY, J.S., WOOLLER, R.D., SKIRA, I.J. & SERVENTY, D.L. 1989. Age-dependent survival of breeding Short-tailed Shearwaters *Puffinus tenuirostris*. *Journal of Animal Ecology* 58: 175–188.
BRESSOU, C. 1931. Le survol de la Camargue par les avions. *Actes de la Réserve Zoologique et Botanique de Camargue: annexe du Bulletin de la Société Nationale d'Acclimatation* 7: 50–52.
BRICHETTI, P. & CHERUBINI, G. 1997. Popolazione di uccelli acquatici nidificanti in Italia. Situazione 1996. *Avocetta* 21: 218–219.
BRITTON, R., GROOT, H. & JOHNSON, A.R. 1985. The daily cycle of feeding activity of the Greater Flamingo in relation to the dispersion of the prey *Artemia*. *Wildfowl* 37: 151–155.
BRITTON, R.H. & JOHNSON, A.R. 1987. An ecological account of a Mediterranean salina: the Salin-de-Giraud, Camargue (S. France). *Biological Conservation* 42: 185–230.
BRODKORB, P. 1963. Catalogue of fossil birds. *Bulletin of the Florida State Museum* 7: 179–293.
BROEKHUYSEN, G.J. 1975. South Africa. Pp.61–64 in Kear, J. & Duplaix-Hall, N (eds), *Flamingos*. Poyser, Berkhamsted.
BROOKE, M. DE L. 1978. Some factors affecting the laying date, incubation and breeding success of the Manx Shearwater, *Puffinus puffinus*. *Journal of Animal Ecology* 47: 477–495.
BROOKE, M. & BIRKHEAD, T. 1991. *The Cambridge Encyclopedia of Ornithology*. Cambridge University Press, Cambridge, New York, Port Chester, Melbourne & Sydney.
BROOKE, R.K. 1984. *South African Red Data Book—Birds*. South African National Science Programmes Report No. 97, South African Foundation for Research Development, Pretoria.
BROSSET, A. 1959. Les oiseaux de l'Embouchure de la Moulouya (Maroc oriental): les migrateurs. *Alauda* 27: 36–60.
BROWN, L.H. 1955. The breeding of Lesser and Greater Flamingoes in East Africa. *Journal of the East African Natural History Society* 22: 159–162.
BROWN, L.H. 1958. The breeding of the Greater Flamingo *Phoenicopterus ruber* at Lake Elmenteita, Kenya Colony. *Ibis* 100: 388–420.
BROWN, L.H. 1959. *The Mystery of the Flamingos*. Country Life, London.
BROWN, L. 1967. Elmenteita. Pp. 244–245 in Pearson (ed.), *Journal of Wildlife and Safari in Kenya*. East African Publishing House, Nairobi.
BROWN, L.H. 1973. *The Mystery of the Flamingos*. Revised edition. East African Publishing House, Nairobi.
BROWN, L.H. 1975. East Africa. Pp. 38–48 in Kear, J. & Duplaix-Hall, N. (eds), *Flamingos*. Poyser, Berkhamsted.
BROWN, L.H. & BRITTON, P.L. 1980. *The Breeding Seasons of East African Birds*. The East African Natural History Society, Nairobi.
BROWN, L.H. & ROOT, A. 1971. The breeding behaviour of the Lesser Flamingo *Phoeniconaias minor*. *Ibis* 113: 147–172.

BROWN, L.H., POWELL-COTTON, D. & HOPCROFT, J.B.D. 1973. The breeding of the Greater Flamingo and Great White Pelican in East Africa. *Ibis* 115: 352–374.
BROWN, L.H., URBAN, E.K. & NEWMAN, K. 1982. *The Birds of Africa*. Academic Press, London, New York, Paris, San Diego, San Francisco, Sao Paulo, Sydney, Tokyo & Toronto.
BROWN, M.J. & PICKERING, S.P.C. 1992. The mortality of captive flamingos at Slimbridge 1975–89. *Wildfowl* 43: 185–190.
BROWN, B. & BOLDT A. 2001. Flight characteristics of birds: 1. Radar measurements of speeds. *Ibis* 143: 178–204.
BUB, H. 1991. *Bird Trapping and Bird Banding*. Cornell University Press, Ithaca.
BUNDY, G. 1976. *The Birds of Libya*. Checklist No. 1. British Ornithologists' Union, London.
BUNDY, G. & WARR, E. 1980. A checklist of the birds of the Arabian Gulf States. *Sandgrouse* 1: 4–49.
BURBRIDGE, A.A. & FULLER, P.J. 1982. Banded Stilt breeding at Lake Barlee, Western Australia. *Emu* 82: 212–216.
BURGER, J. & GOCHFELD, M. 1991. *The Common Tern: its breeding biology and social behaviour*. Columbia University Press, New York.
BURGER, J. & GOCHFELD, M. 1992. Experimental evidence for aggressive antipredator behaviour in black skimmers (*Rhynchops niger*). *Aggressive Behaviour* 18: 241–248.
BURLEY, N. 1983. The meaning of assortative mating. *Ethology and Sociobiology* 4: 191–203.
BURLEY, N. & MORAN, N. 1979. The significance of age and reproductive experience in the mate preferences of feral pigeons, *Columba livia*. *Animal Behaviour* 27: 686–698.
BURNHAM, K.P., ANDERSON, D.R., WHITE, G.C., BROWNIE, C. & POLLOCK, K.H. 1987. *Design and Analysis Methods for Fish Survival Experiments Based on Release-Recapture*. American Fisheries Society Monograph 5.
BUSSE, P. 1980. The new-Euring code. *The Ring* 104–105: 149–156.
CAM, E. & MONNAT, J.Y. 2000. Apparent inferiority in first-time breeders in the kittiwake: the role of heterogeneity among age-classes. *Journal of Animal Ecology* 69: 380–394.
CAMPBELL, B. & LACK, E. (eds). 1985. *A Dictionary of Birds*. Poyser, Berkhamsted.
CAMPREDON, P. 1987. La reproduction des oiseaux d'eau sur le Parc National du Banc d'Arguin (Mauritanie) en 1984–1985. *Alauda* 55: 187–210.
CARP, E. 1980. *A Directory of Western Palearctic Wetlands*. IUCN, Gland.
CARRO, C. & BERNIS, F. 1968. Sobre los nombres de los flamencos (*Phoenicopterus* spp.). *Ardeola* 14: 183–208.
CASLER, C.L. & ESTÉ, E.E. 2000. Caribbean Flamingos feeding at a new solar saltworks in western Venezuela. *Waterbirds* 23 (Special Publication 1): 193–197.
CASTAN, R. 1960. Le Flamant rose en Tunisie (*Phoenicopterus ruber roseus* PALLAS). Nidification dans le Chott Djerid en 1959, et déplacements en cours d'année. *Alauda* 28: 15–19.
CASTAN, R. 1963. Notes de Tunisie (Région de Gabès). *Alauda* 31: 294–303.
CASTRO, H. 1991. Areas de especial interés para el Flamenco y su conservación en Andalucía: las Salinas de Cabo de Gata (Almería). Pp. 209–226 in Junta de Andalucía (ed.), Reunión técnica sobre la situación y problemática del flamenco rosa (*Phoenicopterus ruber roseus*) en el Mediterráneo Occidental y Africa Noroccidental. Agencia de Medio Ambiente, Galan, Seville.
CÉZILLY, F. 1985. Premier bilan de la vague de froid survenue en janvier 1985 pour quelques espèces d'échassiers hivernant en Camargue. *Bulletin of the Association Régionale Provence, Alpes, Côte d'Azur & Corse pour la Protection des Oiseaux et de la Nature* 27: 21–24.
CÉZILLY, F. 1993. Nest desertion in the Greater Flamingo *Phoenicopterus ruber roseus*. *Animal Behaviour* 45: 1038–1040.
CÉZILLY, F. 2004. Assortative mating. Pp. 876–881 in Bekoff, M. (ed.), *The Encyclopedia of Animal Behavior*. Greenwood Press, Westport, Connecticut.

CÉZILLY, F. & JOHNSON, A.R 1992. Exotic Flamingos in the Western Mediterranean region: a case for concern? *Colonial Waterbirds* 15: 261–263.

CÉZILLY, F. & JOHNSON, A.R. 1995. Re-mating between and within breeding seasons in the Greater Flamingo *Phoenicopterus ruber roseus*. *Ibis* 137: 543–546.

CÉZILLY, F. & NAGER, R.G. 1995. Comparative evidence for a positive association between divorce and extra-pair paternity in birds. *Proceedings of the Royal Society* B 262: 7–12.

CÉZILLY, F. & NAGER, R.G. 1996. Age and breeding performance in monogamous birds: the influence of pair stability. *Trends in Ecology and Evolution* 11: 27.

CÉZILLY, F., DUBOIS, F. & PAGEL, M. 2000. Is mate fidelity related to site fidelity? A comparative analysis in Ciconiiformes. *Animal Behaviour* 59: 1143–1152.

CÉZILLY, F., TOURENQ, C. & JOHNSON, A.R. 1994a. Variation in parental care with offspring age in the Greater Flamingo. *Condor* 96: 809–812.

CÉZILLY, F., GOWTHORPE, P., LAMARCHE, B. & JOHNSON, A.R. 1994b. Observations on the breeding of the Greater Flamingo, *Phoenicopterus ruber roseus*, in the Banc d'Arguin National Park, Mauritania. *Colonial Waterbirds* 17: 181–183.

CÉZILLY, F., BOY, V., TOURENQ, C.J. & JOHNSON, A.R. 1997. Age-assortative pairing in the Greater Flamingo *Phoenicopterus ruber roseus*. *Ibis* 139: 331–336.

CÉZILLY, F., VIALLEFONT, A., BOY, V. & JOHNSON, A.R. 1996. Annual variation in survival and breeding probability in Greater Flamingos. *Ecology* 77: 1143–1150.

CÉZILLY, F., BOY, V., GREEN, R.E., HIRONS, G.J.M. & JOHNSON, A.R. 1995. Interannual variation in Greater Flamingo breeding success in relation to water levels. *Ecology* 76: 20–26.

CHAPMAN, A. 1884. Flamingos breeding in Spain. *Ibis* 5th Series 2: 66–99.

CHAPMAN, A. & BUCK, W.J. 1910. *Unexplored Spain*. Incafo, Madrid & London.

CHOUDHURY, S. 1995. Divorce in birds: a review of the hypotheses. *Animal Behaviour* 50: 413–429.

CHOUDHURY, S. & BLACK, J.M. 1993. Mate-selection behaviour and sampling strategies in geese. *Animal Behaviour* 46: 747–757.

CLAY, T. 1974. The Phthiraptera (Insecta) parasitic on flamingoes (Phoenicopteridae: Aves). *Journal of Zoology, London* 172: 483–490.

CLAY, T. 1975. Feather lice. Pp. 159–161 in Kear, J. & Duplaix-Hall, N. (eds), *Flamingos*. Poyser, Berkhamsted.

CLOBERT, J. & LEBRETON, J.-D. 1991. Estimation of demographic parameters in bird populations. Pp. 75–104 in Perrins, C.M., Lebreton, J.-D. & Hirons, G.J.M. (eds), *Bird Population Studies: relevance to conservation and management*. Oxford University Press, Oxford.

CLOBERT, J., LEBRETON, J.-D., ALLAINÉ, D. & GAILLARD, J.M. 1994. The estimation of age-specific breeding probabilities from recaptures or resightings in vertebrate populations. II. Longitudinal models. *Biometrics* 50: 375–385.

CLUTTON-BROCK, T.H. 1988. *Reproductive Success*. University of Chicago Press, Chicago.

COLLAR, N J. & ANDREW, A. 1988. *Birds to Watch: the ICBP world check-list of threatened birds*. International Council for Bird Preservation (Technical Publication No. 8), Cambridge.

COLLAR, N.J. & STUART, S.N. 1985. *Threatened Birds of Africa and Related Islands: the ICBP/IUCN Red Data book*. International Council for Bird Preservation & International Union for Conservation of Nature, Cambridge.

COLLAR, N.J., CROSBY, M.J. & STATTERSFIELD, A.J. 1994. *Birds to Watch 2: the world list of threatened birds*. BirdLife International, Cambridge.

COMIN, F.A., HERRERA-SILVEIRA, J.A. & MARTIN, M. 1997. *Flamingo Footsteps Enhance Nutrient Release from the Sediment to the Water Column*. Hungarian Waterfowl Kozlemenyek, Wetlands International Publication No. 43, Wageningen.

COOPER, J.E. 1975. Capture in Kenya. Pp. 106–108 in Kear, J. & Duplaix-Hall, N. (eds), *Flamingos*. Poyser, Berkhamsted.

COOPER, J.E. & KARSTAD, L. 1975. Tuberculosis in lesser flamingos in Kenya. *Journal of WildlifeDiseases* 11: 32–36.
COPETE, J.L. (ed.). 1998. *Anuari d'ornitologia de Catalunya. 1996.* Grup Catala d'Anellament, Barcelona.
COPETE, J.L. (ed.). 2000. *Anuari d'ornitologia de Catalunya. 1997.* Grup Catala d'Anellament, Barcelona.
CORMACK, R.M. 1964. Estimates of survival from the sighting of marked animals. *Biometrika* 51: 429–438.
COSSON, R.C. & METAYER, C. 1993. Etude de la contamination des flamants de Camargue par quelques éléments traces: Cd, Cu, Hg, Pb, Se et Zn. *Bulletin d'Ecologie* 24: 17–30.
COSSON, R.P., AMIARD, J.C. & AMIARD-TRIQUET, C. 1988. Trace elements in Little Egrets and flamingos of Camargue, France. *Ecotoxicology and Environmental Safety* 15: 107–116.
COSTA, L.T. & RUFINO, R. 1997. Contagens de aves aquáticas em Portugal—Janeiro de 1997. *Airo* 8: 25–32.
COULSON, J.C. 1963. Improved coloured-rings. *Bird Study* 10: 109–111.
COULSON, J.C. 1966. The influence of the pair-bond and age on the breeding biology of the Kittiwake Gull *Rissa tridactyla. Journal of Animal Ecology* 35: 269–279.
CRAMP, S. & SIMMONS, K.E.L. 1977. *The Birds of the Western Palearctic.* Vol. 1. Oxford University Press, Oxford.
CRESPON, J. 1840. *Ornithologie du Gard et des pays circonvoisins.* Castel, Montpellier.
CRESPON, J. 1844. *Faune méridionale.* Vol.1. Nimes Ballivet & Fabre.
CURRY-LINDAHL, K. 1981. *Bird Migration in Africa.* Academic Press, London, NewYork, Toronto, Sydney & San Francisco.
CUTHBERT, F.J. 1985. Mate retention in Caspian Terns. *Condor* 87: 74–78.
DAICKER, B., BRÜCKNER, R., HELDSTAB, A. & PAGAN, O. 1996. Struktur und funktion der Nickhaut beim Rosaflamingo. *Der Ornithologische Beobachter* 93: 59–68.
DALL'ANTONIO, P., BACCETTI, N. & CIANCHI, F. 1996. Origine, fenologia e movimenti dei fenicotteri della Laguna di Orbetello. *Rivista Itialiana di Ornitologia, Milano* 66: 97–117.
DANCHIN, E. & WAGNER, R.H. 1997. The evolution of coloniality: the emergence of new perspectives. *Trends in Ecology and Evolution* 12: 342–347.
DANEEL, A.B.C. & ROBERTSON, H.G. 1982. Two previously undocumented breeding records of the Greater Flamingo in the Orange Free State. *Ostrich* 53: 51–52.
DARLING, F.F. 1938. *Bird Flocks and the Breeding Cycle.* Cambridge University Press, Cambridge.
DARLUC, M. 1782–1786. *Histoire Naturelle de Provence.* Vol. 1. J.J. Niel, Avignon.
DAVIES, N.B. 1991. Mating systems. Pp. 263–294 in Krebs, J.R. & Davies N.B. (eds), *Behavioural Ecology: an evolutionary approach.* Blackwell Scientific Publications, Oxford.
DEAN, W.R.J. 2000. *The Birds of Angola an annotated checklist.* BOU Checklist No. 18. British Ornithologists' Union, Tring.
DE JUANA, A.E. 1996. Noticario ornitológico. *Ardeola* 43: 239–259.
DELACOUR, J. 1961. Anseriformes. Pp. 39–40 in Grey, P. (ed.), *The Encyclopedia of the Biological Sciences.* Chapman & Hall, London.
DELANY, S., REYES, C., HUBERT, E., PIHL, S., REES, E., HAANSTRA, L & VAN STRIEN, A. 1999. *Results from the International Waterbird Census in the Western Palearctic and Southwest Asia 1995 and 1996.* Wetlands International Publication No. 54, Wageningen.
DEL HOYO, J., ELLIOTT, A. & SARGATAL, J. (eds). 1992. *Handbook of the Birds of the World.* Vol. 1. Lynx Edicions, Barcelona.
DE MAROLLES, M. 1836. *La chasse au fusil.* Barrois & Duprat, Paris.
DEMENTE'EV, G.P., MEKLENBURSTEV, R.N., SUDILOVSKAYA, A.M. & SPANGENBERG, E.P. (eds). 1951. *Birds of the Soviet Union.* Vol. II. Demente'ev, G.P. & Gladkov, N.A., Moscow. [In Russian; English translation by Israel Program for Scientific Translations, Jerusalem, 1969]

DERVIEUX, A., LEBRETON, J.-D. & TAMISIER, A. 1980. Technique et fiabilité des dénombrements aeriens de canards et de foulques hivernant en Camargue. *Revue d'Ecologie (La Terre et la Vie)* 34: 69–99.
DI CASTRI, F., GOODALL, D.W. & SPECHT, R.L. (eds). 1981. *Ecosystems of the World 11*: Mediterranean-type shrublands. Elsevier Scientific Publishing Company, Amsterdam, Oxford & New York.
DODMAN, T. & DIAGAMA, C.H. 2003. *African Waterbird Census/Les Dénombrements d'Oiseaux d'Eau en Afrique 1999, 2000 & 2001*. Wetlands International Global Series No. 16, Wageningen.
DODMAN, T. & TAYLOR, V. 1995. *African Waterfowl Census 1995*. IWRB, Slimbridge.
DODMAN, T. & TAYLOR, V. 1996 *African Waterfowl Census 1996*. Wetlands International, Wageningen.
DODMAN, T., BÉIBRO, H.Y., HUBERT, E. & WILLIAMS, E. 1999. *African Waterbird Census 1998*. Les Dénombrements d'Oiseaux d'Eau en Afrique 1998. Wetlands International, Dakar.
DODMAN, T., DE VAAN, C., HUBERT, E. & NIVET, C. 1997. *African Waterfowl Census 1997*. Wetlands International, Wageningen.
DOMERGUE, C. 1949a. Le Chott Djerid, lieu de ponte des Flamants roses. *Bulletin de la Société d'Histoire Naturelle de Tunisie* 2: 32–33.
DOMERGUE, C. 1949b. Le Chott Djerid, station et lieu de ponte du Flamant rose (*Phoenicopterus ruber roseus* LINNE). *Bulletin de la Société des Sciences Naturelles de Tunis* 2: 119–128.
DOMERGUE, C. 1951–52. Les Flamants roses. *Bulletin de la Société des Sciences Naturelles de Tunis* 5: 45–64.
DRAULANS, D. 1988. The importance of heronries for mate attraction. *Ardea* 76: 187–192.
DUBOIS, F., CÉZILLY, F. & PAGEL, M. 1998. Mate fidelity and coloniality in waterbirds: a comparative analysis. *Oecologia* 116: 433–440.
DUGAN, P.J. & JONES, T. 1993. Ecological change in wetlands: a global overview. Pp. 34–38 in Moser, M., Prentice, R.C. & Van Vessem, J. (eds), *Waterfowl and Wetland Conservation in the 1990s: a global perspective*. IWRB Special Publication 26, Slimbridge.
DUNNET, G.M. 1991. Introductory remarks: recruitment in long-lived birds. *Proceedings of the International Ornithological Congress XX*: 1639–1640.
DUNNING, J.B. 1993. *CRC Handbook of Avian Body Masses*. CRC Press, Boca Raton, Ann Arbor, London & Tokyo.
DUPLAIX-HALL, N. & KEAR, J. 1975. Breeding requirements in captivity. Pp. 131–141 in Kear, J. & Duplaix-Hall, N. (eds), *Flamingos*. Poyser, Berkhamsted.
DUPUY, A. 1969. Catalogue ornithologique du Sahara Algérien. *Oiseau et Revue Française d'Ornithologie* 39: 140–160.
DUPUY, A.R. 1976. Reproduction de Flamants roses (*Ph. ruber*) au Sénégal. *Oiseau et Revue Française d'Ornithologie* 46: 294–296.
DUPUY, A.R. 1979. Reproductions de Pélicans blancs et Flamants roses au Sénégal. *Oiseau et Revue Française d'Ornithologie* 49: 323–324.
DUPUY, A. & VERSCHUREN, J. 1978. Note sur les oiseaux, principalement aquatiques, de la région du Parc National du delta du Saloum (Sénégal). *Gerfaut* 68: 321–345.
EKEN, G. 1997. The breeding population of some species of waterbirds at Gediz Delta, Western Turkey. *Zoology in the Middle East* 14: 53–68.
ELLIOTT, C.C.H. & JARVIS, M.J.F. 1973. Fifteenth ringing report. *Ostrich* 44: 34–78.
ELLIS, T.M., BOUSFIELD, R.B., BISSETT, L.A., DYRTING, K.C., LUK, G.S.M., TSIM, S.T., STURM-RAMIRE, K., WEBSTER, G., GUAN, Y. & PEIRIS, J.S.M. 2004. Investigation of outbreaks of highly pathogenic H5N1 avian influenza in waterfowl and wild birds in Hong Kong in late 2002. *Avian Pathology* 33: 492–505.
ENDLER, J.A. 1980. Natural selection on color patterns in *Poecilia reticulata*. *Evolution* 34: 76–91.

ENS, B., CHOUDHURY, S. & BLACK, J. 1996. Mate fidelity and divorce in monogamous birds. Pp. 344–401 in Black, J (ed.). *Partnerships in Birds: the study of monogamy*. Oxford University Press, Oxford.
ENS, B., SAFRIEL, U.N. & HARRIS, M.P. 1993. Divorce in the long-lived and monogamous Oystercatcher *Haematopus ostralegus*: incompatibility or choosing the better option? *Animal Behaviour* 45: 1199–1217.
ERDEM, O. 1989. *Turkey's Bird Paradises*. Green Series 5. Republic of Turkey, Ministry of Environment, Rekmay Ltd, Ankara.
ESPIE, R.H.M., OLIPHAANT, L.W., JAMES, P.C., WARKENTIN, I.G. & LIESKE, D.J. 2000. Age-dependent breeding performance in Merlins (*Falco columbarius*). *Ecology* 81: 3404–3415.
ESTÉ, E.E. & CASLER, C.L. 2000. Abundance of benthic macroinvertebrates in Caribbean Flamingo feeding areas at Los Olivitos Refuge, western Venezuela. *Waterbirds* 23 (Special Publication 1): 87–94.
EVANS, M.E. 1977. Recognising individual Bewick's Swans by bill pattern. *Wildfowl* 23: 153–158.
EVANS, M.E. 1979. The effects of weather on the wintering of Bewick's Swans *Cygnus columbianus bewickii* at Slimbridge, England. *Ornis Scandinavia* 10: 124–132.
EVANS, M.I. 1994. *Important Bird Areas in the Middle East*. BirdLife International, Cambridge.
FAIN, M.G. & HOUDE, P. 2004. Parallel radiations in the primary clades of birds. *Evolution* 58: 2558–2573.
FEDUCCIA, A. 1976. Osteological evidence for shorebird affinities of the flamingos. *Auk* 93: 587–601.
FEDUCCIA, A. 1977. Hypothetical stages in the evolution of modern ducks and flamingos. *Journal of Theoretical Biology* 67: 715–721.
FEDUCCIA, A. 1980. *The Age of Birds*. Harvard University Press, Cambridge.
FEDUCCIA, A. 1996. *The Origin and Evolution of Birds*. Yale University Press, New Haven.
FERNANDEZ-CRUZ, M. 1970. Actividad del Centro de Migración de la Sociedad Española de Ornitológia, Bienio 1969–1970. *Ardeola* 16: 5–29.
FERNANDEZ-CRUZ, M., MARTIN-NOVELLA, C., PARIS, M., IZQUIERDO, E., CAMACHO, M., RENDÓN, M. & RUBIO, J.C. 1988. Revisión y puesta al dia de la invernada del flamenco (*Phoenicopterus ruber roseus*) en la Peninsula Ibérica. Pp. 23–53 in Telleria, J.L. (ed.), *Invernada de Aves en la Peninsula Ibérica*. Monografias de la SEO No. 1.
FERNANDEZ-CRUZ, M., MARTIN-NOVELLA, C., PARIS, M., FERNANDEZ-ALCAZAR, G., SANCHEZ, E.G., NEVADO, J.C., RENDÓN, M. & RUBIO, J.C. 1991. Dinámica de la Población del Flamenco (*Phoenicopterus ruber roseus* Pallas) en España. Pp. 11–45 in *Reunión técnica sobre la situación y problemática del Flamenco rosa* Phoenicopterus ruber roseus *en el Mediterráneo Occidental y Africa Noroccidental*. Junta de Andalucía, Sevilla.
FERRER, M. 1990. Hematological studies in birds. *Condor* 92: 1085–1086.
FERRER, M. 1992. Natal dispersal in relation to nutritional condition in Spanish Imperial Eagles. *Ornis Scandinavica* 23: 104–107.
FERRER, M., GARCIA-RODRIGUEZ, T., CARRILLO, J.C., & CASTROVIEJO, J. 1987. Hematocrit and blood chemistry values in captive raptors. *Comparative Biochemistry and Physiology* 87A: 1123–1127.
FERRER, X., GARCIA, L. & PURROY, F.J. 1976. Sobre el Flamenco en España y su población en 1974. *Bulletin Estación Central Ecológia, Madrid*, Fifth yr. 9: 55–72.
FINNEY, G. & COOKE, F. 1978. Reproductive habits in the Snow Goose: the influence of female age. *Condor* 80: 147–158.
FISHER, H. 1972. The nutrition of birds. Pp. 431–469 in D.S Farner & J.R. King (eds), *Avian Biology* Vol. 2. Academic Press, New York.
FISHER, J. & FISHER, M.L. 1969. The visits of Laysan Albatrosses to the breeding colony. *Micronesica* 5: 173–201.

FISHPOOL, L.D.C. & EVANS, M.I. 2001. Important Bird Areas in Africa and Associated Islands: priority sites for conservation. BirdLife Conservation Series No. 11. UK Pisces Publications & BirdLife International, Newbury & Cambridge.

FJELDSÅ, J. 1977. *Guide to the Young of European Precocial Birds*. Skarv Nature Publications, Strandgården.

FJELDSÅ, J. & KRABBE, N. 1990. *Birds of the High Andes*. Copenhagen Zoological Museum, University of Copenhagen.

FLAMANT, R. 1994. Aperçu des programmes de marquage d'oiseaux à l'aide de bagues de couleur, colliers et marques alaires en Europe. *Aves* 31: 65–186.

FLINT, P.R. & STEWART, P.F. 1992. *The Birds of Cyprus*. Checklist No. 6. Second edition. British Ornithologists' Union, Tring.

FOLLESTAD, A., LARSEN, B.H., NYGARD, T. & ROV, N. 1988. Estimating numbers of moulting Eiders *Somateria mollissima* with different flock size and flock structure. *Fauna norv. Ser. C. Cinclus* 11: 3.

FORSLUND, P. & LARSSON, K. 1991. The effect of mate change and new partner's age on reproductive success in the Barnacle Goose, *Branta leucopsis*. *Behavioural Ecology* 2: 116–122.

FORSLUND, P. & PÄRT, T. 1995. Age and reproduction in birds—hypotheses and tests. *Trends in Ecology and Evolution* 10: 374–378.

FOX, D.L. 1975. Carotenoids and pigmentation. Pp. 162–182 in Kear, J. &. Duplaix-Hall, N. (eds), *Flamingos*. Poyser, Berkhamsted.

FOX, D.L., ELLIOT SMITH, V. & WOLFSON, A.A. 1967. Carotenoid selectivity in blood and feathers of Lesser (African), Chilean and Greater (European) Flamingos. *Comparative Biochemistry and Physiology* 23: 225–232.

FOX, V.E., LINDEQUE, P.M., SIMMONS, R.E., BERRY, H.H., BRAIN, C. & BRABY, R. 1997. Flamingo 'rescue' in Etosha Park, 1994: technical, conservation and economic considerations. *Ostrich* 68: 72–76.

FRAZIER, S. 1996. *Directory of Wetlands of International Importance: an update*. Ramsar Convention Bureau, Gland.

FREDERICK, P.C. 1987. Extrapair copulations in the mating system of White Ibis (*Eudocimus albus*). *Behaviour* 100: 170–201.

FREEMAN, S.N. & NORTH, P.M. 1990. Estimation of survival rates of British, Irish and French Grey Herons. *The Ring* 13: 139–166.

FRETWELL, S.D. & LUCAS, H.L. 1970. On territorial behaviour and other factors influencing habitat distribution in birds. I. Theoretical development. *Acta Biotheoretica* XIX: 16–36.

GABRION, C., MACDONALD-CRIVELLI, G. & BOY, V. 1982. Dynamique des populations larvaires du cestode *Flamingolepis liguloides* dans une population d'*Artemia* en Camargue. *Acta Oecologica, Oecologica Generalis* 3: 273–293.

GADOW, H. 1877. Anatomie des *Phoenicopterus ruber* Pall. und seine Stellung in System. *Journal für Ornithologie* 37: 236–245.

GADOW, H. 1892. On the classification of birds. *Proceedings of the Zoologocal Society of London* 1892: 229–256.

GALLET, E. 1949. Les Flamants roses de Camargue. Payot, Lausanne. (English edition 1952)

GARCIA-ORCOYEN, T.C., VALLECILLO, C.G. & VALLADARES, M.A. 1992. How many inland Mediterranean wetlands will there be in the year 2000? Pp. 28–31 in Finlayson, C.M., Hollis, G.E. & Davies, T.J. (eds), *Managing Mediterranean Wetlands and their Birds. Proceedings of Symposium, Grado, Italy, 1991*. IWRB Special Publication 20, IWRB Slimbridge.

GARCIA RODRIGUEZ, L., CASTRO NOGUEIRA, L., MIRALLES GARCIA, J.M. & CASTRO NOGUEIRA, H. 1982. *Cabo de Gata*. Editorial Everest, León.

GARCIA-RODRIGUEZ, T., FERRER, M., CARILLO, J.C. & CASTROVIEJO, J. 1987. Metabolic responses of *Buteo buteo* to long-term fasting and refeeding. *Comparative Biochemistry and Physiology* 87A: 381–386.

GAVRILOV, E. 1986. Individually marked flamingos from Kazakhstan. *The Ring* 128–129: 236.
GERHARTS, L.D. & VOOUS, K.M. 1968. Natural catastrophes in the flamingo colony of Bonaire, (Netherlands Antilles). *Ardea* 56: 188–192.
GILBERTSON, M., MORRIS, R.D. & HUNTER, R.A. 1976. Abnormal chicks and PCB residue levels in eggs of colonial birds on the lower Great Lakes (1971–73). *Auk* 93: 434–442.
GILISSEN, N., HAANSTRA, L., DELANY, S., BOERE, G. & HAGEMEIJER, W. 2002. *Numbers and Distribution of Wintering Waterbirds in the Western Palearctic and Southwest Asia in 1997, 1998, and 1999: results from the International Waterbird Census.* Wetlands International Global Series No. 11, Wageningen.
GILL, F.B. 1995. *Ornithology.* Second edition. W.H. Freeman, New York.
GISTSOV, A.P. 1994. Flamingo v severo-vostochnom Prikaspii [Flamingos in the north-east part of the Caspian Basin]. *Selevinia* 3: 89–92. [In Russian with English summary]
GLASSOM, D. & BRANCH, G.M. 1997a. Impact of predation by greater flamingos *Phoenicopterus ruber* on the macrofauna of two southern African lagoons. *Marine Ecology Progress Series* 149: 1–12.
GLASSOM, D. & BRANCH, G.M. 1997b. Impact of predation by greater flamingos *Phoenicopterus ruber* on the meiofauna, microflora and sediment properties of two southern African lagoons. *Marine Ecology Progress Series* 150: 1–10.
GLOUTNEY, M.L. & CLARK, R.G. 1991. The significance of body mass to female dabbling ducks during late incubation. *Condor* 93: 811–816.
GOCHFELD, M. 1980. Mechanisms and adaptive value of reproduction synchrony in colonial seabirds. Pp. 207–270 in Burger, J., Olla, B.L., & Winn, H.E. (eds), *Behavior of Marine Animals*, Vol. 4. Plenum Press, New York.
GOODMAN, S.M. & MEININGER, P.L. 1989. *The Birds of Egypt.* Oxford University Press, Oxford.
GOWTHORPE, P., LAMARCHE, B., BINAUX, R., GUEYE, A., LEHLOU, S.M., SALL, M.A. & SAKHO, A.C. 1996. Les oiseaux nicheurs et les principaux limicoles Paléarctiques du Parc National du Banc d'Arguin (Mauritanie). *Alauda* 64: 81–126.
GRAFEN, A. 1990. Biological signals as handicaps. *Journal of Theoretical Biology* 144: 517–546.
GRAVES, J., HAY, R.T., SCOLLAN, M. & ROWES, S. 1991. Extra-pair paternity in the shag *Phalacrocorax aristotelis* as determined by DNA fingerprinting. *Journal of Zoology, London* 226: 399–408.
GREEN, R.E., HIRONS, G.J.M., & JOHNSON, A.R. 1989. The origin of long-term cohort differences in the distribution of Greater Flamingos *Phoenicopterus ruber roseus* in winter. *Journal of Animal Ecology* 58: 543–555.
GRIMMETT, R., INSKIP, C. & INSKIP, T. 1998. *Birds of the Indian Subcontinent.* Christopher Helm, London.
GRIMMETT, R.F.A. & JONES, T.A. 1989. *Important Bird Areas in Europe.* ICBP Technical Publication No. 9. ICBP, IWRB, RSPB, Cambridge.
GRUSSU, M. 1999. La nidificazione del Fenicottero *Phoenicopterus ruber roseus* in Sardegna. *Aves Ichnusae (Bulletin of the Sardinian Ornithological Society)* 2: 3–46.
GUEDES, R.S. & TEIXERA, A.M. 1991. O Flamingos em Portugal. Pp. 53–61 in *Reunión técnica sobre la situación y problematica del Flamenco rosa* Phoenicopterus ruber roseus *en el Mediterráneo Occidental y Africa Noroccidental.* Junta de Andalucía Sevilla.
GUICHARD, G. 1951. Les Flamants de Camargue. *Oiseau et Revue Française d'Ornithologie* 21: 48–54.
GUITART, R., CLAVERO, R., MATEO, R. & MÁÑEZ, M. 2005. Levels of persistent organochlorine residues in eggs of Greater Flamingos from the Guadalquivir marshes (Doñana), Spain. *Journal of Environmental Science and Health* B 40: 753–760.
GURNEY, J.H. 1921. *Early Annals of Ornithology.* H.F. & G. Witherby, London.

GUSTAFSSON, L. & PÄRT, T. 1990. Acceleration of senescence in the collared flycatcher (*Ficedula albicollis*) by reproductive costs. *Nature* 347: 279–281.

GWINNER, E. 1996. Circannual clocks in avian reproduction and migration. *Ibis* 138: 47–63.

HAFNER, H., JOHNSON, A.R. & WALMSLEY, J.G. 1979. Compte rendu ornithologique camarguais pour les années 1976 et 1977. *Revue d'Ecologie (La Terre et la Vie)* 33: 307–324.

HAFNER, H., JOHNSON, A.R. & WALMSLEY, J.G. 1980. Compte rendu ornithologique camarguais pour les années 1978 et 1979. *Revue d'Ecologie (La Terre et la Vie)* 34: 621–647.

HAFNER, H., JOHNSON, A.R. & WALMSLEY, J.G. 1982. Compte rendu ornithologique camarguais pour les années 1980 et 1981. *Revue d'Ecologie (La Terre et la Vie)* 36: 573–601.

HAFNER, H., JOHNSON, A.R. & WALMSLEY, J.G. 1985. Compte rendu ornithologique camarguais pour les années 1982 et 1983. *Revue d'Ecologie (La Terre et la Vie)* 40: 87–112.

HAFNER, H., PINEAU, O. & KAYSER, Y. 1994. Ecological determinants of annual fluctuations in numbers of breeding Little Egrets *Egretta garzetta* in the Camargue, S. France. *Revue d'Ecologie (La Terre et la Vie)* 49: 53–62.

HAFNER, H., PINEAU, O. & WALLACE, J.P. 1992. The effect of winter climate on the size of the Cattle Egret (*Bubulcus ibis* L.) population in the Camargue. *Revue Ecologie (La Terre et la Vie)* 47: 403–410.

HAFNER, H., KAYSER, Y., BOY, V., FASOLA, M., JULLIARD, A.C, PRADEL, R. & CÉZILLY, F. 1998. Local survival, natal dispersal, and recruitment in Little Egrets *Egretta garzetta*. *Journal of Avian Biology* 29: 216–227.

HAGEMEIJER, W.J.M. & BLAIR, M.J. 1997. *The EBCC Atlas of European Breeding Birds: their distribution and abundance.* Poyser, London.

HANCOCK, J.A., KUSHLAN, J.A. & KAHL, M.P. 1992. *Storks, Ibises and Spoonbills of the World.* Academic Press, London.

HANCOCK, P. 1990. The flamingos of Sowa. *Bushcall* 1: 16–21.

HANDRINOS, G. & AKRIOTIS, T. 1997. *The Birds of Greece.* Christopher Helm, A&C Black, London.

HANSEN, T.F. & PRICE, D.K. 1995. Good genes and old age: do old mates provide superior genes? *Journal of Evolutionary Biology* 8: 759–778.

HARRIS, M.P. 1981. Age determination and first breeding of British Puffins. *British Birds* 74: 246–256.

HARRIS, M.P. 1984. *The Puffin.* Poyser, Calton.

HATCH, S.A. 1987. Copulations and mate-guarding in the Northern Fulmar. *Auk* 104: 450–461.

HAWKEY, C., HART, M.G., & SAMOUR, H.J. (1984a). Age-related haematological changes and haematopathological responses in Chilean flamingos (*Phoenicopterus chilensis*). *Avian Pathology* 13: 223–229.

HAWKEY, C., HART, M.G., SAMOUR, H.J., KNIGHT, J.A., & HUTTON, R.E. (1984b). Haematological findings in healthy and sick captive rosy flamingos (*Phoenicopterus ruber ruber*). *Avian Pathology* 13: 163–172.

HEIM DE BALSAC, H. & MAYAUD, N. 1962. *Oiseaux du Nord-Ouest de l'Afrique.* Encyclopédie ornithologique. Lechevalier, Paris.

HENDERSON, M. 1921. Flamingos on migration in Mediterranean. *British Birds* 14: 234–235.

HENRY, G.M. 1971. *A Guide to the Birds of Ceylon.* Oxford University Press, London, New York & Bombay.

HIDALGO DE TRUCIOS, S.J. & CARRANZA, J. 1991. Timing, structure and functions of the courtship display in male Great Bustard. *Ornis Scandinavica* 22: 360–366.

HILL, G.E. 1990. Female house finch prefer colourful males: sexual selection for a condition-dependent trait. *Animal Behaviour* 40: 563–572.
HILL, G.E. 1992. Proximate basis of variation in carotenoid pigmentation in male house finches. *Auk* 109: 1–12.
HIMMATSINHJI, M.K. 1991. The 'flamingo city' in the Rann of Kutch. *Newsletter for Birdwatchers (India)* 31: 3–4.
HOBBS, R.J., RICHARDSON, D.M., & DAVIS, G.W. 1995. Mediterraneran-type ecosystems: opportunities and constraints for studying the function of biodiversity. Pp. 1–42 in Davis, G.W. & Richardson, D.M. (eds), *Ecological Studies* Vol. 109. Springer-Verlag, Berlin Heidelberg.
HOFFMAN, R.H. 1985. An evaluation of banding Sandhill Cranes with colored leg bands. *North American Bird Bander* 10: 46–49.
HOFFMANN, L. 1954. Essais de baguage de Flamants en Camargue. *La Terre et la Vie* 101: 39–43.
HOFFMANN, L. 1955. La nidification des Flamants en 1955. *La Terre et la Vie* 9: 327–328.
HOFFMANN, L. 1957a. La nidification des Flamants en 1956. *La Terre et la Vie* 11: 179–181.
HOFFMANN, L. 1957b. Les effets de la vague de froid de février 1956 sur la faune des vertébrés de Camargue. *La Terre et la Vie* 11: 186–197.
HOFFMANN, L. 1959. La nidification des Flamants en 1957. *La Terre et la Vie* 13: 74–76.
HOFFMANN, L. 1962. La nidification des Flamants en 1959. *La Terre et la Vie* 109: 78–79.
HOFFMANN, L. 1963. La nidification des Flamants en 1960 et 1961. *La Terre et la Vie* 110: 289–297.
HOFFMANN, L. 1964a. La nidification des Flamants en 1962 et 1963. *La Terre et la Vie* 18: 331–333.
HOFFMANN, L. 1964b. La valeur des salins comme milieux biologiques. *IUCN Publication, New Series* 3: 55–69.
HOFFMANN, L. & JOHNSON, A.R. 1991. Extent and control of flamingo damage to rice crops in the Camargue (Rhône delta, S. France). Pp. 119–127 in *Reunion técnica sobre la situación y problemática del Flamenco rosa* Phoenicopterus ruber roseus *en el Mediterraneo Occidental y Africa Noroccidental*. Junta de Andalucia, Sevilla.
HOFFMANN, L., HAFNER, H. & SALATHÉ, T. 1996. The contribution of colonial waterbird research to wetland conservation in the Mediterranean region. *Colonial Waterbirds* 19 (Special Publication 1): 12–30.
HOLLIDAY, R. 1995. *Understanding Ageing*. Cambridge University Press, Cambridge.
HOUDE, P., SHELDON, F.H. & KREITMAN, M. 1995. A comparison of solution and membrane-bound DNA × DNA hybridization, as used to infer phylogeny. *Journal of Molecular Evolution* 40: 678–688.
HOULIHAN, P.F. & GOODMAN, S.M. 1986. *The Birds of Ancient Egypt*. Aris & Phillips, Warminster.
HUGHES, A. 1932. Observations Zoologiques. Pp. 70–74 in *Actes de la Réserve Zoologique et Botanique, Annexe du Bulletin de la Société Nationale d'Acclimatation* No. 8.
HUGHES, B. 1996. The ruddy duck and the threat to the white-headed duck. In Proceedings of the Anatidae 2000 Conference, Strasbourg, France, 5–9 December 1994. Birkan, M., van Vessem, J., Havet, P., Madsen, J., Trolliet, B. & Moser, M. (eds), *Gibier Faune Sauvage, Game and Wildlife 13*, Wetlands International Publication No. 40: 1127–1141.
HUGHES, R.H. & HUGHES, J.S. 1992. *A Directory of African Wetlands*. IUCN, Gland, UNEP, Nairobi, WCMC, Cambridge.
HUNTER, F.M., BURKE, T. & WATTS, S.E. 1992. Frequent copulation as a method of paternity assurance in the Northern Fulmar. *Animal Behaviour* 44: 149–156.
HURLBERT, S.H. & CHANG, C.C.Y. 1983. Ornitholimnology: effects of grazing by the Andean Flamingo (*Phoenicoparrus andinus*). *Ecology* 80: 4766–4769.
HURLBERT, S.H., LOAYZA, W. & MORENO, T. 1986. Fish-flamingo-plankton interactions in the Peruvian Andes. *Limnology and Oceanography* 3: 457–468.

IAPICHINO, C. & MASSA, B. 1989. *The Birds of Sicily*. BOU Checklist No. 11. British Ornithologists' Union, Tring.
IBAÑEZ GONZALEZ, J.M., LOPEZ JURADO, L.F., MCIVOR, J. & TALAVERA TORRALBA, P.A. 1974. Primer dato de reproducción de flamenco (*Phoenicopterus ruber*) en Alicante. *Ardeola* 20: 328–330.
IMBER, M.J. 1976. Breeding biology of the grey-faced petrel *Pterodroma macroptera gouldi*. *Ibis* 118: 51–64.
ISENMANN, P. 1993. *Oiseaux de Camargue*. Société d'Etudes Ornithologiques, Brunoy.
ISENMANN, P. & MOALI, A. 2000. *Oiseaux d'Algérie*. [*Birds of Algeria*.] Société d'Etudes Ornithologiques de France, Paris.
JACOB, J.-P. & JACOB, A. 1980. Nouvelles données sur l'avifaune du lac de Boughzoul (Algérie). *Alauda* 48: 209–219.
JADHAV, A. & PARASHARYA, B.M. 2004. Counts of flamingos at some sites in Gujarat State, India. *Waterbirds* 27(2): 141–146.
JENKIN, P.M. 1957. The filter-feeding and food of flamingoes (*Phoenicopteri*) *Philosophical Transactions of the Royal Society, London* B 240: 410–493.
JOHNSON, A. 1992. The western Mediterranean population of flamingos: is it at risk? Pp. 215–219 in Finlayson, C.M., Hollis, G.E. & Davies, T. J. (eds), *Managing Mediterranean Wetlands and their Birds*. Proceedings of Symposium, Grado, Italy, 1991. IWRB Special Publication No. 20, IWRB, Slimbridge.
JOHNSON, A. 1994. Greater Flamingo *Phoenicopterus ruber*. Pp. 106–107 in Tucker, G.M. & Heath, M. F. *Birds in Europe: their conservation status*. BirdLife Conservation Series No. 3, BirdLife International, Cambridge.
JOHNSON, A. 1997a. *Phoenicopterus ruber* Greater Flamingo. In *BWP Update* 1: 15–23.
JOHNSON, A. 1997b. Long-term studies and conservation of Greater Flamingos in the Camargue and Mediterranean. *Colonial Waterbirds* 20: 306–315.
JOHNSON, A., MESLÉARD, F. & RIOLS, C. 1997. Deux espèces à valeur patrimoniale: le Flamant rose et la Grue cendrée. Pp. 53–68 in Clergeau, P. (ed.), *Oiseaux à risque en ville et en campagne*. INRA, Paris.
JOHNSON, A.R. 1966. Les Flamants en 1964 et 1965. *La Terre et la Vie* 20: 255–257.
JOHNSON, A.R. 1970. La nidification des Flamants en 1968 et 1969. *La Terre et la Vie* 24: 594–603.
JOHNSON, A.R. 1973. La nidification des flamants de Camargue en 1970 et 1971. *La Terre et la Vie* 27: 95–101.
JOHNSON, A.R. 1975a. La nidification des flamants de Camargue en 1972 et 1973. *La Terre et la Vie* 29: 113–115.
JOHNSON; A.R. 1975b. Camargue flamingos. Pp. 17–25 in Kear, J. &. Duplaix-Hall, N. (eds), *Flamingos*. Poyser, Berkhamsted.
JOHNSON, A.R. 1976. La nidification des flamants de Camargue en 1974 et 1975. *La Terre et la Vie* 30: 593–598.
JOHNSON, A.R. 1979. Greater Flamingo (*Phoenicopterus ruber roseus*) ringing in the Camargue and an analysis of recoveries. *The Ring* 100: 53–58.
JOHNSON, A.R. 1982. Construction of a breeding island for flamingos in the Camargue. Pp. 204–208 in Scott, D.A. (ed.), *Managing Wetlands and their Birds*. International Waterfowl Research Bureau, Slimbridge:
JOHNSON, A.R. 1983. Etho-écologie du Flamant rose (*Phoenicopterus ruber roseus* Pallas) en Camargue et dans l'ouest Paléarctique. Unpublished PhD thesis, Université Paul Sabatier, Toulouse.
JOHNSON, A.R. 1985. *Les effets de la vague de froid de janvier 1985 sur la population de flamants roses hivernant en France*. IWRB/ICBP Flamingo Working Group Special Publication No. 2.
JOHNSON, A.R. 1989a. Movements of Greater Flamingos *Phoenicopterus ruber roseus* in the western Palearctic. *La Terre et la Vie* 44: 75–94.

JOHNSON, A.R. 1989b. Population studies and conservation of Greater Flamingos in the Camargue. Pp. 49–63 in Spaans, A.L. (ed.), *Wetlands en Watervogels*. Pudoc, Wageningen.
JOHNSON, A.R. 1990. Taking a closer look at the flamingos on Cyprus in winter. *Gli Uccelli d'Italia* 15: 5–10.
JOHNSON, A.R. 1991. Conservation of breeding flamingos in the Camargue (Southern France). *Species* 17: 33–34.
JOHNSON, A.R. 1995 (ed.). *IWRB Flamingo Specialist Group Newsletter* No. 7.
JOHNSON, A.R. 1999. Flamant rose *Phoenicopterus ruber roseus*. Pp. 332–333 in Rocamora, G., & Yeatman-Berthelot, D. *Oiseaux menacés et à surveiller en France. Listes rouges et recherche de priorités. Populations. Tendances. Menaces. Conservation.* Société d'Etudes Ornithologique de France/Ligue pour la Protection des Oiseaux, Paris.
JOHNSON, A.R. 2000a. An overview of the Greater Flamingo ringing program in the Camargue (southern France) and some aspects of the species' breeding biology studied using marked individuals. *Waterbirds* 23 (Special Publication 1): 2–8.
JOHNSON, A.R. 2000b. Flamingo Specialist Group: past, present, and future activities. *Waterbirds* 23 (Special Publication 1): 200–205.
JOHNSON, A.R. & GREEN, R.E. 1990. Survival and breeding of Greater Flamingos *Phoenicopterus ruber roseus* in the wild after a period of care in captivity. *Wildfowl* 41: 117–121.
JOHNSON, A.R., GREEN, R.E. & HIRONS, G.J.M. 1991. Survival rates of Greater Flamingos in the west Mediterranean region. Pp. 250–271 in Perrins, C.M., Lebreton, J.-D. & Hirons, G.J.M. (eds), *Bird Population Studies: relevance to conservation and management*. Oxford University Press, Oxford.
JOHNSON, A.R., CÉZILLY, F. & BOY, V. 1993. Plumage development and maturation in the Greater Flamingo *Phoenicopterus ruber roseus*. *Ardea* 81: 25–34.
JOHNSON, K. 1988. Sexual selection in pinion jays. I: female choice and male–male competition. *Animal Behaviour* 36: 1038–1047.
JOHNSON, K.P., KENNEDY, M. & MCCRACKEN, K. 2006. Reinterpreting the origins of flamingo lice: cospeciation or host-switching? *Biology Letters* 2: 275–278.
JOHNSTON, V.H. & RYDER, J.P. 1987. Divorce in larids: a review. *Colonial Waterbirds* 10: 16–26.
JOLLY, G.M. 1965. Explicit estimates from capture-recapture data with both death and immigration-stochastic models. *Biometrika* 52: 225–247.
JUNTA DE ANDALUCÍA (ed.) 1991. Reunión técnica sobre la situación y problemática del flamenco rosa (*Phoenicopterus ruber roseus*) en el Mediterráneo Occidental y Africa Noroccidental. Agencia de Medio Ambiente, Galan, Seville.
KAHL, M.P. 1975a. Distribution and numbers: a summary. Pp. 93–102 in Kear, J. & Duplaix-Hall, N. (eds), *Flamingos*. Poyser, Berkhamsted.
KAHL, M.P. 1975b. Ritualized displays. Pp. 142–149 in Kear, J. & Duplaix-Hall, N. (eds), *Flamingos*. Poyser, Berkhamsted.
KAHL, M.P. 1975c. Flamingo Group. *Bulletin ICBP* 12: 220–222.
KALETA, E.F. & MARSCHALL, H.J. 1981. Newcastle disease in a zoo affecting demoiselle cranes (*Anthropoides virgo*), greater flamingos (*Phoenicopterus ruber*) and a pied imperial pigeon (*Ducula bicolor*). *Avian Pathology* 10: 395–401.
KANYANIBWA, S., SCHIERER, A., PRADEL, R. & LEBRETON, J.-D. 1990. Changes in adult survival in a western European population of White Stork (*Ciconia ciconia*) *Ibis* 132: 27–35.
KASPAREK, M. 1985. Die Sultanssumpfe, Naturgeschichte eines Vogelparadieses in Anatolien. Max Kasparek Verlag, Heidelberg.
KATONDO, J. 1997. Aperçu des dénombrements nationaux d'oiseaux d'eau de janvier 1995 en Tanzanie. In Dodman, T. (ed.), Stratégie préliminaire pour le suivi des Oiseaux d'eau en Afrique. *Wetlands International—AEME Publication* 43: 125–129.

KAYSER, Y., GIRARD, C., MASSEZ, G., CHÉRAIN, Y., COHEZ, D., HAFNER, H., JOHNSON, A., SADOUL, N., TAMISIER, A. & ISENMANN, P. 2003. Compte rendu ornithologique camarguais pour les années 1995–2000. *Revue d'Ecologie (Terre Vie)* 58: 5–75.

KEAR, J. & DUPLAIX-HALL, N. (eds). 1975. *Flamingos*. Poyser, Berkhamsted.

KEBEDE, E. & HILLMAN, J.C. 1989. First recorded breeding of Greater Flamingo *Phoenicopterus ruber roseus* in Ethiopia. *IWRB–ICBP Flamingo Research Group Newsletter* No.5.

KHOKHLOV, A.N. 1995. Ornithological observations in Western Turkmenia. Stavropol University Press, Stavropol. [In Russian]

KHROKOV, V.V. 1996. *The Red Data book of Kazakhstan*. Vol. 1 Animals, Part 1 Vertebrates. Third edition. Almaty, Istanbul.

KHUMAR, S. & BHARGAVA, R.N. 1996. Sambha Lake—a new breeding ground for flamingos in India. *Sanctuary Asia* 16: 59.

KILIC, A. 1988. The Eregli Marshes: a new nesting site for the Greater Flamingo, *Phoenicopterus ruber roseus*, in Turkey. *Zoology in the Middle East* 2: 39–42.

KING, C.E. 1994. Management and research implications of selected behaviours in a mixed colony of flamingos at Rotterdam Zoo. *International Zoo Yearbook* 33: 103–113.

KIRWAN, G. 1992. A freshwater breeding record of Greater Flamingo *Phoenicopterus ruber* in Turkey. *Sandgrouse* 14: 56–57.

KLOCKENHOFF, H. & MADEL, G. 1970. Uber die Flamingos (*Phoenicopterus ruber roseus*) der Dasht-e-Nawar in Afghanistan. *Journal für Ornithologie* 111: 78–84.

KNOX, A.G., COLLINSON, M., HELBIG, A.J., PARKIN, D.T. & SANGSTER, G. 2002. Taxonomic recommendations for British birds. *Ibis* 144: 707–710.

KOMDEUR, J., BERTELSEN, J. & CRACKNELL, G. 1992. *Manual for Aeroplane and Ship Surveys of Waterfowl and Seabirds*. IWRB Special Publication No.19, Slimbridge.

KONING, F.J. & ROOTH, J. 1975. Greater Flamingos in Asia. Pp. 33–34 in Kear, J. & Duplaix-Hall, N. (eds), *Flamingos*. Poyser, Berkhamsted.

KUMERLOEVE, H. 1962. Der Flamingo, *Phoenicopterus ruber*, in Kleinasien und Syrien. *Vogelwelt* 83: 177–181.

KUMERLOEVE, H. 1966. Le Lac Djabboul à l'est d'Alep (Syrie), lieu d'hivernage des Flamants. *Alauda* 34: 39–44.

KUSHLAN, J.A. 1987. Recovery plan for the US breeding population of the Wood Stork. *Colonial Waterbirds* 10(2): 259–262.

KUSHLAN, J.A. 1993. Colonial waterbirds as bioindicators of environmental change. *Colonial Waterbirds* 16(2): 223–251.

LACK, D. 1954. *The Natural Regulation of Animal Numbers*. Clarendon, Oxford.

LACK, D. 1968a. Bird migration and natural selection. *Oikos* 19: 1–9.

LACK, D. 1968b. *Ecological Adaptations for Breeding in Birds*. Methuen & Co. Ltd, London.

LANG, E.M. 1963. Flamingoes raise their young on a liquid containing blood. *Experientia* 19: 532–533.

LANG, E.M., THIERSCH, A., THOMMEN, H. & WACKERNAGEL, H. 1962. Was füttern die Flamingo (*Phoenicopterus ruber*) ihren Jungen? *Der Ornithologische Beobachter* 59: 173–176.

LAVAUDAN, L. 1924. *Voyage de Guy Babault en Tunisie*. Résultat scientifique Oiseaux, Paris.

LEBRETON, J.-D. & CLOBERT, J. 1991. Bird population dynamics, management and conservation: the role of mathematical modelling. Pp. 105–125 in Perrins, C.M., Lebreton, J.-D. & Hirons, G.J.M. (eds), *Bird Population Studies: relevance to conservation and management*. Oxford University Press, Oxford.

LEBRETON, J.-D., PRADEL, R. & CLOBERT, J. 1993. The statistical analysis of survival in animal populations. *Trends in Ecology and Evolution* 8: 91–95.

LEBRETON, J.-D., HEMERY, G., CLOBERT, J. & COQUILLART, H. 1990. The estimation of age-specific breeding probabilities from recaptures or resightings in vertebrate populations. I. Transversal models. *Biometrics* 46: 609–622.

LEBRETON, J.-D., BURNHAM, K.P., CLOBERT, J. & ANDERSON, D.R. 1992. Modeling survival and testing biological hypotheses using marked animals: a unified approach with case studies. *Ecological Monographs* 62: 67–118.
LEDANT, J.-P. & VAN DIJK, G. 1977. Situation des zones humides algériennes et de leur avifaune. *Aves* 14: 217–232.
LEFTWICH, A.W. 1968. *A Dictionary of Zoology*. Cox & Wyman Ltd, London, Reading & Fakenham.
LI, Z.W.D. & MUNDKUR, T. 2004. *Numbers and Distribution of Waterbirds and Wetlands in the Asia-Pacific Region. Results of the Asian Waterbird Census: 1997–2001.* Wetlands International, Kuala Lumpur.
LIECHTI, F. & BRUDERER, B. 1998. The relevance of wind for optimal migration theory. *Journal of Avian Biology* 29: 561–568.
LINDGREN, C.J. & PICKERING, S.P.C. 1997. Ritualised displays and display frequencies of Andean Flamingos *Phoenicopterus andinus*. *Wildfowl* 48: 194–201.
LITVINENKO, N.M. & SHIBAEV, YU.V. 1999. Some new ornithological records and observations from the extreme south-west part of Primorye. *The Russian Journal of Ornithology* 71: 9–16.
LIVERSEDGE, R. 1962. Further notes on the wildfowl of the O.F.S. goldfields. *Ostrich* 29: 29–32.
LLOYD, C.S. & PERRINS, C.M. 1977. Survival and age of first breeding in the Razorbill (*Alca torda*). *BIRD-BANDING* 48: 239–252.
LOGAN, C.A. 1991. Mate switching and mate choice in female Northern Mockingbirds: facultative monogamy. *Wilson Bulletin* 103: 277–281.
LOMONT, H. 1938. Rapport ornithologique. *Actes de la Réserve de Camargue* 21: 13–20.
LOMONT, H. 1953. Sur le comportement nourricier de *Phoenicopterus ruber roseus* PALLAS. *Vie et Milieu* 4: 713–717.
LOMONT, H. 1954a. Observations ornithologiques sur les Flamants. *La Terre et la Vie* 8: 28–38.
LOMONT, H. 1954b. Rapport sur le baguage des jeunes flamants en 1953. *La Terre et la Vie* 8: 44–46.
LOPEZ, A. & MUNDKUR, T. (eds). 1997. *The Asian Waterfowl Census 1994–1996. Results of the coordinated waterbird census and an overview of the status of wetlands in Asia.* Wetlands International, Kuala Lumpur.
LOTT, D., SCHOLZ, S.D. & LEHRMAN, D.S. 1967. Exteroceptive stimulation of the reproductive system of the female Ring Dove (*Streptopelia risoria*) by the male and by the colony milieu. *Animal Behaviour* 15: 433–437.
LOZANO, G.A. 1994. Carotenoids, parasites, and sexual selection. *Oikos* 70: 309–311.
LOZANO, G.A. 2001. Carotenoids, immunity, and sexual selection: comparing apples and oranges? *American Naturalist* 158: 200–203.
LUKE, A. 1992. Spanish pilots force flamingos to flee. *New Scientist* 5 September 1992, No. 1837.
LUKOVTSEV, YU. S. 1990. Flamingo. Short communication, p. 81 in *Materials for Red Data Book*. Central Scientific Research Laboratory, Main Department of Hunting Industry, Moscow. [In Russian]
LYON, B.E. 1993. Conspecific brood parasitism as a flexible female reproductive tactic in American coots. *Animal Behaviour* 46: 911–928.
MACDONALD, G.H. 1980. The use of *Artemia* cysts as food by the flamingo (*Phoenicopterus ruber roseus*) and shelduck (*Tadorna tadorna*). *The Brine Shrimp Artemia* 3: 97–104.
MACNAE, W. 1960. Greater Flamingoes eating crabs. *Ibis* 102: 325–326.
MADON, P. 1932. Contribution à l'étude du régime alimentaire du Flamant rose. *Alauda* 4: 37–40.
MADRIGAL DIAZ, L. 1999. Flamencos y manchegos. *Quercus* 155: 7.

MAGNIN, G. & YARAR, M. 1994. Some notes on the breeding of the Greater Flamingo *Phoenicopterus ruber* and White Pelican *Pelecanus onocrotalus* in Turkey. *Bulletin Ornithological Society of the Middle East* 32: 28–30.
MAGNIN, G. & YARAR, M. 1997. *Important Bird Areas in Turkey*. Dogal Hayati Koruma Dernegi, Istanbul.
MAHÉ, E. 1985. Contribution à l'étude scientifique de la région du Banc d'Arguin (littoral mauritanien: 21°20′N/19°20′W). Peuplements avifaunistiques. Unpublished PhD thesis, Université des Sciences et Techniques du Languedoc, Montpellier.
MAINARDI, D. 1962. Immunological data on the philogenetic relationship and taxonomic position of flamingoes (Phoenicopteridae). *Ibis* 104: 426–428.
MAÑEZ, M. 1991. Sobre la reproducción del Flamenco (*Phoenicopterus ruber roseus*) en las Marismas del Guadalquivir (SW de España) con especial referencia al año 1988. Pp. 111–117 in *Reunión técnica sobre la situación y problemática del Flamenco rosa Phoenicopterus ruber roseus en el Mediterráneo Occidental y Africa Noroccidental*. Junta de Andalucía, Sevilla.
MANNING, J.T. 1985. Choosy females and correlates of male age. *Journal of Theoretical Biology* 116: 349–354.
MARC, H., NAUDOT, C. & QUENIN, V. 1948. *Terre de Camargue (Terro Camarguenco)*. Arthaud, Grenoble & Paris.
MARION, L. 1988. Evolution des stratégies demographiques, alimentaire et utilisation de l'espace chez le Héron cendré en France: importance des contraintes énergétiques et humaines. PhD thesis, University of Rennes.
MASCITTI, V. & KRAVETZ, F.O. 2002. Bill morphology of South American flamingos. *Condor* 104: 73–83.
MAS CORNELLÀ, M. 2000. *Las manifestaciones rupestres prehistóricas de la zona Gaditana. Arqueología Monografías*. Junta de Andalucía, Consejeria de Cultura, Seville.
MATHEVON, N. 1997. Individuality of contact calls in the Greater Flamingo *Phoenicopterus ruber* and the problem of backgound noise in a colony. *Ibis* 139: 513–517.
MAYR, G. 2004. Morphological evidence for sister group relationship between flamingos (Aves: Phoenicopteridae) and grebes (Podicipedidae). *Zoological Journal of the Linnean Society* 140(2): 157–169.
MCCULLOCH, G. 2002. Salt lake survivors, flamingos of the Makgadikgadi. *Africa—Birds & Birding* 17: 24–30.
MCCULLOCH, G. & IRVINE, K. 2004. Breeding of Greater and Lesser Flamingos at Sua Pan, Botswana, 1998–2001. *Ostrich* 75: 236–242.
MCCULLOCH, G., AEBISCHER, A. & IRVINE, K. 2003. Satellite tracking of flamingos in southern Africa: the importance of small wetlands for management and conservation. *Oryx* 37: 480–483.
MCDONALD, D.B., FITZPATRICK, J.W. & WOOLFENDEN, G.E. 1996. Actuarial senescence and demographic heterogeneity in the Florida scrub jay. *Ecology* 77: 2373–2381.
MEINERTZHAGEN, R., 1958. Greater Flamingoes in Kenya. *Ibis* 100: 624.
MEININGER, P.L. & ATTA, G.A.M. 1994. *Ornithological Studies in Egyptian Wetlands 1989/90*. FORE Report No. 94–01, WIWO Report No. 40, Zeist.
MEININGER, P.L., SORENSEN, U.G. & ATTA, G.A.M. 1986. Breeding birds of the lakes of the Nile Delta, Egypt. *Sandgrouse* 7: 1–20.
MENDELSSOHN, H. 1975. Report on the status of some bird species in Israel in 1974. *XII Bulletin of the ICBP*: 265–270.
MIDDLEMISS, E. 1961. Notes on the Greater Flamingo. *Bokmakierie* 13: 9–14.
MILLIMAN, J.D., JEFTIC, L. & SESTINI, G. 1992. *Climatic Change in the Mediterranean*. UNEP.
MILLS, J.A. 1973. The influence of age and pair-bond on the breeding biology of the Red-billed Gull *Larus novaehollandiae scopulinus*. *Journal of Animal Ecology* 42: 147–162.

MILON, P., PETTER, J.J. & RANDRIANASOLO, G. 1973. *Faune de Madagascar, XXXV Oiseaux.* ORSTOM, Tananarive, & CNRS, Paris.

MITCHELL, C., BLACK, J.M., OWEN, M. & WEST, J. 1988. On renesting in semi-captive Barnacle Geese. *Wildfowl* 39: 133–135.

MOHAMED, S.A. 1991. On the movement and distribution of the Greater Flamingo *Phoenicopterus ruber* in Bahrein, Arabian Gulf. *Arab Gulf Journal of Scientific Research* 9: 133–142.

MØLLER, A.P. & DE LOPE, F. 1999. Senescence in a short-lived migratory bird: age-dependent morphology, migration, reproduction and parasitism. *Journal of Animal Ecology* 68: 163–171.

MONTES, C. & BERNUÉS, M. 1991. Incidencia del Flamenco rosa (*Phoenicopterus ruber roseus*) en el funcionamiento de los ecosistemas acuáticos de la Marisma del Parque Nacional de Doñana (SW España). Pp. 103–110 in *Reunión técnica sobre la situación y problemática del Flamenco rosa* Phoenicopterus ruber roseus *en el Mediterráneo Occidental y Africa Noroccidental.* Junta de Andalucía, Sevilla.

MORGAN, N.C. 1982. An ecological survey of standing waters in North West Africa: II. Site descriptions for Tunisia and Algeria. *Biological Conservation* 24: 83–113.

MORRISON, T. 1975. Conservation in South America. Pp. 80–83 in Kear, J. & Duplaix-Hall, N. (eds). *Flamingos.* Poyser, Berkhamsted.

MULLIÉ, W.C. & MEININGER, P.L. 1983. Waterbird trapping and hunting in Lake Manzala, Egypt, with an outline of its economic significance. *Biological Conservation* 27: 23–43.

MUNDKUR, T. & TAYLOR, V. 1993. *Asian Waterfowl Census 1993.* AWB, Kuala Lumpur & IWRB, Slimbridge.

MUNDKUR, T., PRAVEZ, R., SHIVRAJKUMAR KHACHER & NAIK, R.M. 1989. Hitherto unreported nest site of Lesser Flamingo *Phoeniconaias minor* in the Little Rann of Kutch, Gujarat. *Journal of the Bombay Natural History Society* 86: 281–285.

MUÑOZ, F. & GARCIA, A.R. 1983. Historia de Fuente de Piedra. Ajalvir, Madrid.

MURTON, R.K. & WESTWOOD, N.J. (eds). 1977. *Avian Breeding Cycles.* Clarendon Press, Oxford.

NAGER, R.G., JOHNSON, A.R., BOY, V., RENDÓN-MARTOS, M., CALDERON, J. & CÉZILLY, F. 1996. Temporal and spatial variation in dispersal in the Greater Flamingo (*Phoenicopterus ruber roseus*). *Oecologia* 107: 204–211.

NAUROIS, R. de 1965. Une colonie reproductrice du Petit Flamant Rose, *Phoeniconaias minor* (GEOFFROY) dans l'Aftout es Sahel (sud-ouest Mauritanien). *Alauda* 33: 166–175.

NAUROIS, R. de 1969a. Peuplements et cycles de reproduction des oiseaux de la Côte Occidentale d'Afrique. *Mémoires du Muséum Nat. D'Histoire Naturel.* Nouvelle, Série A: Zoologie. Paris.

NAUROIS, R. de 1969b. Le Flamant Rose (*Phoenicopterus ruber*) a-t-il niché en nombre et régulièrement dans l'Archipel du Cap-Vert? *Oiseau et Revue Française d'Ornithologie* 39: 28–37.

NAUROIS, R. de. 1994. *Les Oiseaux de l'Archipel du Cap Vert.* Instituto de Investigação Cientifica Tropical, Lisbon.

NAUROIS, R. de & BONNAFFOUX, D. 1969. L'avifaune de l'île du Sel (Ilha do Sal, Archipel du Cap Vert). *Alauda* 57: 93–113.

NELSON, B. 1978. *The Gannet.* Poyser, Berkhamsted.

NELSON, J.B. 2005. Pelicans, Cormorants and their relatives. Oxford; Oxford University Press.

NIETHAMMER, J. 1970. Die Flamingos am Ab-i-Istada in Afghanistan. *Natur und Museum* 100: 201–210.

NISBET, I.C.T. 2001. Detecting and measuring senescence in wild birds: experience with long-lived seabirds. *Experimental Gerontology* 36: 833–843.

NISBET, I.C.T., WINCHELL, J.M. & HEISE, A.E. 1984. Influence of age on the breeding biology of Common Terns. *Colonial Waterbirds* 7: 117–126.
NOGGE, G. 1974. Beobachtungen an den Flamingobrutplätzen Afghanistans. *Journal für Ornithologie* 115: 142–151.
NUR, N. & HASSON, O. 1984. Phenotypic plasticity and the handicap principle. *Journal of Theoretical Biology* 110: 275–297.
OGILVIE, M.A. 1972. Large numbered leg bands for individual identification of swans. *Journal of Wildlife Management* 36: 1261–1265.
OGILVIE, M. & OGILVIE, C. 1986. *Flamingos*. Alan Sutton Publishing Ltd, Gloucester.
OKUMURA, J. & TASAKI, I. 1969. Effects of fasting, refeeding and dietary protein levels on uric acid and ammonia content of blood, liver and kidney in chickens. *Journal of Nutrition* 97: 316–320.
OLIVER, G. 1980. Les Flamants roses *Phoenicopterus ruber roseus* de l'Etang de Canet (Pyrénées Orientales). *Alauda* 48: 255–259.
OLSON, S.L. 1985. The fossil record of birds. Pp.79–252 in Farner, D.S., King, J.R. & Parkes, K.C. (eds), *Avian Biology*, Vol. 8. New York, Academic Press.
OLSON, S.L. & FEDUCCIA, A. 1980. Relationships and evolution of flamingos. *Smithsonian Contributions in Zoology* 316.
OLSON, V.A. & OWENS, I.P.F. 1998. Costly sexual signals: are carotenoids rare, risky or required? *Trends in Ecology and Evolution* 13: 510–514.
ORIANS, G.H. 1969. Age and hunting success in the Brown Pelican (*Pelecanus occidentalis*). *Animal Behaviour* 17: 316–319.
ORING, L.W. 1982. Avian mating systems. Pp. 1–92 in Farner, D.S., King, J.R. & Parkes, K.C. (eds), *Avian Biology*, Vol. 6. Academic Press, New York.
ORNITHOLOGICAL SOCIETY OF TURKEY. 1975. *Ornithological Society of Turkey Bird Report* No. 3, 1970–1973. Records and Editorial Committee, Turkey.
ORSINI, P. 1994. *Les oiseaux du Var*. Association pour le Muséum de Toulon.
PALMA, R.C., JOHNSON, A.R., CÉZILLY, F., THOMAS, F. & RENAUD, F. 2002. Diversity and distribution of feather lice on Greater Flamingoes (*Phoenicopterus ruber roseus*) in the Camargue, southern France. *New Zealand Entomologist* 25: 87–89.
PALUDAN, K. 1959. On the Birds of Afghanistan. Reprinted from Vidensk. *Medd.Dansk naturh. For.*, Vol. 122.
PANOUSE, J.B. 1958. Nidification des Flamants roses au Maroc. *Compte-rendu de la Société des Sciences Naturelles de Maroc* 24: 110.
PAPAYANNIS, TH. & SALATHÉ, T. 1999. *Les zones humides méditerranéennes à l'aube du 21ᵉ siècle*. Medwet, Tour du Valat, Arles.
PARKER, G.A. & STUART, R.A. 1976. Animal behaviour as a strategy optimizer: evolution of resource assessment strategies and optimal emigration thresholds. *American Naturalist* 110: 1055–1076.
PAZ, U. 1987. *The Birds of Israel*. The Stephen Greene Press, Lexington.
PEINADO, V.I., POLO, F.J., VISCOR, G. & PALOMEQUE, J. 1992. Hematologic and blood chemistry values for several flamingo species. *Avian Pathology* 21: 55–64.
PENNY, M. 1974. *The Birds of Seychelles and the Outlying Islands*. Collins, London.
PENOT, J. 1963. Rapport ornithologique pour 1960 et 1961. *La Terre et la Vie*: 280–288.
PERENNOU, C. 1991. *African Waterfowl Census 1991*. IWRB, Slimbridge.
PERENNOU, C. 1992. *African Waterfowl Census 1992*. IWRB, Slimbridge.
PERENNOU, C. & MUNDKUR, T. 1991. *Asian Waterfowl Census 1991*. IWRB, Slimbridge.
PERENNOU, C. & MUNDKUR, T. 1992. *Asian and Australian Waterfowl Census 1992*. IWRB, Slimbridge.
PERENNOU, C., MUNDKUR, T., SCOTT, D.A., FOLLESTAD, A. & KVENILD, L. 1994. *The Asian Waterfowl Census 1987–9: distribution and status of Asian Waterfowl*. AWB Publication No. 86, Kuala Lumpur & IWRB Publication No. 24, Slimbridge.

PERENNOU, C., SADOUL, N., PINEAU, O., JOHNSON, A. & HAFNER, H. 1996. *Management of Nest Sites for Colonial Waterbirds*. Medwet Publication No. 4, Tour du Valat.

PERKINS, L.E.L. & SWAYNE, D.E. 2003. Comparative susceptibility of selected avian and mammalian species to Hong Kong-origin H5N1 high-pathogenicity avian influenza virus. *Avian Diseases* 47: 956–967.

PERRINS, C.M. 1996. Eggs, egg formation and the timing of breeding. *Ibis* 138: 2–15.

PERRINS, C.M. & MCCLEERY, R.H. 1985. The effect of age and pair bond on the breeding success of Great Tits *Parus major*. *Ibis* 127: 306–315.

PÉTÉTIN, M. & TROTIGNON, J. 1972. Prospection hivernale au Banc d'Arguin (Mauritanie). *Alauda* 40: 195–213.

PHILLIPS, N.R. 1982. Observations on the birds of North Yemen in 1979. *Sandgrouse* 4: 37–59.

PICAZO, J. 1999. Flamenco Comun. Pp. 306–307 in de la Puente, J., Pinilla, J. & Lorenzo, J.A. Noticiario Ornitologico 1999. *Ardeola* 46: 305–314.

PICKERING, S. 1992. The comparative biology of flamingos Phoenicopteridae at the Wildfowl and Wetlands Trust Centre, Slimbridge. *International Zoo Yearbook* 31: 139–146.

PICKERING, S. & DUVERGE, L. 1992. The influence of visual stimuli provided by mirrors on the marching displays of lesser flamingos *Phoeniconaias minor*. *Animal Behaviour* 43: 1048–1050.

PICKERING, S., CREIGHTON, E. & STEVENS-WOOD, B. 1992. Flock size and breeding success in flamingoes. *Zoo Biology* 11: 229–234.

PINEAU, J. & GIRAUD-AUDINE, M. 1979. *Les Oiseaux de la Peninsule Tingitane*. Institut Scientifique Charia Ibn Batouta, Rabat.

PLATT, J.B. 1994. Greater Flamingo *Phoenicopterus ruber* at Khor Dubai, United Arab Emirates. *Sandgrouse* 14: 72–80.

POMEROY, D.E. 1962. Birds with abnormal bills. *British Birds* 55(2): 49–72.

PORTER, R.F., MARTINS, R.P., SHAW, K.D. & SØRENSEN, U. 1996. The status of non-passerines in southern Yemen and the records of the OSME survey in spring 1993. *Sandgrouse* 17: 22–53.

PORTER, R.N. & FORREST, G.W. 1974. First successful breeding of Greater Flamingo in Natal, SA. *Lammergeyer* 21: 26–33.

POULSEN, H. 1975. Copenhagen Zoo. Pp. 117–120 in Kear, J. & Duplaix-Hall, N. (eds), *Flamingos*. Poyser, Berkhamsted.

POWER, H.W. 1998. Quality control and the important questions in avian conspecific brood parasitism. In Rothstein, S.I. & Robinson, S.K. (eds), *Parasitic Birds and their Hosts: studies in coevolution*. Oxford University Press, Oxford.

PRADEL, R. 1992. Estimation et comparaison de probabilités de survie par suivi individuel et utilisation en biologie des populations. Doctoral thesis. University of Montpellier II, Montpellier, France.

PRADEL, R. 1996. Utilization of capture-mark-recapture for the study of recruitment and population growth. *Biometrics* 52: 703–709.

PRADEL, R. & LEBRETON, J.-D. 1991. *User's manual for Program SURGE, version 4.2*. CEFE/CNRS, Montpellier.

PRADEL, R., JOHNSON, A.R., VIALLEFONT, A., NAGER, R.G. & CÉZILLY, F. 1997. Local recruitment in the Greater Flamingo: a new approach using capture-mark-recapture data. *Ecology* 78: 1431–1445.

PRATER, A.J. 1979. Trends in accuracy of counting birds. *Bird Study* 26: 198–200.

PROBST, J.-M. 1997. *Animaux de la Réunion*. Ile de la Réunion: Azalées editions, Saint-Marie.

PUERTA, M.L., HUECAS, V., & GARCIA DEL CAMPO, A.L. 1989. Hematology and blood chemistry of the Chilean flamingo. *Comparative Biochemistry and Physiology* 94A: 623–625.

PUERTA, M.L., GARCIA DEL CAMPO, A.L., ABELENDA, M., FERNANDEZ, A., HUECAS, V. & NAVA, M.P. 1992. Hematological trends in flamingos, *PHOENICOPTERUS RUBER*. *Comparative Biochemistry and Physiology and Comparative Physiology* 102A: 683–686.
PURROY, F.J. (ed.). 1997. *Atlas des las Aves de España (1975–1995)*. Sociedad Española de Ornitologia/BirdLife, Lynx Edicions Barcelona.
QUIQUERAN DE BEAUJEU, P. 1551. *De Laudibus Provinciae Paris*. 3 Vols. French translation Arles: M. Claret. 1614. La Provence.
RAINBOLT, R.E., AUGERI, D.M., PIERCE, S.M. & BERGESON, M.T. 1997. Greater Flamingos breed on Aldabra. *Wilson Bulletin* 109: 351–353.
RAMADAN-JARADI, G. & RAMADAN-JARADI, M. 1999. An updated checklist of the birds of Lebanon. *Sandgrouse* 21: 132–170.
RAMO, C., SANCHEZ, C. & ST AUBIN, L.H. 1992. Lead poisoning of Greater Flamingos *Phoenicopterus ruber*. *Wildfowl* 43: 220–222.
RECHER, H.F. & RECHER, J.A. 1969. Comparative foraging efficiency of adult and immature little blue heron (*Florida caerulea*). *Animal Behaviour* 17: 320–322.
REES, E.C. 1981. The recording and retrieval of bill pattern variations in *Cygnus columbianus bewickii*. Pp. 105–119 in Matthews, G.V.T. & Smart, M. (eds), *Proceedings IWRB Symposium Sapporo*. International Waterowl Research Bureau, Slimbridge.
REES, E.C., OWEN, M., GITAY, H. & WARREN, S. 1990. The fate of plastic leg rings used on geese and swans. *Wildfowl* 41: 43–52.
REID, J.M., BIGNAL, E.M., BIGNAL, S., MCCRACKEN, D.I. & MONAGHAN, P. 2003. Age specific reproductive performance in red-billed chough *Pyrrhocorax pyrrhocorax*: patterns and processes in a natural population. *Journal of Animal Ecology* 72: 765–776.
REID, W.V. 1988. Age correlations within pairs of breeding birds. *Auk* 105: 278–285.
REILLY, P.N. & BALMFORD, P. 1975. A breeding study of the little penguin *Eudyptula minor* in Australia. Pp. 161–187 in Stonehouse, B (ed.), *The Biology of Penguins*. MacMillan, London.
RENDÓN, M. 1997a. Anillamiento para lectura a distancia: Los flamencos de Fuente de Piedra. *La Garcilla* 100: 31–33.
RENDÓN, M. 1997b. El Flamenco comun (*Phoenicopterus ruber roseus*). Pp. 62–63 in Purroy, F.J. (ed.), *Atlas des las Aves de España* (1975–1995). Sociedad Española de Ornitologia/BirdLife, Lynx Edicions, Barcelona.
RENDÓN, M., GARRIDO, A., RAMÍREZ, J.M., RENDÓN-MARTOS, M. & AMAT, J.A. 2001. Despotic establishment of breeding colonies of greater flamingos, *Phoenicopterus ruber*, in southern Spain. *Behavioral Ecology and Sociobiology* 50: 55–60.
RENDÓN-MARTOS, M. 1996. La laguna de Fuente de Piedra en la dinámica de la población de flamencos (*Phoenicopterus ruber roseus*) del Mediterráneo Occidental. PhD thesis, University of Málaga.
RENDÓN-MARTOS, M. & JOHNSON A.R. 1996. Management of nesting sites for Greater Flamingos. *Colonial Waterbirds* Special Publication 19: 167–183.
RENDÓN, M., VARGAS, J.M. & RAMÍREZ, J.M. 1991. Dinámica temporal y reproducción del Flamenco común (*Phoenicopterus ruber roseus*) en la laguna de Fuente de Piedra (Sur de España). Pp. 135–153 in Junta de Andalucía (ed.), *Reunión técnica sobre la situación y problemática del flamenco rosa* (Phoenicopterus ruber roseus) *en el Mediterráneo Occidental y Africa Noroccidental*. Agencia de Medio Ambiente, Galan, Seville.
RENDÓN-MARTOS, M., VARGAS, J.M., RENDÓN, M.A., GARRIDO, A. & RAMÍREZ, J.M. 2000. Nocturnal movements of breeding Greater Flamingos in southern Spain. *Waterbirds* 23 Special Publication (1): 9–19.
RÉSERVE NATIONALE DE CAMARGUE. 1987. Compte rendu ornithologique camarguais pour les années 1984–1985. *Revue d'Ecologie (La Terre et la Vie)* 42: 167–191.
RICHTER, N.A. & BOURNE, G.F. 1990. Sexing Greater Flamingos by weight and linear Measurements. *Zoo Biology* 9: 317–323.

RIDLEY, M.W. 1954. Observations on the diet of flamingos. *Journal of the Bombay Natural History Society* 52: 5–7.
ROBERT, F. & GABRION, C. 1991. Cestodes de l'avifaune camarguaise. Rôle d'*Artemia* (Crustacea, Anostraca) et stratégies de rencontre hôte-parasite. *Annales de Parasitologie Humaine et Comparée* 66: 226–235.
ROBERTS, T.J. 1991. *The Birds of Pakistan, Vol. 1. Non-passeriformes*. Oxford University Press, Karachi.
ROBERTSON, H.G. & JOHNSON, P.G. 1979. First record of Greater and Lesser Flamingos breeding in Botswana. *Botswana Notes and Records* 11: 115–119.
ROBERTSON, R.J. & RENDELL, W.B. 2001. A long-term study of reproductive performance in tree swallows: the influence of age and senescence on output. *Journal of Animal Ecology* 70: 1014–1031.
ROBIN, A.P. 1966. Nidifications sur l'Iriki, daya temporaire du sud Marocain en 1965. *Alauda* 34: 81–101.
ROBIN, A.P. 1968. L'avifaune de l'Iriki (sud Marocain). *Alauda* 36: 237–253.
ROBLEDANO, F. & CALVO, J.F. 1991. Flamenco *Phoenicopterus ruber*. Pp. 84–85 in Urios, V., Escobar, J.V., Pardo, R. & Gomez, J.A. (eds). *Atlas de las Aves Nidificantes de la Comunidad Valenciana*. Conselleria d'Agricultura i Pesca, Picanya, Valencia.
ROCAMORA, G. & YEATMAN-BERTHELOT, D. 1999. *Oiseaux menacés et à surveiller en France*. Société Ornithologique de France/Ligue pour la Protection des Oiseaux, Paris.
ROLLAND, C., DANCHIN, E. & DE FRAIPONT, M. 1998. The evolution of coloniality in birds in relation to food, habitat, predation, and life-history traits: a comparative analysis. *American Naturalist* 151: 514–529.
ROOTH, J. 1965. The Flamingoes on Bonaire (Netherlands Antilles), habitat, diet and reproduction of *Phoenicopterus ruber ruber*. *Uitgaven Natuurwetenschappelijke Studiekring voor Suriname en de Nederlandse Antillen, Utrecht* 41: 1–151.
ROOTH, J. 1975. Caribbean Flamingos in a man-made habitat. Pp. 75–79 in Kear, J. & Duplaix-Hall, N. (eds), *Flamingos*. Poyser, Berkhamsted.
ROOTH, J. 1982. A man-made breeding sanctuary for flamingos on Bonaire, Netherlands Antilles. Pp. 172–176 in Scott, D.A. (ed.), *Managing Wetlands and their Birds*. International Waterfowl Research Bureau, Slimbridge.
ROSE, P.M. (ed.) 1992. *Western Palearctic Waterfowl Census 1992*. IWRB, Slimbridge.
ROSE, P.M. (ed.). 1995. *Western Palearctic and South West Asia Waterfowl Census 1994*. IWRB Publication 35, Slimbridge.
ROSE, P.M. & SCOTT, D.A. 1994. *Waterfowl Population Estimates*. IWRB Publication 29, IWRB, Slimbridge.
ROSE, P.M. & SCOTT, D.A. 1997. *Waterfowl Population Estimates*. Wetlands International Publication 44, Wetlands International, Wageningen.
ROSE, P.M. & TAYLOR, V. 1993. *Western Palearctic and South West Asia Waterfowl Census 1993*. IWRB, Slimbridge.
ROWLEY, I. 1983. Re-mating in birds. Pp. 311–360 in Bateson, P. (ed.), *Mate Choice*. Cambridge University Press, Cambridge.
RUSTAMOV, E.A. 1994. The wintering waterfowl of Turkmenistan. *Wildfowl* 45: 242–247.
SABAT, P., FERNANDO NOVOA, F. & PARADA, M. 2001. Digestive constraints and nutrient hydrolysis in nestlings of two flamingo species. *Condor* 103: 396–399.
SADOUL, N., WALMSLEY, J. & CHARPENTIER, B. 1998. *Salinas and Nature Conservation*. Medwet Publication No. 9. Tour du Valat, Arles.
SADOUL, N., JOHNSON, A.R., WALMSLEY, J.G. & LÉVÊQUE, R. 1996. Changes in the numbers and the distribution of colonial Charadriiformes breeding in the Camargue, southern France. *Colonial Waterbirds* 19 (Special Publication 1): 46–58.
SAETHER, B.E. 1990. Age-specific variation in reproductive performance in birds. *Current Ornithology* 7: 251–283.

SAINO, N., AMBROSINI, R., MARTINELLI, R. & MØLLER, A.P. 2002. Mate fidelity, senescence in breeding performance and reproductive trade-offs in the barn swallow. *Journal of Animal Ecology* 71: 309–319.
SALATHÉ, T. 1983. La prédation du Flamant rose *Phoenicopterus ruber roseus* par le Goéland leucophé *Larus cachinnans* en Camargue. *Revue Ecologique (La Terre et la Vie)* 37: 87–115.
SANGSTER, G. 1997. Trends in systematics: species limits in flamingos, with comments on lack of concensus in taxonomy. *Dutch Birding* 19: 193–198.
SANGSTER, G. 2005. A name for the flamingo–grebe clade. *Ibis* 147: 612–615.
SCHENK, H., MURGIA, P.-F. & NISSARDI, S. 1995. Prima nidificazione del Fenicottero rosa (*Phoenicopterus ruber roseus*) in Sardegna e problemi di conservazione delle specie coloniali nello stagno di Molentargius. *Supplement Ricerche Biologia della Selvaggina* 22: 313–321.
SCHENKER, A. 1978. Höchstalter europäischer vögel im zoologischen Garten Basel. *Der Ornithologische Beobachter* 75: 96–97.
SCHMITZ, R.A. & BALDASSARRE, G. 1992. Contest asymmetry and multiple bird conflicts during foraging among non-breeding American Flamingos in Yucatan, Mexico. *Condor* 94: 254–259.
SCHMITZ, R.A., ALONSO AGUIRRE, A., COOK, R.S. & BALDASSARRE, G.A. 1990. Lead poisoning of Caribbean Flamingos in Yucatan, Mexico. *Wildlife Society Bulletin* 18: 399–404.
SCOTT, D.A. 1975. Iran. Pp. 28–32 in Kear, J. & Duplaix-Hall, N. (eds), *Flamingos*. Poyser, Berkhamsted.
SCOTT, D.A. (ed.). 1989. *A Directory of Asian Wetlands*. IUCN, Gland, Cambridge.
SCOTT, D.A. (ed.). 1995. *A Directory of Wetlands in the Middle East*. IUCN, Gland & IWRB, Slimbridge.
SCOTT, D.A. & ROSE, P.M. 1989. *Asian Waterfowl Census 1989*. IWRB, Slimbridge.
SEBER, G.A.F. 1965. A note on the multiple recapture census. *Biometrika* 52: 249–259.
SERLE, W., MOREL, G.J. & HARTWIG, W. 1977. *A Field Guide to the Birds of West Africa*. William Collins Sons & Co. Ltd, London.
SERRA, L., MAGNANI, A., DALL'ANTONIA, P. & BACCETTI, N. 1997. Risultati dei censimenti degli acquatici svernanti in Italia, 1991–1995. *Biologia e Conservazione della Fauna* 101: 1–312.
SERIOT, J. 2000. Les oiseaux nicheurs rares et menacés en France en 1998. *Ornithos* 7: 1–18.
SERIOT, J. 2001. Les oiseaux nicheurs rares et menacés en France en 1999. *Ornithos* 8: 121–135.
SERIOT, J. 2002. Les oiseaux nicheurs rares et menacés en France en 2000. *Ornithos* 9: 225–241.
SHAW, P. 1985. Age-differences within breeding pairs of Blue-eyed Shags *Phalacrocorax atriceps*. *Ibis* 127: 537–543.
SHELDON, F. & SLIKAS, B. 1997. Advances in Ciconiiform systematics 1976–1996. *Colonial Waterbirds* 20: 106–114.
SHIVRAJKUMAR OF JASDAN, NAIK, R.M. & LAVKUMAR, K.S. 1960. A visit to the flamingos in the Great Rann of Kutch. *Journal of the Bombay Natural History Society* 57: 465–478.
SHIRIHAI, H. 1996. *The Birds of Israel*. Academic Press, London.
SIBLEY, C.G. & AHLQHUIST, J.E. 1990. *Phylogeny and Classification of Birds*. Yale University Press, New Haven & London.
SIBLEY, C.G., CORBIN, K.W. & HAAVIE, J.H. 1969. The relationships of the flamingos as indicated by the egg-white proteins and hemoglobins. *Condor* 71: 155–179.
SIEGEL-CAUSEY, D. & KHARITONOV, S.P. 1990. The evolution of coloniality. Pp. 285–330 in Power, D.M. (ed.), *Current Ornithology* Vol. 7. Plenum Press, New York.
SIMMONS, R. 1988. Honest advertising, sexual selection, courtship displays and body condition of polyginous male harriers. *Auk* 105: 303–307.

SIMMONS, R.E. 1996. Population declines, viable breeding areas, and management options for flamingos in Southern Africa. *Conservation Biology* 10: 504–514.
SIMMONS, R.E. 1997. The Lesser Flamingo in Southern Africa: a summary. Pp. 50–61 in Howard, G. (ed.). *Conservation of the Lesser Flamingo in Eastern Africa and Beyond*. Proceedings of workshop at Lake Bogoria, IUCN, Kenya.
SIMMONS, R.E., BARNARD, P. & JAMIESON, I.G. 1999. What precipitates influxes of wetland birds to ephemeral pans in arid landscapes? Observations from Namibia. *Ostrich* 70: 145–148.
SIMPSON, K., SMITH, J.N.M. & KELSALL, J.P. 1987. Correlates and consequences of coloniality in great blue herons. *Canadian Journal of Zoology* 65: 572–577.
SLATER, C.A. 1990. First arrival dates at two Fulmar *Fulmaris glacialis* colonies in Norfolk. *Bird Study* 37: 1–4.
SOUTHERN, W.E. 1974. Copulatory wing-flagging: a synchronising stimulus for nesting ring-billed gulls. *Bird-Banding* 45: 210–216.
STANFORD, J.K. 1954. A survey of the ornithology of Northern Libya. *Ibis* 96: 449–473.
STEVENS, E.F. 1991. Flamingo breeding: the role of group displays. *Zoo Biology* 10: 53–63.
STEVENS, E.F., BEAUMONT, J.F., CUSSON, E.W. & FOWLER, J. 1992. Nesting behaviour in a flock of Chilean Flamingos. *Zoo Biology* 11: 209–214.
STORER, R.W. 1971. Classification of birds. Pp. 1–18 in Farner, D.S. & King, J.R. (eds), *Avian Biology*, Vol. 1. Academic Press, New York.
STUDER-THIERSCH, A. 1966. Altes und Neues über das Fütterungssekret der Flamingos *Phoenicopterus ruber*. *Der Ornitologische Beobachter* 63: 85–89.
STUDER-THIERSCH, A. 1974. Die Balz der Flamingogattung, unter besonderer Berücksichtigung von *Ph. ruber roseus*. *Zeitschrift für Tierpsychologie* 36: 212–266.
STUDER-THIERSCH, A. 1975a. Basle Zoo. Pp. 121–130 in Kear, J. & Duplaix-Hall, N. (eds). *Flamingos*. Poyser, Berkhamsted.
STUDER-THIERSCH, A. 1975b. Group displays in *Phoenicopterus*. Pp. 150–158 in Kear, J. & Duplaix-Hall, N. (eds). *Flamingos*. Poyser, Berkhamsted.
STUDER-THIERSCH, A. 1986. Tarsus length as an indication of sex in the flamingo genus *Phoenicopterus*. *International Zoo Yearbook* 24/25: 240–243.
SUCHANTKE, A. 1959. Die paarung beim Flamingo. *Der Ornithologische Beobachter* 56: 94–97.
SULLIVAN, M.S. 1994. Mate choice as an information gathering process under time constraint: implications for behaviour and signal design. *Animal Behaviour* 47: 141–151.
SWIFT, J.J. 1960. Densité des nids et notions de territoire chez le flamant de Camargue. *Alauda* 28: 1–14.
TAVECCHIA, G., PRADEL, R., BOY, V., JOHNSON, A.R. & CÉZILLY, F. 2001. Sex- and age-related variation in survival probability and cost of first reproduction in Greater Flamingos. *Ecology* 82: 165–174.
TAYLOR, V. 1993. *African Waterfowl Census 1993*. IWRB, Slimbridge.
TAYLOR, V. & ROSE, P.M. 1994. *African Waterfowl Census 1994*. IWRB, Slimbridge.
TELLERIA, J.L. 1981. *La migración de las aves en el Estrecho de Gibraltar, Vol. II: aves no planeadoras*. Universidad Complutense, Madrid.
THAKKER, P.S. 1982. Flamingos breeding in Thol Lake Sanctuary near Ahmedabad. *Journal of the Bombay Natural History Society* 79: 668.
THÉVENOT, M., VERNON, R. & BERGIER, P. 2003. *The Birds of Morocco*. BOU Checklist No. 20. British Ornithologists' Union & British Ornithologists' Club, Tring.
THIBAULT, J.C. 1983. *Les oiseaux de la Corse*. Gerfau Impression, Paris.
THIBAULT, M., KAYSER, Y., TAMISIER, A., SADOUL, N., CHERAIN, Y., HAFNER, H., JOHNSON, A. & ISENMANN, P. 1997. Compte rendu ornithologique camarguais pour les années 1990–1994. *La Terre et la Vie, Rev. Ecol.* 52: 261–315.

THIÉRY, A., ROBERT, F. & GABRION, C. 1990. Distribution des populations d'*Artemia* et de leur parasite *Flamingolepsis liguloides* (Cestode, Cyclophyllidea), dans les salins du littoral méditerranéen français. *Canadian Journal of Zoology* 68: 2199–2204.

THOMPSON, P.S., BAINES, D., COULSON, J.C. & LONGRIGG, G. 1994. Age at first breeding, philopatry and breeding site-fidelity in the Lapwing *Vanellus vanellus*. *Ibis* 136: 474–484.

TIANA, V. (ed.). 2000. *Nidificazione e inanellamento dei Fenicotteri in Sardegna*. Associazione per il Parco Molentargius Saline Poetto, Cagliari.

TICEHURST, C.B., COX, P. & CHEESMAN, R.E. 1926. Additional notes on the avifauna of Iraq. *Journal of the Bombay Natural History Society* 31: 91–119.

TOPPER, U. & TOPPER, U. 1988. *Arte rupestre en la Provincia de Cadiz*. Diputación Provincial de Cadiz, Chiclana de la Frontera, Cadiz.

TOURENQ, C., JOHNSON, A.R. & GALLO, A. 1995. Adult aggressiveness and crèching behaviour in the Greater Flamingo, *Phoenicopterus ruber roseus*. *Colonial Waterbirds* 18: 216–221.

TOURENQ, C., AULAGNIER, S., MESLÉARD, F., DURIEUX, L., JOHNSON, A., GONZALEV, G. & LEK, S. 1999. Use of artificial neural networks for predicting rice crop damage by greater flamingos in the Camargue, France. *Ecological Modelling* 120: 349–358.

TREEP, J.M. 2000. Flamingos presumably escaped from captivity find suitable habitat in western Europe. *Waterbirds* 23 Special Publication (1): 32–37.

TRIVERS, R.L. 1972. Parental investment and sexual selection. Pp. 136–179 in Campbell, B. (ed.), *Sexual Selection and the Descent of Man, 1871–1971*. Aldine-Atherton, Chicago.

TROTIGNON, J. 1975. Mauritania. Pp. 35–37 in Kear, J. &. Duplaix-Hall, N. (eds), *Flamingos*. Poyser, Berkhamsted.

TROTIGNON, J. 1976. La nidification sur le Banc d'Arguin (Mauritanie) au printemps 1974. *Alauda* 44: 119–133.

TROTIGNON, E. & TROTIGNON, J. 1981. Recensement hivernal 1979–1980 des Spatules, des Flamants et des Pélicans blancs sur le Banc d'Arguin (Mauritanie). *Alauda* 49: 203–215.

TROTIGNON, E. & J., BAILLOU, M., DEJONGHE, F., DUHAUTOIS, L. & LECOMTE, M. 1980. Recensement hivernal des limicoles et autres oiseaux aquatiques sur le Banc d'Arguin (Mauritanie) Hiver 1978–1979. *Oiseau et Revue Française d'Ornithologie* 50: 323–343.

TROUCHE, L. 1938. Le Flamant rose de Camargue, erratique? sédentaire? nicheurs? *Alauda* 10: 159–187.

TSOUGRAKIS, Y & KARDAKARI, N. (eds). 1996. *Birds of the Aegean*. Hellenic Ornithological Society, Athens.

TUCKER, G.M. & HEATH, M.F. 1994. *Birds in Europe: their conservation status*. BirdLife Conservation Series No. 3, BirdLife International, Cambridge.

TUITE, C. 1981a. Flamingos in East Africa. *Swara* 4: 36–38.

TUITE, C.H. 1981b. Standing crop densities and distribution of *Spirulina* and benthic diatoms in East African alkaline saline lakes. *Freshwater Biology* 11: 345–360.

TUITE, C.H. 2000. The distribution and density of Lesser Flamingos in East Africa in relation to food availability and productivity. *Waterbirds* 23 (Special Publication 1): 2–63.

UNDERHILL, L.G., TREE, A.J., OSCHADLEUS, H.D. & PARKER, V. 1999. *Review of Ring Recoveries of Waterbirds in Southern Africa*. Avian Demography Unit, University of Cape Town, Cape Town.

UYS, C.J. & MACLEOD, J.G.R. 1967. The birds of the De Hoop Vlei region, Bredasdorp, and the effect of the 1957 inundation over a 10-year period (1957–1966) on the distribution of species, bird numbers and breeding. *Ostrich* 38: 233–254.

UYS, C.J. & MARTIN, J. 1961. The breeding of the Greater Flamingo in the Bredasdorp District. *African Wildlife* 15: 97–105.

UYS, C.J., BROEKHUYSEN, G.J., MARTIN, J. & MACLEOD, J.G. 1963. Observations on the breeding of the Greater Flamingo *Phoenicopterus ruber* LINNAEUS in the Bredasdorp District, South Africa. *Ostrich* 34: 129–154.

VALLE, C.A. & COULTER, M.C. 1987. Present status of the Flightless Cormorant, Galapagos Penguin and Greater Flamingo in the Galapagos Islands, Ecuador, after the 1982–83 El Niño. *Condor* 89: 276–281.
VALVERDE, J.A. 1957. *Aves del Sahara Español (Estudio ecológico del desierto)*. Instituto de Estudios Africanos, Madrid.
VALVERDE, J.A. 1960. Vertebrados de las Marismas del Guadalquivir. *Archivos del Instituto de Aclimatacion de Almeria* 9: 9–168.
VALVERDE, J.A. 1963. La reproducción de flamencos en Andalucía en el año 1963. *Ardeola* 9: 55–65.
VAN HEERDEN, J. 1974. Botulism in the Orange Free State Goldfields. *Ostrich* 45: 182–184.
VAN TUINEN, M., BUTVILL, D.B., KIRSCH, J.A.W. & HEDGES, S.B. 2001. Convergence and divergence in the evolution of aquatic birds. *Proceedings of the Royal Society London* B 268: 1345–1350.
VARGAS YÁÑEZ, J.M., BLASCO RUIZ, M. & ANTUÑEZ CORRALES, A. 1983. *Los vertebrados de la Laguna de Fuentepiedra (Málaga)*. Monografias 28. Universidad de Málaga.
VARSHAVSKIY, S.N., VARSHAVSKIY, B. & GARBUZOV, V.K. 1977. Some rare and endangered birds of Northern Cis-Aral area. Pp. 146–153 in *Rare and Vanishing Animals and Birds of Kazakhstan*. Alma-Ata.
VASILJEV, V.E. 1986. *The Territory Where Flamingoes Winter*. Ashkhabad, Magarif.
VERE BENSON, S. 1970. *Birds of Lebanon and the Jordan Area*. Warne & Co. Ltd., ICBP, London & New York.
VERHENCAMP, S.L., BRADBURY, J.W. & GIBSON, R.M. 1989. The energetic cost of display in male sage grouse. *Animal Bevaviour* 38: 885–896.
VERHEYEN, R. 1959. Contribution à l'anatomie et à la systématique de base de Ciconiiformes (Parker 1868). *Bulletin Institut Royal des Sciences Naturelles de Belgique* 35: 1–34.
VERNER, W. 1909. *My Life Among the Wild Birds in Spain*. John Bale, Sons & Danielsson Ltd, London.
VERNON, J.D.R. 1973. Observations sur quelques oiseaux nicheurs du Maroc. *Alauda* 41: 101–109.
VIALLEFONT, A. 1995. Robustesse et flexibilité des analyses demographiques par capture-recapture: de l'estimation de la survie à la détection de compromis évolutifs. PhD thesis, Univerité de Montpellier II, Montpellier.
VOLKOV, Ye.N. 1977. Distribution and numbers of the Central Kazakhstan population of flamingos. Pp. 153–167 in Sludskiy, A.A. *Rare and Vanishing Animals and Birds of Kazakhstan*. Nauka, Alma-Ata. [In Russian]
VOLPONI, S. 1996. Bill deformity in a Pygmy Cormorant (*Phalacrocorax pygmaeus*) chick. *Colonial Waterbirds* 19 (1): 147–148.
VOOUS, K.H. 1960. *Atlas of European Birds*. Elsevier, Amsterdam (English translation by Nelson, London).
WAGNER, R.H.J. 1993. The pursuit of extra-pair copulations by female birds: a new hypothesis of colony formation. *Journal of Theoretical Biology* 163: 333–346.
WALMSLEY, J. 2006. Tadorne de Belon *Tadorna tadorna*. Pp. 65–66 in Lascève, M., Crocq, C., Kabouche, B., Flitti, A. & Dhermain, F. *Oiseaux remaquables de Provence: ecologie, statut et conservation*. Delachaux & Niestlé, Paris.
WALMSLEY, J.G. 1991. Feeding association of Slender-billed Gulls with Greater Flamingos. *British Birds* 84: 508.
WALMSLEY, J.G. 1994. An assessment of the Greater Flamingo population in Egypt in winter and spring 1989–1990. In Meininger, P.L. & Atta, G.A.M. (eds), 1994. *Ornithological Studies in Egyptian Wetlands 1989/90*. FORE Report No. 94–01, WIWO Report No. 40, Zeist
WARHAM, J. 1990. *The Petrels: their ecology and breeding systems*. Academic Press, London.
WARNCKE, K. 1970. Beitrag zur Vogelwelt des Zentralanatolischen Beckens. *Die Vogelwelt* 91: 176–184.

WARNCKE, K. 1971. The flamingo: a new breeding bird for Turkey. *Bulletin of the Ornithological Society of Turkey* 7: 4–6.
WATSON, G.E. 1960. Flamingos in Greece. *Ibis* 102: 135–136.
WEGE, M.I. & RAVELING, D.G. 1983. Factors influencing the timing, distance, and paths of migrations of Canada geese. *Wilson Bulletin* 95: 209–221.
WEIMERSKIRCH, H. 1992. Reproductive effort in long-lived birds: age-specific patterns of condition, reproduction and survival in the wandering albatross. *Oikos* 64: 464–473.
WELCH, G. & WELCH, H. 1984. Birds seen on an expedition to Djibouti. *Sandgrouse* 6: 1–23.
WESTERNHAGEN, W. VON 1970. Über der Brutvögel der Banc d'Arguin (Mauritanien). *Journal für Ornithologia* 111: 206–226.
WETLANDS INTERNATIONAL. 2002. *Waterbird Population Estimates*. Third edition. Wetlands International Global Series No. 12, Wageningen.
WETMORE, A. 1960. A classification of the birds of the world. *Smithsonian Miscellaneous Collection* 139: 11.
WETTEN, J. VAN, OULD MBARA, C., BINSBERGEN, M. & SPANJE, T. VAN. 1990–91. *Zones humides du sud de la Mauritanie*. RIN Contributions to Research on Management of Natural Resources, Leersum.
WICKLER, W. & SEIBT, U. 1983. Monogamy: an ambiguous concept. Pp. 33–50 in Bateson, P. (ed.), *Mate Choice*. Cambridge University Press, Cambridge.
WICKRAMASINGHE, R.H. 1997. Greater Flamingos at Bundala, Sri Lanka. *Oriental Bird Club Bulletin* 26: 53.
WILLIAMS, A.J. & VELÁSQUEZ, C. 1997. Greater Flamingo. Pp. 112–113 in Harrison, J.A., Allan, D.G., Underhill, L.G., Herremans, M., Tree, A.J., Parker, V. & Brown, C.J. (eds), *The Atlas of Southern African Birds. Volume 1: Non-passerines*. BirdLife South Africa, Johannesburg.
WILLIAMS, L.E. & JOANEN, T. 1974. Age of first breeding in the Brown Pelican. *Wilson Bulletin* 86: 279–280.
WINKLER, R. 1999. Flamant rose. Pp. 34–35 in Avifaune de Suisse. *Nos Oiseaux Supplement* 3.
WITTENBERGER, J.F. & HUNT, G.J. 1985. The adaptive significance of coloniality in birds. Pp. 1–79 in Farner, D.S., King, J.R. & Parked, K.C. (eds). *Avian Biology*. Vol. 8. Academic Press, New York.
WOLFF, W.J. & SMIT, C.J. 1990. The Banc d'Arguin, Mauritania, as an environment for coastal birds. *Ardea* 78: 17–38.
WOODWORTH, B.L., FARM, B.P., MUFUNGO, C., BORNER, M. & OLE KUWAI, J. 1997. A photographic census of flamingos in the Rift Valley lakes of Tanzania. *African Journal of Ecology* 35: 326–334.
WOOLER, R.D. & COULSON, J.C. 1977. Factors affecting the age of first breeding of the Kittiwake *Rissa tridactyla*. *Ibis* 119: 339–349.
WORLD CONSERVATION MONITORING CENTRE. 1990. Directory of Wetlands of International Importance. Ramsar Convention Bureau. Switzerland, Gland.
YEATES, G.K. 1947. Bird Life in Two Deltas. Faber & Faber, London.
YEATES, G.K. 1948. Some supplementary notes on the birds of the Rhone delta. *Ibis* 90(3): 425–433.
YEATES, G.K. 1950. *Flamingo City*. Country Life Ltd, London & Charles Scribners, New York.
YEATMAN-BERTHELOT, D. 1991. *Atlas des oiseaux de France en hiver*. Société Ornithologique de France, Paris.
YOM-TOV, Y. & OLLASON, J.G. 1976. Sexual dimorphism and sex ratios in wild birds. *Oikos* 27: 81–85.

YOUNG, E. 1967. Leg paralysis in the Greater Flamingo and Lesser Flamingo following capture and transportation. *Zoo Yearbook* 7: 226–227.

ZAHAVI, A. 1975. Mate selection: a selection for a handicap. *Journal of Theoretical Biology* 53: 205–214.

ZUBAKIN, V.A. 1985. Types of coloniality in the family Laridae. *Proceedings of the International Ornithological Congress* 18: 1250–1252.

ZWEERS, G., DE JONG, F., BERKHOUD, H. & VAN DEN BERGE, J.C. 1995. Filter feeding in flamingos. *Condor* 97: 297–324.

Index

Ab-e-Istada Lake, Afghanistan 251, 254
Abu Dhabi 251
Afghanistan 37, 44, 251, 253–4
Africa 14, 36, 43, 45, 47, 48, 55
   breeding sites 231, 232–6, 256–9, 259–62
   distribution 70, 71, 81, 74, 84–5
   flyways 102–4, 104–6, 106–7
African-Eurasian Migratory Waterbirds Agreement (AEWA) 188
agriculture 197–9, 215
aircraft 177, 202–3, 227
Akgöl, Turkey 250
Al Wathba Wetland Reserve, Abu Dhabi 251
albinism 24
Algeria 36, 37, 237
Ali, Salim 42–3, 44, 55, 57
Anaticola phoenicopteri 28, 29
Anatidae 20
Anatoceus pygaspis 28, 29
Anseriformes 19, 20, 21
Asia 47, 55, 57
avian botulism 175, 212
avian influenza virus (AIV) 176
avian mycobacteriosis 176
avian tuberculosis 176
Avocet 208, 216

Badger 194
bands 13–14, 47, 55–7, 62
   band loss 59–60
   France 55–6
   Italy 56
   Spain 56
barnacles 60
Basel Zoo, Switzerland 15, 66, 67, 134, 181–2
Bern Convention 188, 201
bill 10, 18, 22, 23, 110
   bill patterns 28–30, 66
   deformed 179, 191–2
   feeding apparatus 111–12
BirdLife International 15, 48, 49, 223

Boar, Wild 194, 204, 216, 238, 239
Bombay Natural History Society 57
Bonn (CITES) Convention 188, 201
Botswana 36, 45, 259, 260–1
Boughzoul Dam, Algeria 37
breeding 141–2
   captive flocks 66–8
   colony establishment 150–3
   egg-laying 155–7
   egg-laying dates 142–5, 148–9, 150, 151
   environmental factors 165–70
   incubation 157–8
   nest building 153–4
   nest density 155
   nest relief and attentive periods 158
   nestlings 159–64
   new colonies 150
   rainfall and water levels 145–9
breeding dispersal 97–8
breeding ecology 63–6
breeding numbers 80, 81, 83, 84, 85, 87–8
breeding sites 229–31
   Asia 232, 250–6
   East Africa 232, 256–9
   Mediterranean 231, 236–50
   predator control 203
   Southern Africa 232, 259–62
   wardening 201–3
   West Africa 231, 232–6
breeding success 60–1, 165–70
   age-specific breeding success 170–2
Brown, Leslie 43, 44, 47, 256

Cagliari, Sardinia 236, 243–4
Camalti Tuzlasi, Turkey 237, 247–8
Camargue, France 12–14, 16–17, 43–4, 226–7, 236, 237–8, 242–3
   breeding islands 204–11
   Etang du Fangassier 54, 63–5, 205–10, 242

foraging flights 121
pair-bonding 134–5
relevance to other flamingo species 227–8
Salin de Giraud 37–41, 242–3
Tour du Valat Biological Station 12
captive flocks 66–8, 222
capture 51–2
   Africa 55
   France 52–3
   India 55
   Iran 55
   Italy 55
   recoveries and resightings 57–8, 94, 95, 108, 109, 178, 183
   Spain 53–4
Charadriiformes 20, 21
chicks 159
   Camargue 165, 170
   capture and ringing 13
   crèching 65, 159–60
   feeding 161–4
   fledging 164
   Fuente de Piedra 211
   soda 'anklets' 212, 213–14
Chott Boul, Mauritania 36, 233, 234
Chotts Fedjaj, Tunisia 44, 123, 237, 246
Ciconiiformes 19, 20, 21, 181
circadian rhythms 119–20
classification 19
climatic events 100–1
Clostridium botulinum 175
coastal lagoons 36
coastal mudflats 36
Cockatoo, Sulphur-crested 182
collisions 177, 182
Colonial Waterbird Society 15
coloniality 32–4
colony establishment 150–1
   artificial colonies 226–7
   breeding attempts per year 152
   despotism 151–2
   false-nesting' 152–3
Colpocephalum 28, 29
Compagnie des Salins du Midi et des Salines del'Est (CSME) 210, 242

Index

Condor, Andean, 182
conservation 187–8, 223
 agriculture 197–9, 215
 artificial colonies 226–7
 breeding-site management 204–12
 breeding-site protection 201–3
 environmental pollution 189–92
 exotic flamingos and feral populations 192–3
 human disturbance 193–4, 202–3
 legal protection 201
 predators and scavengers 194–7
 rescue and rehabilitation 212–14
 salt industry 215–17
 threats 188–9
 wetlands 189, 199–201
Coot 174
copulation 131–2
 extra-pair copulations 132–3
cormorants 192
 Cormorant 235
counts 48–51
 breeding numbers 80, 81, 83, 84, 85, 87–8
 wintering numbers 81–2, 83, 84–5, 86
cranes 9, 56
 Common 90
 Siberian White 182

Dasht-e-Nawar Lake, Afghanistan 37, 44, 251, 253–4
data storage and analyses 59
Delacour, Jean 22
disease 176, 193, 225
dispersal
 breeding dispersal 97–8
 environmentally induced dispersal 100–1
 natal dispersal 96–7
 post-fledging dispersal 93–5
distribution 69–70
 Asia 73–4, 83–4
 East Africa 74, 84
 Mediterranean 70–3, 81–2
 research 48–51
 Southern Africa 74, 84–5
 West Africa 70, 71, 81
Djerid, Tunisia 43, 44, 237, 246
dogs 216
Domergue, Charles 43
Doñana, Spain 45, 236, 238–9
doves 21
drinking flights 124–5

ducks 18, 19, 21, 51, 90, 174, 176, 198
 Ruddy 193
 White-headed 193
Dunaliela salina 41
dye-marking 62–3

eagles 176
Ebro delta, Spain 37, 236, 241–2
egg-laying 155
 clutch size 157
 dates 142–5, 148–9, 150, 151
 egg characteristics 155–7
Egret, Cattle 174
Egret, Little 174
Egypt 45, 237, 247
Eider, Common 50–1, 91
El Malaha, Egypt 45, 237, 247
Elmenteita Lake, Kenya 37, 43, 44, 195, 257–8
endangered status 223–4
environment 219–20
Eregli Marshes, Turkey 237, 250
estuaries 36
Ethiopia 37, 257
Etosha Pan, Namibia 36, 44, 259, 260, 262
Europe 107
European Union for Bird Ringing (Euring) 59

false-nesting 152–3, 237, 260
feather lice 28–9
feeding apparatus 111–12
feeding behaviour 113–17
 'grubbing' 116, 117
 heron-like 'running' 119
 skimming 116, 118
 stamping 116, 117
 stamping — 'marking time' 116, 117–18
 up-ending 115, 116, 117
 walking 'leaving tracks' 116, 118–19
 walking 116, 117
feeding chicks 161–4
feeding cones 117, 118
feral flamingos 107, 192–3
filter-feeding 112
fish 195, 259
fish farms 37, 197, 198, 224
Flamingo Specialist Group (FSG) 15, 16, 49, 50, 189, 222
Flamingo Supporter Scheme 58
Flamingo, American 21, 67
Flamingo, Andean 22
Flamingo, Caribbean 21, 66, 67, 107, 113, 114, 182, 223

 breeding 145, 192
 feeding 119, 181
 foraging flights 120–1
 plumage 26
Flamingo, Chilean 22, 66, 107, 192, 193
Flamingo, Greater 9, 10, 11
 Camargue 12–14, 16–17
 habitat 34–7
 life history 32–4
 morphology 22–32
 population trends 85–8
 systematics 21, 22
 taxonomy 21
 world population 74–5, 75–80
 world range 46
Flamingo, James' (Puna) 10, 22
Flamingo, Lesser 10, 22, 33, 50, 55, 176, 192, 193, 255
 captive breeding 67
 chick mortality 213
 feeding 118
 Rift Valley lakes 256, 258, 259, 260
Flamingolepis caroli 28
Flamingolepis flamingo 28
Flamingolepis liguloides 28
flamingos 9–10, 18
 ancestry and relationships 19–21
 ancient history 10–11
 recent history 11–12
 research 222–3
fledging 164
flight behaviour 90–2
flightlessness 26–7
flock sizes, Camargue 92
flyways 103, 107
 east and southern Africa 106–7
 east Mediterranean and south-west Asia 104–6
 north-west Europe 107
 west Mediterranean and north-west Africa 102–4
food 112–13
foraging areas 35–7
foraging ecology 110
 circadian rhythms 119–20
 drinking flights 124–5
 feeding apparatus 111–12
 feeding behaviour 113–19
 food 112–13
 foraging flights 120–4
 research 61–3
 vigilance 119
fossil record 19, 20
foxes 202, 216
 Red 182, 194, 204

France 36, 37, 52–3, 55–6, 236, 237–8, 242–3
Fuente de Piedra, Spain 14, 36, 44, 236, 238, 239–40
 breeding islands 211–12

gales 101
Gallet, Etienne 43, 44
Gannet, Northern 94, 187
geese 9, 18, 19, 20, 21, 56, 110, 176, 215
 Barnacle 32, 60, 135, 138
 Bean 91
 Greater Canada 90
 Greylag 238
grebes 21
Greece 36, 237
group displays
 alert posture 127
 display flights 128–9
 function 129–31
 head-flagging 127
 inverted wing-salute 128
 marching 127–8
 scratching 128
 twist-preen 128
 wing-leg stretch 128
 wing-salute 128
gulls 169, 193, 208, 216, 249
 Herring 195
 Lesser Black-backed 197
 Ring-billed 130
 Slender-billed 114
 Yellow-legged 12, 34, 154, 160, 194, 194, 195–6, 241
 culls 203, 210–11
Gynandrotaenia stammeri 28

habitat 34–7
Halobacterium 41
handling 51–2
 Africa 55
 France 52–3
 India 55
 Iran 55
 Italy 55
 Spain 53–4
Harrier, Marsh 197
herons 19, 110, 133, 192
 Great Blue 137
 Grey 235
Hoffman, Luc 12, 13, 14, 17, 44, 182, 205
Hoop Vlei, South Africa 45, 259, 261–2
hunting 176
Hyena, Spotted 176
Hymenolepididae 28

ibises 19
 Scarlet 23
Ilot des Flamants, Mauritania 232, 235–6
impoundments 37
incubation 157–8
India 42–3, 55, 57, 251, 254–6
inland lakes 37
inter-colony movements 98–9
International Council for Bird Preservation (ICBP) 15
International Waterbird Census (IWC) 47, 48–9, 50, 75–80
Iran 37, 44, 55, 57, 251, 252
Iraq 251
Iriki depression, Morocco 43, 44, 237, 238
Italian National Wildlife Institute (INFS) 56
Italy 14, 46, 55, 56, 236, 237, 244–5

Juncitarsus gracillimus 19

Kaolack saltpans, Senegal 45, 233
Kazakhstan 36, 44, 57, 250, 251, 252–3
Kenya 36, 37, 44, 256–9
Kiaone Islands, Mauritania 43, 44, 232, 234–5
Kite, Black 197
Kuwait 251

Laguna Petrola, Spain 236
Larids 192
lead poisoning 176–7, 191
Libya 45
life history 32–4
Linnaeus 21
longevity 181–2, 221–2

Magadi Lake, Kenya 36, 44, 257, 258
Makgadikgadi Pans, Botswana 36, 45, 259, 260–1
Margherita di Savoia, Italy 236, 244–5
marshes 36
mating 126–7
 age-assortative pairing 138–40
 group displays 127–9, 129–31
 pair formation and copulation 131–3
 seasonal monogamy 133–7

Mauritania 43, 44, 45, 232, 233, 234–6
measurements 30–2
Mediterranean 12, 13, 14, 45–7, 55
 control of flamingo population 224–5
Meikel's tract 23
metapopulation dynamics 221, 224
Mexico 177, 223
migration 90, 91, 99–100, 107–9
mistral 90, 208
Mockingbird, Northern 135
Molentargius-Quartu lagoon, Sardinia 243–4
monogamy 133
 divorce 135–7
 mate fidelity 134–5
Morocco 36, 43, 44, 237
morphology 22–32
mortality 174
 avian botulism 175
 epidemics 176
 extreme weather 174–5
 trapped in mud 175, 208
moult 26–7
movements 89, 92–3, 107–9
 breeding dispersal 97–8
 environmentally induced dispersal 100–1
 inter-colony movements 98–9
 natal dispersal 96–7
 post-fledging dispersal 93–5
 spatial patterns 101–7
mud 175, 208

Namibia 36, 44, 259, 260, 262
natal dispersal 96–7
Natron Lake, Tanzania-Kenya 36, 43, 44, 256, 257, 258, 259
neck collars 57
nesting
 'false-nesting' 152–3
 nest density 155
 nest relief and attentive periods 158
 nest-building 153–4
 nesting sites 34–5
Newcastle disease virus (NDV) 176
nighthawks 21
numbers 74–80
 research 48–51

Orbetello, Italy 36, 236
Orthomyxoviridae 176
Owl, Eagle 182

Index

pair bonds 68, 134–7
  age-assortative pairing 138–40
pair formation 131–3
Panouse 43
pans 36
Pantano de El Hondo, Spain 236, 241
parasites 27–8, 193
  feather lice 28–9
pelicans 249
  Great White 88, 188, 194, 195, 250
Penguin, Emperor 32
Penguin, Little 137
Petrel, Grey-faced 132
Phoeniconaias 22
Phoenicoparrus andinus 22
Phoenicoparrus jamesi 22
Phoenicopteriformes 20, 21
Phoenicopterus 10
  antiquorum 21
  chilensis 22
  croizeti 19
  minor 22
  roseus 22
  ruber 21, 22
  ruber glyphorhynchus 21–2
  ruber ruber 21
  rubus roseus 22
phylogenetic affinities 19
Pigeon, Feral 140
playas 36
plumage 23–4
  plumage development 24–6
pollution 37, 189–92
population 173–4
  control 224–5
  estimated world population 74–5, 75–80
  population trends 85–8
  sub-populations 102–7
Portugal 36
post-fledging dispersal 93–5
predators 119, 152, 160, 165, 176, 188, 189, 194–7, 208
  predator control 203
Presbyornis 19, 20
Procellariiformes 221
Progynotaeniidae 28
Puffin 120

Rann of Kutch, India 42–3, 55, 57, 251, 255–6
raptors 197
recoveries 57–8, 94, 95, 108, 109, 178, 183
recruitment 182–3

age-specific probability of recruitment 184–6
age-specific proportions of breeders 183–4
Recurvirostridae 20
Red Data Books 188, 223
rescue 212–14
research 42–5, 218–19
  breeding success 60–1
  captive flocks 66–8
  capture and handling 51–5
  contemporary research 45–8
  data storage and analyses 59
  distribution and numbers 48–51
  foraging ecology 61–3
  other flamingo species 222–3
  recording attentive periods 66
  recording breeding birds 63–5
  recoveries and resightings 57–8, 94, 95, 108, 109, 178, 183
  rings and bands 55–7
  survival rates 59
resightings 57–8, 94, 95, 108, 109, 178
rice fields 13, 37, 197–9, 220, 224, 226
  protection 215
Rift Valley, Africa 43, 256, 258, 259, 260
rings 55–7
  France 56
  Iran 57
  Italy 56
  Kazakhstan 57
  Spain 56

Sahara 43
Salins, Les 210, 242
salt industry 14, 37, 205, 206, 210, 211, 215–17
salt-steppe 36
saltpans 11, 37, 215–17
Sambhar Lake, India 251, 254–5
Santa Gilla, Sardinia 244
Santa Pola saltpans, Spain 236, 241
Sardinia 14, 36, 236, 237, 243–4
Scott, Sir Peter 19
seabirds 133, 136, 221
Senegal 45, 233
senescence 182, 221–2
Seyfe Gölü, Turkey 237, 249
Shalla Lake, Ethiopia 37, 257
Shelduck, Common 216, 226
Sidi el Hani, Tunisia 45, 237, 245
Sidi Mansour, Tunisia 44, 237, 245–6

Slimbridge Flamingo Symposium 22
South Africa 44–5, 259–62
Spain 36, 37, 44–5, 53–4, 56, 176, 236, 237, 238–42
  foraging flights 121–2
  spatial patterns 101–7
Species Survival Commission (SSC) 15, 49, 189
spoonbills 20
  Roseate 23
Sri Lanka 251
St Lucia, South Africa 45, 259, 261
Stilt, Australian Banded 18, 20, 21
storks 18, 19, 20, 21, 133
  Marabou 160, 176, 194, 195
  White 59
Studer-Thiersch, Adelheid 15, 66–7, 134
Sultansazligi, Turkey 45, 237, 249–50
surveys 48–51
survival 59, 177–8, 181
  adult 180–1
  juvenile and immature 178–80
  longevity and senescence 181–2
swans 19, 56
  Bewick's 59, 60, 90
  Black 97, 197
  Whooper 60, 91
swifts 21
systematics 21–2

Tanzania 36, 257
Tengiz, Kazakhstan 36, 44, 251, 252–3
terns 208, 210, 216, 241, 249
  Caspian 137, 235, 236
  Gull-billed 236
  Royal 235, 236
  Sooty 137
tidal mudflats 36
Tilapia alcalica 259
Tilapia grahami 195
Tixerent 43
Tunisia 36, 43, 237, 245–6
  foraging flights 122–3
Turkey 36, 45, 237, 247–50
  foraging flights 123–4
Tuz Gölü, Turkey 45, 237, 248–9

United Nations Environment Programme 48
Uromiyeh, Iran 44, 251, 252

328 Index

Valle di Comacchio saltpans, Italy 236
Valverde, Jose Antonio 44
vigilance 119
vocalisations 32
Vulture, Egyptian 197
Vulture, Turkey 175

waders 936, 0, 216, 249
wadi mouths 36
wanderings 99–100
wardening 201–3
Waterbird 48

Waterbird Society 15
weather 100–1, 174–5, 212, 214, 225
wetlands 100
  loss and degradation 189
Wetlands International (IWRB) 15, 47, 48, 49, 189, 222
wheelies 117, 118
Wild Birds Directive 188, 201
Wildfowl and Wetlands Trust, Slimbridge 15, 22, 67, 134
Wildfowl Trust 15
Wildlife Society of East Africa 213

winds 90, 101
winter 100–1
wintering numbers 81–2, 83, 84–5, 86
wire collisions 177, 182
World Conservation Union (IUCN) 15, 48, 49, 222, 223

Yeates, George 44

Zululand 259